Tributes
Volume 44

A Question is More Illuminating than an Answer
A Festscrift for Paulo A. S. Veloso

Volume 34
Models: Concepts, Theory, Logic, Reasoning, and Semantics. Essays Dedicated to Klaus-Dieter Schewe on the Occasion of his 60th Birthday
Atif Mashkoor, Qing Wang and Bernhrd Thalheim, eds.

Volume 35
Language, Evolution and Mind. Essays in Honour of Anne Reboul
Pierre Saint-Germier, ed.

Volume 36
Logic, Philosophy of Mathematics and their History.
Essays in Honor of W. W. Tait
Erich H. Reck, ed.

Volume 37
Argumentation-based Proofs of Endearment. Essays in Honor of Guillermo R. Simari on the Occasion of his 70th Birthday
Carlos I. Chesñevar, Marcelo A. Falappa, Eduardo Fermé, Alejandro J. García, Ana G. Maguitman, Diego C. Martínez, Maria Vanina Martinez, Ricardo O. Rodríguez, and Gerardo I. Simari, eds.

Volume 38
Logic, Intelligence and Artifices. Tributes to Tarcísio H. C. Pequeno
Jean-Yves Béziau, Francicleber Ferreira, Ana Teresa Martins and Marcelino Pequeno, eds.

Volume 39
Word Recognition, Morphology and Lexical Reading. Essays in Honour of Cristina Burani
Simone Sulpizio, Laura Barca, Silvia Primativo and Lisa S. Arduino, eds

Volume 40
Natural Arguments. A Tribute to John Woods
Dov Gabbay, Lorenzo Magnani, Woosuk Park and Ahti-Veikko Pietarinen, eds.

Volume 41
On Kreisel's Interests. On the Foundations of Logic and Mathematics
Paul Weingartner and Hans-Peter Leeb, eds.

Volume 42
Abstract Consequence and Logics. Essays in Honor of Edelcio G. de Souza
Alexandre Costa-Liete, ed.

Volume 43
Judgements and Truth. Essays in Honour of Jan Woleński
Andrew Schumann, ed.

Volume 44
A Question is More Illuminating than the Answer: A Festschrift for Paulo A.S. Veloso
Edward Hermann Haeusler, Luiz Carlos Pinheiro Dias Pereira and Jorge Petrucio Viana, eds.

Tributes Series Editor
Dov Gabbay   dov.gabbay@kcl.ac.uk

A Question is More Illuminating than an Answer
A Festscrift for Paulo A. S. Veloso

edited by

Edward Hermann Haeusler

Luiz Carlos Pinheiro Dias Pereira

Jorge Petrucio Viana

© Individual authors and College Publications 2020. All rights reserved.

ISBN 978-1-84890-353-1

College Publications
Scientific Director: Dov Gabbay
Managing Director: Jane Spurr

http://www.collegepublications.co.uk

Cover design by Laraine Welch

All rights reserved. No part of this publication may be reproduced, stored in a retrieval system or transmitted in any form, or by any means, electronic, mechanical, photocopying, recording or otherwise without prior permission, in writing, from the publisher.

Contents

Preface · 1

Acknowledgements · 8

Memory Propositional Dynamic Logic
Mario Benevides and Bruno Lopes · 9

Fixed-Parameter Tractability of Decidable Prefix-Vocabulary FO Classes
Luis Henrique Bustamante, Ana Teresa Martins, and Francicleber Martins Ferreira · 28

An Objectual Semantics for First-Order LFI1 with an Application to Free Logics
Walter Carnielli and Henrique Antunes · 58

Remarks on a Nice Theorem of Monsieur Glivenko
Ítala M. Loffredo D'Ottaviano and Evandro Luís Gomes · 92

The Arithmetization-Free Component of Gödel's Proof
Rodrigo A. Freire · 112

A Working Mathematician between Philosophers: On the Logical Analysis of Magnitudes
Abel Lassalle-Casanave and Eduardo N. Giovannini · 120

Boolean Real Semigroups
Francisco Miraglia and Hugo R. O. Ribeiro · 136

Kolgomorov-Veloso Problems and Dialectica Categories
Valeria de Paiva and Samuel G. da Silva · 176

On the Construction of Explosive Relation Algebras
Carlos G. Lopez Pombo, Marcelo F. Frias, and Thomas S. E. Maibaum · 202

Euclidean Machines: General Theory of Problem
Wagner de Campos Sanz · 236

When Databases Roamed Computing: Formal database specification revisited perspective of language
Ionuţ Ţuţu, Claudia E. Chiriţă, and José L. Fiadeiro · 261

Living and working with Paulo A. S. Veloso
   Sheila Regina Murgel Veloso 279

Preface

It is difficult to start a text. It is difficult to find the first sentence, even the very first word, the right tone. To use a well-known image, it is like being in a garden in front of many forking paths. Difficulties multiply (by an inconstant value) when the text is a preface to a volume of tributes; in this case we usually have to face a dilemma (which could easily be generalized to an n-place-lemma for any number n): (1) To make a descriptive text with the important academic contributions and achievements of the honoree and with a brief description of the texts that compose the volume or, which is by no means less tempting, (2) to construct a text exploring different aspects of the life and personality of the honoree. These difficulties and hesitations become much more acute when the honoree is someone of the stature of professor Paulo Augusto Silva Veloso. Professor Veloso is not just a great researcher and a peerless teacher; professor Veloso is one of the most important characters in the history of logic and computer science in our country. When we refer to history, we do not refer only to the strictly academic facts of that history, such as scientific production and the formation of teachers and researchers; behind this history there is another, deeper, and in no way less interesting and important history: the one that lives in the catalogue of Veloso's stories–we are certain that those who contributed with texts to this volume or even anyone who has known Paulo Veloso as a teacher, as a co-author, as a travel companion or at informal gatherings can tell a very good story.

To say that professor Veloso's main areas of activity and interest are Logic and the Theory of Computation is to say very little. Indeed, Veloso's work opens up new horizons in fields as varied as automata and formal languages, abstract data types, the mathematical theory of problems, formal specifications, algebraic logic, diagrammatic reasoning, generalized quantifiers, and the history of logic, among many others. The set of texts that make up this volume serves as an unequivocal testimony to the extent of professor Veloso's scientific contributions.

After accepting the invitation to be part of this enterprise, each author (or group of co-authors) sent us a article that was evaluated by a referee and reviewed by the author(s) for the final publication format. The editors would like to express their gratitude and appreciation for authors and reviewers, whose efforts—made in a period as difficult as the one we are experiencing, due to the pandemic—have made the articles that appear in this volume reach the high quality they have.

We now move on to a brief description of the content of each article.

In "Memory Propositional Dynamic Logic" Mario Benevides and Bruno Lopes generalize their previous works on Dynamic Logics for Petri nets, extending the language—iteration excluded—by equipping the modalities with a memory (sequence of names). Moreover, changing the semantics by imposing that states now have memory too (also a sequence of names). They investigate the role played in this setting by two new basic programs $+a$, that adds name a to the current sequence s, and $-a$ that removes one occurrence of a in s—or fail when a does not occur in s. The article contains a process calculus, an axiomatization, proofs of soundness and completeness—for the intended semantics—and a proof that the finite model property holds, which yields a decidability result.

In "Fixed-Parameter Tractability of Decidable Prefix-Vocabulary FO Classes" Luis Henrique Bustamante, Ana Teresa Martins, and Francicleber Martins Ferreira study the parameterized complexity of the satisfiability problem for some prefix-vocabulary classes of first-order logic. The authors consider the natural parameters emerging from these classes' definition, such as the quantifier rank and number of symbols. Taking the classification of maximal decidable classes, they observe that many fragments have fixed-parameter tractability for the satisfiability concerning some of these parameters when combining with the finite model property. The article contains a review of some previous results for pure predicate fragments and some new unpublished results on some classes with function symbols.

In "An Objectual Semantics for First-Order LFI1 with an Application to Free Logics" Walter Carnielli and Henrique Antunes present an object-based semantics for the first-order version of LFI1, along with a corresponding natural deduction system. The authors prove the completeness of their set of natural deduction rules concerning the proposed semantics. The article also contains an adaptation of LFI1 to yield FLFI1, a free logic of formal inconsistency. This later system results from some slight modifications on the former system, and the new system proves to be sound and complete for a non-inclusive dual-domain semantics. Finally, the authors illustrate how combining the FLFI1 consistency operator with its existence predicate may be used to express certain ontological theses concerning the relationships between the notions of consistency and existence.

In "Remarks on a Nice Theorem of Monsieur Glivenko " Itala M. Loffredo D'Ottaviano and Evandro Luís Gomes make a historical incursion, revisiting Glivenko's article of 1929, in which he introduced his well-known "translation of double negation" from classical into intuitionistic propositional logic. The discussion on Glivenko's interpretation is based on some general concepts of translations between logics that the authors have proposed and studied in a series of previous articles. The article also contains a digression on why Kolmogorov's article of 1925, in which it was proposed the first formalization for intuitionistic logic and the first translation from classical into intuitionistic logic, is not mentioned by Glivenko.

In "The Arithmetization-Free Component of Gödel's Proof" Rodrigo A. Freire presents a weaker version of Gödel's first incompleteness theorem, applicable to any consistent first-order theory, in which it is not necessary to consider an arithmetization. This weaker version is a consequence of a usually neglected general pattern of reasoning in Gödel's proof that is not concerned with arithmetic or computability but belongs to general first-order logic with equality. The author argues that this simple theorem suffices to show that the subject matter defined by the axioms of a formal system for arithmetic cannot be extensionally equivalent to the subject matter of finitary metamathematics.

In "A Working Mathematician between Philosophers: On the Logical Analysis of Magnitudes", Abel Lassalle-Casanave and Eduardo N. Giovannini discuss the exclusion of the general concept of magnitude from Hilbert's "Foundations of Geometry" of 1899. The authors illustrate this aspect of Hilbert's geometrical program by examining the elementary case of straight line segments, their operations, and relations. In particular, they analyze how Hilbert explicitly avoided the use of principles for general magnitudes, while he proved their validity for the relevant geometrical objects from his geometrical axioms. The article ends with the author's testimony about their collaboration with Paulo A. S. Veloso in the production of two articles whose themes are closely related to the theme of this work.

In "Boolean Real Semigroups" Francisco Miraglia and Hugo R. O. Ribeiro employ the languages of special groups and real semigroups (RS) to give new, often conceptually different and more precise, proofs of the characterization of RSs whose space of characters is Boolean in the natural Harrison (or spectral) topology. These are results obtained by Murray A. Marshall. As new results, they give a natural Horn-geometric axiomatization of Boolean RSs (in the language of RSs). Moreover, they establish the closure of this class by specific important constructions as Boolean powers, arbitrary filtered colimits, products, reduced products and RS-sums, and surjective RS-morphisms (in particular, quotients). Finally, the article contains characterizations of both the morphisms between Boolean RSs and the quotients of Boolean RSs. These results extend previous work by many authors.

In "Kolgomorov-Veloso Problems and Dialectica Categories" Valeria de Paiva and Samuel G. da Silva establish a categorical connection between Dialectica models, Kolmogorov's problems, Veloso's problems and Blass' problems. The authors show that the use of categories allows us to connect extremely different mathematical areas, using simple methods. The article contains results showing that while Blass and Kolmogorov's notions of a problem can be investigated using ZF's set-theoretical framework, Veloso's problems commit us to a stronger set-theory. The results reported here contribute to categorical methods in the sense that they show a situation in which the use of categories helps to determine where the fundamental choices underlying the axioms are important. According to the authors, if this requirement of stronger foundations is a bug or a feature depends, perhaps, on personal taste and conviction. They sustain that, like many other questions in Mathematics, as soon as one investigates them a little, the notion of 'problem' points out to grand challenges in the foundations of Mathematical Logic, to wit whether one wants to accept or not the Axiom of Choice within their chosen framework.

In "On the Construction of Explosive Relation Algebras" Carlos G. Lopez Pombo, Marcelo F. Frias, and Thomas S. E. Maibaum present some techniques for constructing explosive relation algebras, i.e., relation algebras admitting many non-isomorphic expansions to a fork algebra. B. Jónsson and A. Tarski introduced relation algebras in 1948 whereas fork algebras were defined by P. A. S. Veloso and A. M. Haeberer in 1991. Fork algebras are expansions of relation algebras, but not all relation algebras can be expanded to a fork algebra. Thus, the problem arises of determining how an extensible relation algebra can be extended to a fork algebra. The authors review Paulo Veloso's work on the topic and generalize some of his techniques, producing new classes of expansive relation algebras. The article ends with a charming statement, written by Thomas S. E. Maibaum, recalling some aspects of his friendship and academic collaboration with Paulo A. S. Veloso.

In "Euclidean Machines: General Theory of Problem" Wagner de Campos Sanz approaches Euclidean Geometry via Turing machines. The author bases his work on analogies between Turing machines and constructions on propositions of plane geometry. Both concepts are connected through Paulo A. S. Veloso's General Theory of Problems. More specifically, the article suggests interpreting Euclidean geometry with an ontology of actions employing problems. The author hopes that the development reported here can contribute so that the notion of a problem has a more prominent role in contemporary epistemology, linked to mathematical logic.

In "When Databases Roamed Computing: Formal database specification revisited." Ionuţ Ţuţu, Claudia E. Chiriţă, and José L. Fiadeiro revisit an early article by Marco A. Casanova, Paulo A. S. Veloso, and Antonio L. Furtado where database specifications are developed by stepwise refinement within a heterogeneous conceptual-design framework that encompass logical, algebraic, programming language, grammatical and denotational formalisms. The authors recast some of Casanova, Veloso and Furtado's original results, using concepts and techniques from institution theory and modern algebraic specification that are best suited to deal with heterogeneity. Three levels of specification are considered, corresponding to information representation, database querying and manipulation, and database procedures–thus gradually progressing from high-level descriptions to database applications' implementations. Each level is grounded in a logical system formalized as an institution, while connections between levels are given in terms of institution semi-morphisms and heterogeneous refinements. The introduction of the article ends with a tribute, which celebrates the academic collaboration and personal friendship of the third author with Paulo A. S. Veloso, through a brief review of the role played by Veloso in a 'dissident' field of studies on formal specifications opened by Casanova, Furtado and Veloso.

In "Living and working with Paulo A. S. Veloso." Sheila Regina Murgel Veloso presents a compelling testimony of her fifty years of living and working together with Paulo A. S. Veloso. Throughout Sheila's views, we enter in contact with facts that testify Paulo Veloso's creativity and enthusiasm for understanding and clarifying issues in many aspects of Computer Science, Logic, Mathematics and Philosophy. Sheila restricts her narrative to facts related to the work she and Paulo did together, but anyone who has experienced working with Paulo would agree that her vision extends to the entire spectrum of Paulo Veloso's work as a researcher.

Almost all authors who contributed texts to this volume (and indeed, many other colleagues) at some point in time and on some of the themes mentioned above were co-authors of some work with professor Veloso. Working with professor Veloso, or rather, having had the chance to see him at work, is something that contributed in a fundamental and indelible way to our formation as researchers. We could certainly make a list of professor Veloso's qualities–generosity, affectionate irony, unusual intelligence, ability to transform difficult problems into interesting and amusing problems, humbleness to reconsider answers,...–, but we will always be left with the impression that the list would be essentially incomplete in the case of this *homme d'esprit*. Maybe four words would suffice to convey the core of our message to Professor Veloso today: thank you very much!

> Edward Hermann Haeusler
> *Department of Informatics*
> *Pontifical Catholic University of Rio de Janeiro (PUC-Rio),*
> *Rio de Janeiro, Brazil*
> hermann@inf.puc-rio.br

> Luiz Carlos Pinheiro Dias Pereira
> *Department of Philosophy*
> *Pontifical Catholic University of Rio de Janeiro (PUC-Rio),*
> *Rio de Janeiro, Brazil*
> luiz@inf.puc-rio.br

> Jorge Petrucio Viana
> *Institute of Mathematics and Statistics*
> *Fluminense Federal University (UFF),*
> *Niterói, Brazil*
> petrucio_viana@id.uff.br

Postscript

Paulo A. S. Veloso died on November 14, 2020, shortly after the completion of this volume, which was in the process of being sent for publication. We intended to organize an event in his honor, in which a copy of the book would be given to him and in which we could show him the importance and the meaning of his work for the Brazilian community in the areas of philosphy, mathematics, logic, and theory of computation. With his passing, the academic community lost a great researcher and an inspiring professor, but, above all, it lost a reference both in the personal and professional fields, which leaves a gap that the Brazilian logical community may never be able to fill.

Acknowledgements

We would like to thank all the authors who joined us in this project and Jane Spurr for all the help during the editorial process. We would like also to thank all the agencies that supported (partially or not) the works in here presented, namely CONICET (Argentina), FWF (Austria), CNPq (Brazil), CAPES (Brazil), ANPCyT (Argentina).

Memory Propositional Dynamic Logic[‡]

Mario Benevides* Bruno Lopes*

* Instituto de Computação
Universidade Federal Fluminense
mario@ic.uff.br
bruno@ic.uff.br

Abstract

This work generalizes our previous works [Benevides et al., 2016, 2018; Lopes et al., 2014] on Dynamic Logics for Petri nets. We have extended the language, equipping the modalities with a memory (sequence of names) and in the semantics, states now have a memory (also a sequence of names). We add to PDL two new basic programs, $+a$, that adds name a to the current sequence s yielding as, and $-a$ that removes one occurrence of a in s; if a does not occur in s, it fails. We provide a simple process calculus for programs. An axiomatization is proposed and soundness and completeness are proved w.r.t. the class of finite models yielding decidability. In this work we do not consider iteration, it is left as future work.

This paper is dedicated to Prof. Paulo Veloso, one of the most brilliant and generous people we had the honor to ever met.

1 Introduction

Propositional Dynamic Logic (PDL) plays an important role in formal specification and reasoning about programs and actions. PDL is a multi-modal logic with one modality for each program π $\langle \pi \rangle$. It has been used in formal specification to reasoning about properties of programs and their behaviour. Correctness, termination, fairness, liveness and equivalence of programs are among the properties usually desired. A Kripke semantics can be provided, with a frame $\mathcal{F} = \langle W, R_\pi \rangle$, where W is a set of possible program states and for each program π, R_π is a binary relation on W such

---

[‡]This work was supported by the Brazilian research agencies CNPq, CAPES and FAPERJ. We also thank Prof. Edward Hermann Haeusler (DI/PUC-Rio) for his valuable insights.

that $(s,t) \in R_\pi$ if and only if there is a computation of π starting in s and terminating in t.

There are a lot of variations of PDL for different approaches [Balbiani & Vakarelov, 2003]. Propositional Algorithmic Logic [Mirkowska, 1981] that analizes properties of programs connectives, the interpretation of Deontic Logic as a variant of Dynamic Logic [Meyer, 1987], applications in linguistics [Kracht, 1995], Multi-Dimensional Dynamic Logic [Petkov, 1987] that allows multi-agent [Khosravifar, 2013] representation, Dynamic Arrow Logic [van Benthem, 1994] to deal with transitions in programs, Data Analysis Logic [del Cerro & Orlowska, 1985], Boolean Modal Logic [Gargov & Passy, 1990], logics for reasoning about knowledge [Fagin et al., 2004], logics for knowledge representation [Lenzerini, 1994] and Dynamic Description Logic [Wolter & Zakharyaschev, 2000].

Our paper falls in the broad category of works that attempt to generalize PDL and build dynamic logics that deal with classes of non-regular programs. As examples of other works in this area, we can mention the works of Harel & Kaminsky [1999]; Harel & Raz [1993] and Löding et al. [2007], that develop decidable dynamic logics for fragments of the class of context-free programs and Abrahamson [1980]; Benevides & Schechter [2008]; Goldblatt [1992]; Peleg [1897], and Peleg [1987], that develop dynamic logics for classes of programs with some sort of concurrency.

This work is inspired in a previous work by the same authors [Benevides et al., 2016; Lopes et al., 2014], which uses Marked Petri net programs as PDL programs. The main idea is to encode Petri nets as PDL programs. So if π is a Petri net program with markup s, then the formula $\langle s, \pi \rangle \varphi$ means that after running this program with the initial markup s in a state u, φ will eventually be true in a state w with a new (updated) markup s' (we have also the dual $[s, \pi]\varphi \leftrightarrow \neg \langle s, \pi \rangle \neg \varphi$). In this work we generalize this idea for PDL programs that have a data structure s which is a sequence of names. We have two special basic programs $+a$, that adds name a to the current sequence s yielding as, and $-a$ that removes one occurrence of a in s, if a does not occur in s it fails. States are equipped with a memory M which is a sequence of names. As an illustration, suppose we want to evaluate in a model $\mathcal{M}$, in a state u, with memory $M(u) = cd$, a formula $\langle cd, (+a \cup -b); -c \rangle \top$ in the following model.

$$u, M(u) = cd \xrightarrow{+a} v, M(v) = acd \xrightarrow{-c} w, M(v) = ad$$
$$\searrow_{-d}$$
$$t, M(v) = c \xrightarrow{-c} t', M(v) = \epsilon$$

The formula $\langle cd, (+a \cup -b); -c\rangle\top$ evaluates to true in the model $\mathcal{M}$, in a a state u, with memory $M(u) = cd$, $\mathcal{M}, u \models \langle cd, (+a \cup -b); -c\rangle\top$.

There are some approaches which extends Dynamic Logic or Modal Logics with some kind of data structure. In Memory Logics [Areces et al., 2008, 2011a,b], models are equipped with a memory M and they enrich the language with operators that can *memorise* and then *retrieve* the current state. Modal Separation Logics [Demri & Deters, 2015] are dynamic logics, inspired in Separation Logic [Reynolds, 2002], which aims to reasoning about data manipulation by programs. We have included the memory in the syntax of the formulae, leading to a simpler modelling and reasoning.

2 Background

2.1 Propositional Dynamic Logic

In this section, we present the syntax and semantics of the most used dynamic logic called PDL for regular programs. We deliberately omit the test and the iteration operator once we do not use it in this work.

Definition 1. *The PDL language consists of a set Φ of countably many proposition symbols, a set Π of countably many basic programs, the boolean connectives $\neg$ and $\wedge$, the program constructors ; (sequential composition), and $\cup$ (non-deterministic choice). The formulas are defined as follows:*

$$\varphi ::= p \mid \top \mid \neg\varphi \mid \varphi_1 \wedge \varphi_2 \mid \langle\pi\rangle\varphi, \text{ with } \pi ::= a \mid \pi_1; \pi_2 \mid \pi_1 \cup \pi_2,$$

where $p \in \Phi$ and $a \in \Pi$.

In all the logics that appear in this paper, we use the standard abbreviations $\bot \equiv \neg\top$, $\varphi \vee \phi \equiv \neg(\neg\varphi \wedge \neg\phi)$, $\varphi \to \phi \equiv \neg(\varphi \wedge \neg\phi)$ and $[\pi]\varphi \equiv \neg\langle\pi\rangle\neg\varphi$.

Each program π corresponds to a modality $\langle\pi\rangle$, where a formula $\langle\pi\rangle\alpha$ means that after the running of π, α eventually is true, considering that π halts. There is also the possibility of using $[\pi]\alpha$ (as an abbreviation for $\neg\langle\pi\rangle\neg\alpha$) indicating that the property denoted by α holds after every possible run of π.

The semantics of PDL is normally given using a transition diagram, which consists of a set of states and binary relations (one for each program) indicating the possible execution of each program at each state. In PDL literature a transition diagram is called a frame.

Definition 2. *A frame for PDL is a tuple $\mathcal{F} = \langle W, R_\pi\rangle$ where*

- W is a non-empty set of states;
- R_a is a binary relation over W, for each basic program $a \in \Pi$;
- We can inductively define a binary relation R_π, for each non-basic program π, as follows
  - $R_{\pi_1;\pi_2} = R_{\pi_1} \circ R_{\pi_2}$,
  - $R_{\pi_1 \cup \pi_2} = R_{\pi_1} \cup R_{\pi_2}$.

Definition 3. *A model for PDL is a pair $\mathcal{M} = \langle \mathcal{F}, \mathbf{V} \rangle$, where $\mathcal{F}$ is a PDL frame and $\mathbf{V}$ is a valuation function $\mathbf{V} : \Phi \to 2^W$.*

The semantical notion of satisfaction for PDL is defined as follows:

Definition 4. *Let $\mathcal{M} = \langle \mathcal{F}, \mathbf{V} \rangle$ be a model. The notion of satisfaction of a formula φ in a model $\mathcal{M}$ at a state w, notation $\mathcal{M}, w \Vdash \varphi$, can be inductively defined as follows:*

- $\mathcal{M}, w \Vdash p$ iff $w \in \mathbf{V}(p)$;
- $\mathcal{M}, w \Vdash \top$ always;
- $\mathcal{M}, w \Vdash \neg\varphi$ iff $\mathcal{M}, w \nVdash \varphi$;
- $\mathcal{M}, w \Vdash \varphi_1 \wedge \varphi_2$ iff $\mathcal{M}, w \Vdash \varphi_1$ and $\mathcal{M}, w \Vdash \varphi_2$;
- $\mathcal{M}, w \Vdash \langle \pi \rangle \varphi$ iff there is $w' \in W$ such that $wR_\pi w'$ and $\mathcal{M}, w' \Vdash \varphi$.

For more details on PDL see [Harel et al., 2000].

3 Small Memory Process Calculus

In this section, we propose a very small process (program) calculus for PDL memory programs.

Let $\mathcal{N} = \{a, b, c, \ldots\}$ be a set of names or running actions, denoted by $\alpha, \beta, \ldots$. Also, let $S = \{\epsilon, s_1, s_2, \ldots\}$ be a sequence of names, where ϵ is the empty sequence. We use the notation $a \in s$ to denote that name a occurs in s. Let $\#(s, a)$ be the number of occurrences of name a in s. We say that sequence r is a sub-seguence of s, $r \preceq s$, if for any name a, if $a \in r$ implies $a \in s$ and $\#(r, a) \leq \#(s, a)$.

The process language is defined as follows

$$\pi ::= -\alpha \mid +\alpha \mid \pi_1; \pi_2 \mid \pi_1 \cup \pi_2 \mid \pi^*,$$

A Question is More Illuminating than an Answer

$s_1 a s_2, -a; \pi \xrightarrow{-a} s_1 s_2, \pi$	$s_1 s_2, +a; \pi \xrightarrow{+a} s_1 a s_2, \pi$	$\pi \xrightarrow{\alpha} \pi'$
		$\pi; \tau \xrightarrow{\alpha} \pi'; \tau$
$\pi \xrightarrow{\alpha} \pi'$	$\tau \xrightarrow{\beta} \tau'$	$\pi \xrightarrow{\alpha} \pi'$
$\pi + \tau \xrightarrow{\alpha} \pi'$	$\pi + \tau \xrightarrow{\beta} \tau'$	$\pi^* \xrightarrow{\alpha} \pi'; \pi^*$
$s_1 a s_2, -a \xrightarrow{-a} s_1 s_2, \sqrt{}$	$s_1 s_2, +a \xrightarrow{+a} s_1 a s_2, \sqrt{}$	

Table 1: Transition Relation

where α ranges over $\mathcal{N}$

The *process π with memory s* is the expression s, π.

The *nondeterministic choice* operator (+) denotes that the process will make a nondeterministic choice to behave as either π_1 or π_2. The *sequential composition* operator (;) denotes that the process will first behave as π_1, and when it finishes, it behaves as π_2. The processes $-\alpha$ deletes the name α of the memory and process $+\alpha$ adds the name α into the memory. Finally, The *iteration* operator (*) denotes the process that can successfully terminate or behaves as π a finite number of time and then finishes.

In order to avoid too many parenthesis we assume the following priority for the programs operators: $+ < . <; < * < - < +$.

We write $s, \pi \xrightarrow{\alpha} s', \pi'$ to express that the process π with memory s can perform the action α and after that behave as π' with memory s'. We use the symbol $\sqrt{}$ to denote successful termination. We write $\pi \xrightarrow{\alpha} \sqrt{}$ to express that the process π successfully finishes after performing the action α. When a process finishes inside a parallel composition or sequential composition we write π instead of $\pi | \sqrt{}$, $\sqrt{} | \pi$, $\sqrt{}; \pi$. We also write $\sqrt{}$ instead of $\sqrt{} | \sqrt{}$. In table 1, we present the semantics for the operators based on this notation. It is important to notice that the semantics of processes are given by Labeled transition Systems and these rules can be used to build a LTS, for a given process, where the nodes are processes and the arrows a labeled with actions.. In this table, π, τ are process specifications, while π' and τ' are process specifications or $\sqrt{}$.

Next, the definition of bisimulation is presented. It is an equivalence relation between processes which have mutually similar behavior. The intuition is that two bisimilar processes cannot be distinguished by an external observer.

Definition 5. *Let $\mathcal{P}$ be the set of all processes. A set $Z \subseteq \mathcal{P} \times \mathcal{P}$ is a strong bisimulation if $(\pi, \tau) \in Z$ implies the following, for all α:*

- *If $\pi \xrightarrow{\alpha} \pi'$, then there is $\tau' \in \mathcal{P}$ such that $\tau \xrightarrow{\alpha} \tau'$ and $(\pi', \tau') \in Z$;*

- *If $\tau \xrightarrow{\alpha} \tau'$, then there is $\pi' \in \mathcal{P}$ such that $\pi \xrightarrow{\alpha} \pi'$ and $(\pi', \tau') \in Z$;*

- $\pi \stackrel{END}{\to} \sqrt{}$ *if and only if* $\tau \stackrel{END}{\to} \sqrt{}$.

Definition 6. *Two processes π and τ are strongly bisimilar (or simply bisimilar), denoted by $\pi \sim \tau$, if there is a strong bisimulation Z such that $(\pi, \tau) \in Z$.*

We are assuming, from now on that all processes, in our process calculus, at any state in their execution, can only perform a finite set of actions, i.e., they are *image finite*.

4 Memory Propositional Dynamic Logic without Iteration (M-PDL)

This section presents a the syntax and semantics of M-PDL.

4.1 Language and Semantics

The language of M-PDL* consists of

Propositional symbols: p, q,...

Names: $\mathcal{A} = \{a, b, c, d, \ldots\}$

PDL operator: $\cup$ (non-deterministic choice), and ; (sequential composition)

Sequence of names: $S = \{\epsilon, s_1, s_2, \ldots\}$, where ϵ is the empty sequence. We use the notation $a \in s$ to denote that name a occurs in s. Let $\#(s, a)$ be the number of occurrences of name a in s. We say that sequence r is a sub-seguence of s, $r \preceq s$, if for any name a, if $a \in r$ implies $a \in s$ and $\#(r, a) \leq \#(s, a)$.

Definition 7. *M-PDL Programs*

Basic programs: $-\alpha$ *and* $+\alpha$.

Programs:
$$\pi ::= -\alpha \mid +\alpha \mid \pi_1; \pi_2 \mid \pi_1 \cup \pi_2,$$
where $\alpha \in \mathcal{A}$.

Definition 8. *A formula is defined as*
$$\varphi ::= p \mid \top \mid \neg\varphi \mid \varphi \wedge \varphi \mid \langle s, \pi \rangle \varphi.$$

We use the standard abbreviations $\bot \equiv \neg \top$, $\varphi \vee \phi \equiv \neg(\neg\varphi \wedge \neg\phi)$, $\varphi \to \phi \equiv \neg(\varphi \wedge \neg\phi)$ and $[s, \pi]\varphi \equiv \neg\langle s, \pi\rangle\neg\varphi$.

The definition below introduces the *firing* relation. It defines how the sequence is updated after a basic action is performed. .

Definition 9. *We define the update relation $f : S \times \Pi \to S$, as follows*

- $f(s, +a) = sa$

- $f(s, -a) = \begin{cases} s_1 s_2, & if \quad s = s_1 a s_2 \\ \epsilon, & if \quad a \notin s \end{cases}$

We can extend the update relation for basic programs to complex programs as follows.

Definition 10. *Let f be an update relation as defined in definition 9. We inductively define the relation $F : S \times \Pi \mapsto 2^S$ as follows:*

- $\pi = +a, -a$: $F(s, \pi) = \{f(s, \pi)\}$;

- $\pi = \pi_1 \cup \pi_2$: $F(s, \pi) = F(s, \pi_1) \cup F(s, \pi_2)$;

- $\pi = \pi_1; \pi_2$: : $F(s, \pi) = \{r \in S \mid \exists t \in F(s, \pi_1) \ \& \ r \in F(t, \pi_2)\}$

The definitions that follow of frame, model and satisfaction are similar to the one presented in Section 2.1 for PDL. Now, we have to adapt them to deal with the update performed by programs.

Definition 11. *A frame for M-PDL is a 3-tuple $\langle W, R_\pi, M\rangle$, where*

- *W is a non-empty set of states;*

- *$M : W \to S$;*

- *R_α is a binary relation over W, for each basic program α, satisfying the following condition. Let $s = M(w)$*

    - $R_{+a} = \{(w, v) \mid f(M(w), +a) \preceq M(v)\}$
    - $R_{-a} = \{(w, v) \mid M(v) \preceq f(M(w), -a)\}$

- *We can inductively define a binary relation R_π, for each non-basic program π, as follows*

    - $R_\pi = \{(u, v) \mid F(M(u), \pi) \preceq M(v)\}$
    - $R_{\pi_1; \pi_2} = R_{\pi_1} \circ R_{\pi_2}$,

- $R_{\pi_1 \cup \pi_2} = R_{\pi_1} \cup R_{\pi_2}$.

Definition 12. *A model for M-PDL is a pair $\mathcal{M} = \langle \mathcal{F}, \mathbf{V} \rangle$, where $\mathcal{F}$ is a M-PDL frame and $\mathbf{V}$ is a valuation function $\mathbf{V}: \Phi \to 2^W$.*

The semantical notion of satisfaction for M-PDL is defined as follows.

Definition 13. *Let $\mathcal{M} = \langle \mathcal{F}, \mathbf{V} \rangle$ be a model. The notion of satisfaction of a formula φ in a model $\mathcal{M}$ at a state w, notation $\mathcal{M}, w \Vdash \varphi$, can be inductively defined as follows:*

- $\mathcal{M}, w \Vdash p$ *iff* $w \in \mathbf{V}(p)$;
- $\mathcal{M}, w \Vdash \top$ *always;*
- $\mathcal{M}, w \Vdash \neg \varphi$ *iff* $\mathcal{M}, w \nVdash \varphi$;
- $\mathcal{M}, w \Vdash \varphi_1 \wedge \varphi_2$ *iff* $\mathcal{M}, w \Vdash \varphi_1$ *and* $\mathcal{M}, w \Vdash \varphi_2$;
- $\mathcal{M}, w \Vdash \langle s, \pi \rangle \varphi$ *if there exists $v \in W$, $w R_\pi v$, $s \preceq M(w)$ and $\mathcal{M}, v \Vdash \varphi$.*

If $\mathcal{M}, v \Vdash A$ for every state v, we say that A is *valid in the model* $\mathcal{M}$, notation $\mathcal{M} \Vdash A$. And if A is M-PDL valid in all $\mathcal{M}$ we say that A is *valid*, notation $\Vdash A$.

4.2 Axiomatic System

(PL) Enough propositional logic tautologies

(K) $[s, \pi](p \to q) \to ([s, \pi]p \to [s, \pi]q)$

(ND) $\langle s, \pi_1 \cup \pi_2 \rangle \varphi \leftrightarrow \langle s, \pi_1 \rangle \varphi \vee \langle s, \pi_1 \rangle \varphi$

(SC) $\langle s, \pi_1; \pi_2 \rangle \varphi \leftrightarrow \langle s, \pi_1 \rangle \langle r, \pi_2 \rangle \varphi$, such that $r = F(s, \pi_1)$

(SE) $[s, -a] \bot$, if $a \notin s$

(AD) $\langle s_1 s_2, +a; \pi \rangle \varphi \leftrightarrow \langle s_1 a s_2, \pi \rangle \varphi$

(DL) $\langle s_1 a s_2, -a; \pi \rangle \varphi \leftrightarrow \langle s_1 s_2, \pi \rangle \varphi$,

(Sub) If $\Vdash \varphi$, then $\Vdash \varphi^\sigma$, where σ uniformly substitutes proposition symbols by arbitrary formulas.

(MP) If $\Vdash \varphi$ and $\Vdash \varphi \to \psi$, then $\Vdash \psi$.

(Gen) If $\Vdash \varphi$, then $\Vdash [s, \eta] \varphi$.

5 Soundness

The axioms **(PL)**, **(K)**, and **(ND)** and the rules **(Sub)**, **(MP)** and **(Gen)** are standard in the modal logic literature.

Lemma 1. *Validity of M-PDL axioms*

Proof.

1. **SC:** $\Vdash \langle s, \pi_1; \pi_2 \rangle \varphi \leftrightarrow \langle s, \pi_1 \rangle \langle r, \pi_2 \rangle \varphi$, where for some $r \in S$

   - $\Rightarrow$ Suppose $\mathcal{M}, u \Vdash \langle s, \pi_1; \pi_2 \rangle \varphi$ (1). Then there exists v, $uR_{\pi_1;\pi_2}v$, $s \preceq M(u)$ (1) and $\mathcal{M}, v \Vdash \varphi$ (2).
     But, $uR_{\pi_1;\pi_2}v$, then there exists w, $uR_{\pi_1}w$ (3), and $wR_{\pi_2}v$.
     By Definition 11, $uR_{\pi_1}w$ implies that $F(M(u), \pi_1) \preceq M(w)$.
     Taking $r = F(M(u), \pi_1)$, then we have that $wR_{\pi_2}v$ and $r \preceq M(w)$ and $\mathcal{M}, v \Vdash \varphi$. Then,
     $\mathcal{M}, w \Vdash \langle r, \pi_2 \rangle \varphi$. Using (1) and (3) we have $\mathcal{M}, u \Vdash \langle s, \pi_1; \pi_2 \rangle \varphi$.
   - (a) Suppose $\mathcal{M}, u \Vdash \langle s, \pi_1 \rangle \langle r, \pi_2 \rangle \varphi$. Then there exists a w, $uR_{\pi_1}w$ (1), $s \preceq M(u)$ (*) and $\mathcal{M}, w \Vdash \langle r, \pi_2 \rangle \varphi$.
     But, $\mathcal{M}, w \Vdash \langle r, \pi_2 \rangle \varphi$. Then there exists a v, $wR_{\pi_2}v$ (2), $r \preceq M(w)$ and $\mathcal{M}, v \Vdash \varphi$ (**).
     From (1) and (2), we have $uR_{\pi_1;\pi_2}v$, and using this and (*) and (**) we obtain
     $\mathcal{M}, u \Vdash \langle s, \pi_1; \pi_2 \rangle \varphi$.

   So, SC is valid.

2. $\Vdash$ **SE**
   Suppose $f(s, -a) = \epsilon$ and $\nVdash [s, -a]\bot$, so there exists a model $\mathcal{M}$ and a state w such that $\mathcal{M}, w \nVdash [s, -a]\bot$ iff $\mathcal{M}, w \Vdash \langle s, -a \rangle \top$ (1)

   From (1), there is a v such that $wR_{-a}v$, $s \preceq M(w)$ and $\mathcal{M}, v \Vdash \top$

   But $f(s, -a) = \epsilon$ and thus by Definition 11 $(w, v) \notin R_{-a}$, which is a contradiction.

3. $\vdash$ **AD**
   Suppose that there is a world w from a model $\mathcal{M} = \langle W, R_\pi, \mathbf{V}, M \rangle$ where AD is false. For AD to be false in w, there are two cases:

(a) Suppose $\mathcal{M},w \Vdash \langle s, +a; \pi \rangle \varphi$ (1)
and $\mathcal{M},w \nVdash \langle s', \pi \rangle \varphi$ (2)
Definitions 13, and 11 there exists $v \in W$ such that $wR_{+a}v$ and $\mathcal{M},v \Vdash \langle as, \pi \rangle \varphi$ (3)
Thus, by Definitions 11, and 9 (3) constradicts (2).

(b) Suppose $\mathcal{M},w \nVdash \langle s, +a; \pi \rangle \varphi$ (1)
and $\mathcal{M},w \Vdash \langle s', \pi \rangle \varphi$ (2)
By Definition 9 in (2), $f(s, +a) = s'$ thus by Definitions 11 and 13 we have a contradiction with (1).

4. $\vdash$ **DL**
Suppose that there is a world w from a model $\mathcal{M} = \langle W, R_\pi, \mathbf{V}, M \rangle$ where DL is false. For DL to be false in w, there are two cases:

(a) Suppose $\mathcal{M},w \Vdash \langle s_1 a s_2, -a; \pi \rangle \varphi$ (1)
and $\mathcal{M},w \nVdash \langle s_1 s_2, \pi \rangle \varphi$ (2)
By Definition 9 in (1) we have that $f(s_1 a s_2, -a) = s_1 s_2$. Thus using Definition 13 it contradicts (2).

(b) Suppose $\mathcal{M},w \nVdash \langle s_1 a s_2, -a; \pi \rangle \varphi$ (1)
and $\mathcal{M},w \Vdash \langle s_1 s_2, \pi \rangle \varphi$ (2)
By Definition 9 in (2), $f(s_1 s_2, -a) = s_1 s_2$ thus by Definitions 11 and 13 we have a contradiction with (1).

Hence, the axiomatic system is consistent.

$\square$

6 Completeness

Definition 14. *(Fischer and Ladner Closure): Let Γ be a set of formulas. The **closure** of Γ, notation $C_{FL}(\Gamma)$, is the smallest set of formulas satisfying the following conditions:*

1. *$C_{FL}(\Gamma)$ is closed under subformulas,*

2. *if $\langle s, \pi_1; \pi_2 \rangle \varphi \in C_{FL}(\Gamma)$, then $\langle s, \pi_1 \rangle \langle r, \pi_2 \rangle \varphi \in C_{FL}(\Gamma)$,*

3. *if $\langle s, \pi_1 \cup \pi_2 \rangle \varphi \in C_{FL}(\Gamma)$, then $\langle s, \pi_1 \rangle \varphi \in C_{FL}(\Gamma)$ and $\langle s, \pi_2 \rangle \varphi \in C_{FL}(\Gamma)$,*

4. *if $\langle s_1 s_2, +a; \pi \rangle \varphi \in C_{FL}(\Gamma)$, then $\langle s_1 a s_2, \pi \rangle \varphi \in C_{FL}(\Gamma)$,*

5. *if $\langle s_1 a s_2, -a; \pi \rangle \varphi \in C_{FL}(\Gamma)$, then $\langle s_1 s_2, \pi \rangle \varphi \in C_{FL}(\Gamma)$,*

6. if $\varphi \in C_{FL}(\Gamma)$ and φ is not of the form $\neg \psi$, then $\neg \varphi \in C_{FL}(\Gamma)$.

We prove that if Γ is a finite set of formulas, then the closure $C_{FL}(\Gamma)$ of Γ is also finite. We assume Γ to be finite from now on.

Lemma 2. *If Γ is a finite set of formulas, then $C_{FL}(\Gamma)$ is also finite.*

Proof. : This proof is standard in PDL literature [Blackburn et al., 2001]. It should be noticed that the restriction of not using an iteration operator guarantees, in conditions 4 and 5, that $+a$ and $-a$ are applied at most only once by occurence and so the closure is finite.

□

Definition 15. *Let Γ be a set of formulas. A set of formulas $\mathcal{A}$ is said to be an **atom** of Γ if it is a maximal consistent subset of $C_{FL}(\Gamma)$. The set of all atoms of Γ is denoted by $At(\Gamma)$.*

Lemma 3. *Let Γ be a set of formulas. If $\varphi \in C_{FL}(\Gamma)$ and φ is consistent then there exists an atom $\mathcal{A} \in At(\Gamma)$ such that $\varphi \in \mathcal{A}$.*

Proof. We can construct the atom $\mathcal{A}$ as follows. First, we enumerate the elements of $C_{FL}(\Gamma)$ as $\phi_1, \cdots, \phi_n$. We start the construction making $\mathcal{A}_1 = \{\varphi\}$, then for $1 < i < n$, we know that $\vdash \bigwedge \mathcal{A}_i \leftrightarrow (\bigwedge \mathcal{A}_i \wedge \phi_{i+1}) \vee (\bigwedge \mathcal{A}_i \wedge \neg \phi_{i+1})$ is a tautology and therefore either $\mathcal{A}_i \wedge \phi_{i+1}$ or $\mathcal{A}_i \wedge \neg \phi_{i+1}$ is consistent. We take $\mathcal{A}_{i+1}$ as the union of $\mathcal{A}_i$ with the consistent member of the previous disjunction. At the end, we make $\mathcal{A} = \mathcal{A}_n$.

□

Definition 16. *Let Γ be a set of formulas and $\langle s, \pi \rangle \varphi \in At(\Gamma)$. The **canonical relations** over Γ S_π^Γ on $At(\Gamma)$ are defined as follows:*
$\mathcal{A} S_\pi^\Gamma \mathcal{B}$ iff $\bigwedge \mathcal{A} \wedge \langle s, \pi \rangle \bigwedge \mathcal{B}$ is consistent.

Definition 17. *Let $\{\langle s_1, \pi_1 \rangle \varphi_1, ..., \langle s_n, \pi_n \rangle \varphi_n\}$ be the set of all diamond formulas occurring in one atom $\mathcal{A}$. We define the **canonical marking** of $\mathcal{A}$ $M(\mathcal{A})$ as follows*

1. $M(\mathcal{A}) := s_1; s_2; ...; s_n;$

2. *for all basic programs α, if $\mathcal{A} S_\alpha^\Gamma \mathcal{B}$ and $f(M(\mathcal{A}), \alpha) \not\preceq M(\mathcal{B})$, then add to $M(\mathcal{B})$ as few as possible names to make $f(M(\mathcal{A}), \alpha) \preceq M(\mathcal{B})$.*

Definition 18. *Let Γ be a set of formulas. The **canonical model** over Γ is a tuple $\mathcal{M}^\Gamma = \langle At(\Gamma), S_\pi^\Gamma, M^\Gamma, \mathbf{V}^\Gamma \rangle$, where for all propositional symbols p and for all atoms $\mathcal{A} \in At(\Gamma)$ we have*

- $M^\Gamma : At(\Gamma) \mapsto S$, called canonical marking;
- $\mathbf{V}^\Gamma(p) = \{\mathcal{A} \in At(\Gamma) \mid p \in \mathcal{A}\}$ is called canonical valuation;
- S_π^Γ are the canonical relations[1].

Lemma 4. *For all basic programs α, let $s = M(\mathcal{A})$, S_α satisfies*

1. *if $f(s, \alpha) \neq \epsilon$, if $\mathcal{A}S_\alpha\mathcal{B}$ then $f(s, \alpha) \preceq M(\mathcal{B})$*
2. *if $f(s, \alpha) = \epsilon$, then $(\mathcal{A}, \mathcal{B}) \notin S_\alpha$*

Proof. The proof of 1. is straightforward from the definition of canonical marking (definition 17). The proof of 2. follows from axiom R_ϵ. □

Lemma 5 (Existence Lemma for Canonical Models). *Let $\mathcal{A} \in At(\Gamma)$ and $\langle s, \pi \rangle \varphi \in C_{FL}$. Then,*
  $\langle s, \pi \rangle \varphi \in \mathcal{A}$ iff there exists $\mathcal{B} \in At(\Gamma)$ such that $\mathcal{A}S_\pi\mathcal{B}$, $s \preceq M(\mathcal{A})$ and $\varphi \in \mathcal{B}$.

Proof. ⇒: Suppose $\langle s, \pi \rangle \varphi \in \mathcal{A}$. By definition 17 we know that $s \preceq M(\mathcal{A})$. By definition 15, we have that $\bigwedge \mathcal{A} \wedge \langle s, \pi \rangle \varphi$ is consistent. Using the tautology $\vdash \varphi \leftrightarrow ((\varphi \wedge \phi) \vee (\varphi \wedge \neg \phi))$, we have that either $\bigwedge \mathcal{A} \wedge \langle s, \pi \rangle (\varphi \wedge \phi)$ is consistent or $\bigwedge \mathcal{A} \wedge \langle s, \pi \rangle (\varphi \wedge \neg \phi)$ is consistent. So, by the appropriate choice of ϕ, for all formulas $\phi \in C_{FL}$, we can construct an atom $\mathcal{B}$ such that $\varphi \in \mathcal{B}$ and $\bigwedge \mathcal{A} \wedge \langle s, \pi \rangle (\varphi \wedge \bigwedge \mathcal{B})$ is consistent and by definition 16 $\mathcal{A}S_\pi\mathcal{B}$.

⇐: Suppose there is $\mathcal{B}$ such that $\varphi \in \mathcal{B}$ and $\mathcal{A}S_\pi\mathcal{B}$ and $s \preceq M(\mathcal{A})$. Then $\bigwedge \mathcal{A} \wedge \langle s, \pi \rangle \bigwedge \mathcal{B}$ is consistent and also $\bigwedge \mathcal{A} \wedge \langle s, \pi \rangle \varphi$ is consistent. But $\langle s, \pi \rangle \varphi \in C_{FL}$ and by maximality $\langle s, \pi \rangle \varphi \in \mathcal{A}$. □

Lemma 6 (Truth Lemma for Canonical Models). *Let $\mathcal{M} = (W, S_\pi, M, \mathbf{V})$ be a finite canonical model constructed over a formula ϕ. For all atoms $\mathcal{A}$ and all $\varphi \in C_{FL}(\phi)$, $\mathcal{M}, \mathcal{A} \models \varphi$ iff $\varphi \in \mathcal{A}$.*

Proof. : The proof is by induction on the construction of φ.

- Atomic formulas and Boolean operators: the proof is straightforward from the definition of $\mathbf{V}$.

- Modality $\langle s, x \rangle$.

---
[1] For the sake of clarity we avoid using the Γ subscripts

⇒: Suppose $\mathcal{M}, \mathcal{A} \models \langle s, x \rangle \varphi$, then there exists $\mathcal{A}'$ such that $\mathcal{A} S_x \mathcal{A}'$, $s \preceq M(\mathcal{A})$ and $\mathcal{M}, \mathcal{A}' \models \varphi$. By the induction hypothesis we know that $\varphi \in \mathcal{A}'$, and by lemma 5 we have $\langle s, x \rangle \varphi \in \mathcal{A}$.

⇐: Suppose $\mathcal{M}, \mathcal{A} \not\models \langle s, x \rangle \varphi$, by the definition of satisfaction we have $\mathcal{M}, \mathcal{A} \models \neg \langle s, x \rangle \varphi$. Then for all $\mathcal{A}'$, $\mathcal{A} S_x \mathcal{A}'$ and $s \preceq M(\mathcal{A})$ implies $\mathcal{M}, \mathcal{A}' \not\models \varphi$. By the induction hypothesis we know that $\varphi \notin \mathcal{A}'$, and by lemma 5 we have $\langle s, x \rangle \varphi \notin \mathcal{A}$.

□

Definition 19. *Let Γ be a set of formulas. The **proper canonical model** over Γ is a tuple $\mathcal{N}^\Gamma = \langle At(\Gamma), R_\pi^\Gamma, M^\Gamma, \mathbf{V}^\Gamma \rangle$, where for all propositional symbols p and for all atoms $\mathcal{A} \in At(\Gamma)$ we have*
- *$\mathbf{V}^\Gamma(p) = \{\mathcal{A} \in At(\Gamma) \mid p \in \mathcal{A}\}$ is called canonical valuation;*
- *M^Γ is the canonical marking;*
- *$R_\alpha^\Gamma := S_\alpha^\Gamma$, for every basic program π;*
- *we inductively define a binary relation R_π is inductively defined as follows,[2].*

- $R_\pi = \{(u, v) \mid F(M(u), \pi) \preceq M(v)\}$
- $R_{\pi_1;\pi_2} = R_{\pi_1} \circ R_{\pi_2}$,
- $R_{\pi_1 \cup \pi_2} = R_{\pi_1} \cup R_{\pi_2}$,

Lemma 7. *For all programs π, $S_\pi \subseteq R_\pi$.*

Proof. By induction on the length of programs π

- For basic programs α, $S_\alpha = R_\alpha$ (Definition 19)
- $\pi = \pi_1; \pi_2$. We have that $R_\pi = (R_{\pi_1} \circ R_{\pi_2})$. By the induction hypothesis $S_{\pi_i} \subseteq R_{\pi_i}$ (0).

  Suppose $\mathcal{A} S_{\pi_1;\pi_2} \mathcal{B}$, iff $\bigwedge \mathcal{A} \wedge \langle s, \pi_1; \pi_2 \rangle \bigwedge \mathcal{B}$ is consistent, where $s = M(\mathcal{A})$.

  Using axiom (SC), $\bigwedge \mathcal{A} \wedge \langle s, \pi_1 \rangle \langle r, \pi_2 \rangle \varphi \bigwedge \mathcal{B}$ is consistent.

  We can construct a $\mathcal{C}$ such that

  $\bigwedge \mathcal{A} \wedge \langle s, \pi_1 \rangle \bigwedge \mathcal{C}$ is consistent (2) and

  $\bigwedge \mathcal{C} \wedge \langle r, \pi_2 \rangle \varphi \bigwedge \mathcal{B}$ is consistent (3), where $r = M(\mathcal{C})$. As $s_i \preceq s'$, then

  From (2) and (3) we have $\mathcal{A} S_{\pi_1} \mathcal{C}$ and $\mathcal{C} S_{\pi_2} \mathcal{B}$. By the I.H., $\mathcal{A} R_{\pi_1} \mathcal{C}$ and $\mathcal{C} R_{\pi_2} \mathcal{B}$.

  We have that $\mathcal{A}(R_{\pi_1} \circ R_{\pi_2}) \mathcal{B}$. And thus, $R_{\pi_1;\pi_2}$

---
[2] For the sake of clarity we avoid using the Γ superscripts

- $\pi = \pi_1 \cup \pi_2$. It is analogous to the previous case.

$\square$

Lemma 8 (Existence Lemma for Proper Canonical Models). *Let $\mathcal{A} \in At(\Gamma)$ and $\langle s, \pi \rangle \varphi \in C_{FL}$. Then,*
$\langle s, \pi \rangle \varphi \in \mathcal{A}$ iff there exists $\mathcal{B} \in At(\Gamma)$ such that $\mathcal{A} R_\pi \mathcal{B}$, $s \preceq M(\mathcal{A})$ and $\varphi \in \mathcal{B}$.

Proof.

$\Rightarrow$: Suppose $\langle s, \pi \rangle \varphi \in \mathcal{A}$. By the Existence Lemma for Canonical Models, lemma 5, we have then there exists $\mathcal{B} \in At(\Gamma)$ such that $\mathcal{A} S_\pi \mathcal{B}$ and $\varphi \in \mathcal{B}$. As by lemma 7, $S_\pi \subseteq R_\pi$. Thus, there exists $\mathcal{B} \in At(\Gamma)$ such that $\mathcal{A} R_\pi \mathcal{B}$ and $\varphi \in \mathcal{B}$.

$\Leftarrow$: Programs x.

Suppose there exists $\mathcal{B} \in At(\Gamma)$ such that $\mathcal{A} R_x \mathcal{B}$, $s \preceq M(\mathcal{A})$ and $\varphi \in \mathcal{B}$. The proof follows by induction on the structure of x.

- $x = \alpha$ (Basic programs): this is straightforward once $R_\alpha = S_\alpha$ and by the existence lemma 5 for canonical models $\langle s, \alpha \rangle \varphi \in \mathcal{A}$.
- $x = \pi_1; \pi_2$.
  We have $\mathcal{A} R_{\pi_1;\pi_2} \mathcal{B}$ and $\varphi \in \mathcal{B}$. As $R_{\pi_1;\pi_2} = (R_{\pi_1} \circ R_{\pi_2})$, there exists a $\mathcal{C}$ such that $\mathcal{A} R_{\pi_1} \mathcal{C}$ and $\mathcal{C} R_{\pi_2} \mathcal{B}$. By I.H., $\langle r, \pi_2 \rangle \varphi \in \mathcal{C}$ and $\langle s, \pi_1 \rangle \langle r, \pi_2 \rangle \varphi \in \mathcal{A}$. Using axiom (SC), $\langle s, \pi_1; \pi_2 \rangle \varphi \in \mathcal{A}$.
- $\pi = \pi_1 \cup \pi_2$. It is analogous to the previous case.

$\square$

Lemma 9 (Truth Lemma for Proper Canonical Models). *Let $\mathcal{M} = (W, R_\pi, \mathbf{V})$ be a finite proper canonical model constructed over a formula ϕ. For all atoms $\mathcal{A}$ and all $\varphi \in C_{FL}(\phi)$, $\mathcal{M}, \mathcal{A} \models \varphi$ iff $\varphi \in \mathcal{A}$.*

Proof. : The proof is by induction on the construction of φ.

- Atomic formulas and Boolean operators: the proof is straightforward from the definition of $\mathbf{V}$.
- Modality $\langle s, x \rangle$.

⇒: Suppose $\mathcal{M}, \mathcal{A} \models \langle s, x \rangle \varphi$, then there exists $\mathcal{A}'$ such that $\mathcal{A} R_x \mathcal{A}'$ and $\mathcal{M}, \mathcal{A}' \models \varphi$. By the induction hypothesis we know that $\varphi \in \mathcal{A}'$, and by lemma 8 we have $\langle s, x \rangle \varphi \in \mathcal{A}$.

⇐: Suppose $\mathcal{M}, \mathcal{A} \not\models \langle s, x \rangle \varphi$, by the definition of satisfaction we have $\mathcal{M}, \mathcal{A} \models \neg \langle s, x \rangle \varphi$. Then for all $\mathcal{A}'$, $\mathcal{A} R_x \mathcal{A}'$ implies $\mathcal{M}, \mathcal{A}' \not\models \varphi$. By the induction hypothesis we know that $\varphi \notin \mathcal{A}'$, and by lemma 8 we have $\langle s, x \rangle \varphi \notin \mathcal{A}$.

□

Theorem 1 (Completeness for Proper Canonical Models). *M-PDL is complete with respect to the class of Proper Canonical Models.*

Proof. For every consistent formula φ we can build a finite proper canonical model $\mathcal{M}$. By lemma 3, there exist an atom $\mathcal{A} \in At(\varphi)$ such that $\varphi \in \mathcal{A}$, and by the truth lemma 9 $\mathcal{M}, \mathcal{A} \models \varphi$. Therefore, our modal system is complete with respect to the class of finite proper canonical models. Hence, if $\Vdash \varphi$ then $\vdash \varphi$. □

So, as a consequence of the Finite Model Property and Completeness we have the following corollary.

Corollary 1 (Decidability). *The satisfiability problem for M-PDL is decidable.*

7 Usage example

As a usage example, let a coffee machine with three levels of sugar intensity: (i) none, (ii) a teaspoon, and (iii) two teaspoons. An occurrence of c in the input sequence means that a coin was inserted and the amount of occurrence of s means the amount of desired sugar. So, for a user that desires exactly a teaspoon of sugar, given an input sequence ℓ, the M-PDL formula $\langle \ell, -c; -s \rangle \top \wedge [\ell, -c; -s; -s] \bot$ formalizes this scenario.

8 Conclusions and further work

Inspired on Petri-PDL [Benevides et al., 2016; Lopes et al., 2014], a Dynamic Logic tailored to reason about Petri nets where the markup is denoted by a sequence of names, this work proposed M-PDL, an extension of Propositional Dynamic Logic with memory. Such extension increases the expressivity of PDL leading to the possibility to deal with program data, storing and retrieving data from the memory. Using such a memory it is possible to deal with "if-then-else" instructions and we believe that is a powerful tool to reasoning about data manipulation in PDL.

We present a process calculus, propose an axiomatization and a semantics and prove its soundness and completeness and also show that the finite model property holds, which together yields decidability. This version does not use iteration operator, but such inclusion is a natural extension of this work.

Further work also includes other ways to deal with memory, such as instead of a sequence of names using a sequence of boolean variables or a stack structure (which would have interesting applications to the certification and generation of parsers). We also would like to investigate the computational complexity of the validity problem and of the model checking problem. The development of deductive systems for M-PDL and their automatization are also future directions.

References

Abrahamson, K. R. (1980), Decidability and Expressiveness of Logics of Processes, PhD thesis, Department of Computer Science, University of Washington.

Areces, C., Carreiro, F., Figueira, S. & Mera, S. (2011a), Basic model theory for memory logics, in L. D. Beklemishev & R. J. G. B. de Queiroz, eds, 'Logic, Language, Information and Computation - 18th International Workshop, WoLLIC 2011, Philadelphia, PA, USA, May 18-20, 2011. Proceedings', Vol. 6642 of *Lecture Notes in Computer Science*, Springer, pp. 20–34. DOI 10.1007/978-3-642-20920-8_8. URL https://doi.org/10.1007/978-3-642-20920-8_8.

Areces, C., Figueira, D., Figueira, S. & Mera, S. (2008), Expressive power and decidability for memory logics, in W. Hodges & R. J. G. B. de Queiroz, eds, 'Logic, Language, Information and Computation, 15th International Workshop, WoLLIC 2008, Edinburgh, UK, July 1-4, 2008, Proceedings', Vol. 5110 of *Lecture Notes in Computer Science*, Springer, pp. 56–68. DOI 10.1007/978-3-540-69937-8_7. URL https://doi.org/10.1007/978-3-540-69937-8_7.

Areces, C., Figueira, D., Figueira, S. & Mera, S. (2011b), 'The expressive power of memory logics', *Rev. Symb. Log.* 4(2), 290–318. DOI 10.1017/S1755020310000389. URL https://doi.org/10.1017/S1755020310000389.

Balbiani, P. & Vakarelov, D. (2003), 'PDL with intersection of programs: a complete axiomatization', *Journal of Applied Non-Classical Logics* **13**(3-4), 231–276. DOI 10.3166/jancl.13.231-276.

Benevides, M., Lopes, B. & Haeusler, E. H. (2016), Propositional dynamic logic for petri nets with iteration, in A. Sampaio & F. Wang, eds, 'Theoretical Aspects of Computing — ICTAC 2016', Vol. 9965 of *Lecture Notes in Computer Science*, Springer International Publishing, pp. 441–456.

Benevides, M. R. F. & Schechter, L. M. (2008), A Propositional Dynamic Logic for CCS programs, in 'Proceedings of the XV Workshop on Logic, Language', Vol. 5110, pp. 83–97.

Benevides, M. R. F., Lopes, B. & Haeusler, E. H. (2018), 'Towards reasoning about petri nets: A propositional dynamic logic based approach', *Theor. Comput. Sci.* **744**, 22–36. DOI 10.1016/j.tcs.2018.01.007. URL https://doi.org/10.1016/j.tcs.2018.01.007.

Blackburn, P., de Rijke, M. & Venema, Y. (2001), *Modal Logic*, Theoretical Tracts in Computer Science, Cambridge University Press.

del Cerro, L. F. & Orlowska, E. (1985), 'DAL – a logic for data analysis', *Theoretical Computer Science* **36**, 251–264. DOI 10.1016/0304-3975(85)90046-5.

Demri, S. & Deters, M. (2015), 'Separation logics and modalities: a survey', *J. Appl. Non Class. Logics* **25**(1), 50–99. DOI 10.1080/11663081.2015.1018801. URL https://doi.org/10.1080/11663081.2015.1018801.

Fagin, R., Halpern, J. Y., Moses, Y. & Vardi, M. (2004), *Reasoning About Knowledge*, MIT Press.

Gargov, G. & Passy, S. (1990), A note on boolean modal logic, in P. Petkov, ed., 'Mathematical Logic', Springer US, pp. 299–309.

Goldblatt, R. (1992), 'Parallel action: Concurrent Dynamic Logic with independent modalities', *Studia Logica* **51**, 551–558.

Harel, D. & Kaminsky, M. (1999), Strengthened results on nonregular PDL, Technical Report MCS99-13, Faculty of Mathematics and Computer Science, Weizmann Institute of Science.

Harel, D. & Raz, D. (1993), 'Deciding properties of nonregular programs', *SIAM Journal on Computing* **22**(4), 857–874.

Harel, D., Kozen, D. & Tiuryn, J. (2000), *Dynamic Logic*, Foundations of Computing Series, MIT Press.

Khosravifar, S. (2013), 'Modeling multi agent communication activities with Petri Nets', *International Journal of Information and Education Technology* 3(3), 310–314. DOI 10.7763/IJIET.2013.V3.287.

Kracht, M. (1995), 'Synctatic codes and grammar refinement', *Journal of Logic, Language and Information* 4(1), 41–60.
DOI 10.1007/BF01048404.

Lenzerini, M. (1994), Boosting the correspondence between Description Logics and Propositional Dynamic Logics, *in* 'Proceedings of the Twelfth National Conference on Artificial Intelligence', AAAI Press, pp. 205–212.

Löding, C., Lutz, C. & Serre, O. (2007), 'Propositional Dynamic Logic with recursive programs', *Journal of Logic and Algebraic Programming* 73(1-2), 51–69.

Lopes, B., Benevides, M. & Haeusler, H. (2014), 'Propositional dynamic logic for Petri nets', *Logic Journal of the IGPL* 22, 721–736.

Meyer, J.-J. C. (1987), 'A different approach to deontic logic: Deontic logic viewed as a variant of dynamic logic', *Notre Dame Journal of Formal Logic* 29(1), 109–136. DOI 10.1305/ndjfl/1093637776.

Mirkowska, G. (1981), 'PAL – Propositional Algorithmic Logic', *Fundamenta Informaticæ* 4, 675–760. DOI 10.1007/3-540-11160-3_3.

Peleg, D. (1897), 'Concurrent Dynamic Logic', *Journal of the Association for Computing Machinery* 34(2), 450–479.

Peleg, D. (1987), 'Communication in Concurrent Dynamic Logic', *Journal of Computer and System Sciences* 35(1), 23–58.

Petkov, A. (1987), *Propositional Dynamic Logic in two and more dimensions*, Mathematical Logic and its Applications, Plenum Press.

Reynolds, J. C. (2002), Separation logic: A logic for shared mutable data structures, *in* '17th IEEE Symposium on Logic in Computer Science (LICS 2002), 22-25 July 2002, Copenhagen, Denmark, Proceedings', IEEE Computer Society, pp. 55–74. DOI 10.1109/LICS.2002.1029817.
URL https://doi.org/10.1109/LICS.2002.1029817.

van Benthem, J. (1994), *Logic and information flow*, Foundations of Computing, MIT Press.

Wolter, F. & Zakharyaschev, M. (2000), Dynamic Description Logics, *in* 'Proceedings of AiML'98', CSLI Publications, pp. 290–300.

Fixed-Parameter Tractability of Decidable Prefix-Vocabulary FO Classes[‡]

Luis Henrique Bustamante*   Ana Teresa Martins*
Francicleber Martins Ferreira*

* Departamento de Computação
Universidade Federal do Ceará,
lhbusta@lia.ufc.br   ana@dc.ufc.br   fran@lia.ufc.br

Abstract

Parameterized complexity theory is a subarea of computational complexity theory in which the run-time analysis of a computational problem considers, besides the input size, an additional term called a parameter. This idea allows us to recognize "some kind of tractability" for many previously intractable problems. Many problems from Logic have been received attention by some parameterized analysis. We study the parameterized complexity of the satisfiability problem for some prefix-vocabulary classes of first-order logic. We consider the natural parameters emerging from the definition of these classes, such as the quantifier rank and number of symbols. Taking the classification of maximal decidable classes, we observed that, when combining with finite model property, many fragments have fixed-parameter tractability for the satisfiability concerning some of these parameters. We review some previous results for pure predicate fragments and consider some classes with function symbols.

In honor of professor Paulo Veloso for his seminal work, an inspiration for the Brazilian Theoretical Computer Science community.

1 Introduction

Many questions have leveraged the importance of Mathematical Logic for the development of Computing and, in particular, the Computational Complexity Theory. Initially, the starting point was the problem postulated

---

[‡]This research was suported by the Coordenação de Aperfeiçoamento de Pessoal de Nível Superior – Brasil (CAPES) – Finance Code 001, and by the Brazilian National Council for Scientific and Technological Development (CNPq) under the grant number 424188/2016-3.

by Hilbert and Ackerman [1928], the *Entsheidungsproblem* ('the decision problem'), in terms of the problem of validity. Here, we adopt the *satisfiability* version of the problem that asks for an algorithm that answers whether a mathematical sentence in the first-order language (FO) can be satisfiable by some model. In this case, for a given first-order formula φ, the question considers the existence of a model that satisfies φ. Unfortunately, for Hilbert, the *Entsheidungsproblem* was solved negatively and independently by Turing [1936] and Church [1936b,a].

However, some fragments of FO have proved to be decidable for the satisfiability problem. Löwenheim [1915] showed that the satisfiability problem for the monadic fragment of FO is decidable. He also proved that binary predicate formulas are as hard to satisfiability as the class of all first-order formulas implying, on the other hand, an undecidability case. It means that there is an algorithm for converting arbitrary sentences in FO to sentences in this fragment that preserves the satisfiability. Such fragment corresponds to a *reduction class* for the satisfiability problem. These concepts were explored promptly after the papers of Turing and Church, resulting in many decidable and undecidable cases by a restriction on the vocabulary and the prefix of formulas in the prenex normal form. For example, $\forall^*\exists^*$ sentences in pure predicate logic (formulas in the prenex normal form with prefix $\forall^i \exists^j$ for some non-negative integers without function and equality symbols) were proved to be a reduction class, as a corollary of Skolem's normal form in [Skolem, 1920], and it was later improved to $\forall^3 \exists^*$ in [Gödel, 1933]. Then the decision problem turns out to be a *classification problem* where classes of formulas are classified into decidable or reduction classes.

To handle this classification problem, since there exist infinitely many fragments defined by the restrictions on the prefix and vocabulary, Gurevich [1966, 1969, 1976] developed a criterion by which some standard classes can be classified into decidable or undecidable ones. A *prefix-vocabulary class* is a set of first-order logic formulas in the prenex normal form restricted by a prefix set Π, i.e, a set of prefixes denoted by some string in the vocabulary of $\{\forall, \exists, \forall^*, \exists^*\}$ or, for the set of all prefixes (see Definition 1), a vocabulary τ, possibly using the equality symbol $=$. Then, Gurevich came up with a finite number of minimal undecidable fragments (see Table 1) in the sense that any standard undecidable classes contain one of these minimal classes. The definition of prefix-vocabulary classes allowed an incremental and discrete exploration of the classifiability problem for the decidability.

Considering standard prefix classes, it is possible to describe the maximal decidable cases. In [Börger et al., 2001], the authors present the results

Table 1: Minimal Undecidable Prefix-Vocabulary classes [Börger et al., 2001].

Prefix-Vocabulary Class	Reference
Pure predicate logic (without functions and equality =)	
(1) $[\forall\exists\forall,(\omega,1),(0)]$	[Kahr, 1962]
(2) $[\forall^3\exists,(\omega,1),(0)]$	[Surányi, 1959]
(3) $[\forall^*\exists,(0,1),(0)]$	[Kalmár & Surányi, 1950]
(4) $[\forall\exists\forall^*,(0,1),(0)]$	[Denton, 1963]
(5) $[\forall\exists\forall\exists^*,(0,1),(0)]$	[Gurevich, 1966]
(6) $[\forall^3\exists^*,(0,1),(0)]$	[Kalmár & Surányi, 1947]
(7) $[\forall\exists^*\forall,(0,1),(0)]$	[Kostyrko, 1964; Genenz, 1965]
(8) $[\exists^*\forall\exists\forall,(0,1),(0)]$	[Surányi, 1959]
(9) $[\exists^*\forall^3\exists,(0,1),(0)]$	[Surányi, 1959]
Classes with functions or equality	
(10) $[\forall,(0),(2)]_=$	[Gurevich, 1976]
(11) $[\forall,(0),(0,1)]_=$	[Gurevich, 1976]
(12) $[\forall^2,(0,1),(1)]$	[Gurevich, 1969]
(13) $[\forall^2,(1),(0,1)]$	[Gurevich, 1969]
(14) $[\forall^2\exists,(\omega,1),(0)]_=$	[Goldfarb, 1984]
(15) $[\exists^*\forall^2\exists,(0,1),(0)]_=$	[Goldfarb, 1984]
(16) $[\forall^2\exists^*,(0,1),(1)]_=$	[Goldfarb, 1984]

of decidability and the complexity for all *Maximal decidable fragments*, depicted in Table 2. An essential tool for some decidability results is the *finite model property* that guarantees a finite model for all satisfiable formulas within the considered class. However, the classes [all, $(\omega),(1)]_=$ [Rabin, 1969] and $[\exists^*\forall\exists^*,$ all, $(1)]_=$ [Shelah, 1977] are decidable but do not have the finite model property. These classes exhibit the axioms of infinity, which are formulas satisfiable only by infinite models.

We may also classify the decidable classes by the computational complexity of the satisfiability problem. The study of the computational complexity was initiated by Lewis [1980]; Fürer [1981] proving also lower bounds for some relational decidable classes. For most maximal classes, the satisfiability has high computational complexity, non-deterministic and deter-

Table 2: Maximal Prefix-Vocabulary classes [Börger et al., 2001].

Prefix-Vocabulary Class	Reference
(1) [∃*∀*, all]$_=$	[Bernays & Schönfinkel, 1928]
(2) [∃*∀²∃*, all]	[Gödel, 1932; Kalmár, 1933; Schütte, 1934]
(3) [all, (ω), (ω)]	[Löb, 1967; Gurevich, 1969]
(4) [∃*∀∃*, all, all]	[Gurevich, 1973]
(5) [∃*, all, all]$_=$	[Gurevich, 1976]
(6) [all, (ω), (1)]$_=$	[Rabin, 1969]
(7) [∃*∀∃*, all, (1)]$_=$	[Shelah, 1977]

ministic exponential. For example, SAT([all, (ω), (ω)]) and SAT([∃*∀*, all]$_=$) are in NTIME($2^{O(n)}$), and SAT([∃*∀²∃*, all]) is in NTIME($2^{n/\log n}$) where n is the length of the input formula (for some uniform size of symbol representation). In [Börger et al., 2001], some other subclasses are considered as "modest complexity classes". The computational complexity of these classes are restricted to P, NP, co-NP, PSPACE and the second level of the Exponential Hierarchy.

We intend to explore a new classifiability question related to the computational complexity of those intractable classes for the satisfiability using parameterized complexity theory.

Parameterized complexity theory [Downey & Fellows, 2012; Flum & Grohe, 2006] is a branch of computational complexity theory dedicated to the analysis of computational problems regarding an additional factor called *parameter*. This parameter arises from some structural aspects of the problem.

The central notion of parameterized complexity theory is the *fixed-parameter tractability*, which corresponds to a relaxed version of classical tractability where the "intractability" is restricted to some factor in terms of the parameter. A parameterized problem is fixed-parameter tractable if there exists an algorithm that runs in time $f(k) \cdot |x|^c$, where $|x|$ is the input size, c is some constant, k is the parameter, and f is some arbitrary computable function. These definitions briefly presented in this introduction will be more precisely presented in the following sections.

The area of parameterized complexity was introduced as a research field by Downey and Fellows [1992b,a], and it has been applied extensively to many logical problems [Gottlob et al., 2002; Szeider, 2004; Achilleos et al., 2012; Pfandler et al., 2015; de Haan & Szeider, 2016; Lück et al., 2017;

Meier et al., 2019].

Many applications justify the effort in the area of parameterized complexity. Many intractable problems had polynomial behavior when someone restricted to a particular kind of instance. For example, the model checking problem for FO is decidable in PSPACE in the general case. However, when we consider the problem over graphs with bounded degree limiting the size of the formula, the problem can be solved in linear time [Seese, 1996] (For a graph G with degree $d \geq 3$ and a first-order formula φ, the problem $G \models \varphi$ can be solved in $2^{2^{2^{O(k)}}} \cdot n$ time where $k = |\varphi|$ and n is the size of G). A similar result holds for Monadic Second-Order (MSO). In the general case, model checking in MSO is in PSPACE. Courcelle's theorem [Courcelle, 1990] says that it is possible to decide in linear time whether an MSO definable property holds for a given graph when restricted to the class of bounded tree-width graphs.

Observing the decidable cases of the satisfiability problem for FO, we consider a parameterized analysis of the problem based on parameters selected from the prefix-vocabulary class definition. These parameters are, for example, the quantifier rank and the size of the vocabulary. In Section 4, we address the parameterized analysis of the satisfiability of relational classes from Table 4 (1-4) and the *classes with modest complexity*. For these classes, we could establish fixed-parameter tractability with respect to some parameters (Theorems (5-6)-(8-9)). Additionally, we express a lower bound for the satisfiability problem of $[all, (\omega)]$, the class of unary first-order formulas. When parameterized by the quantifier rank only, the problem is unlikely to be fixed-parameter tractable (Proposition 7).

The strategy applied in the previous results is to define a set of fixed-parameter reductions from p-κ-SAT(X) to the propositional parameterized satisfiability p-SAT. This method will imply that the problem is fixed-parameter tractable due to the closure of the class FPT under this kind of reduction. The results depend on the classical conversion of first-order sentences into propositional sentences for a finite domain. We summarize them in Tables 6 and 7.

To close our contributions in this topic, in Section 5, we extend the analysis of parameterized complexity to the functional classes $[all, (\omega), (\omega)]$ and $[\exists^*, all, all]_=$.

The first fixed-parameter tractability results concerning some relational classes, given in Section 4, were published in [Bustamante et al., 2018], and the remaining cases, classes with function symbols, were presented at 19th Brazilian Logic Conference [Bustamante et al., 2019] and appeared in [Bustamante, 2019], but have not been published yet.

A Question is More Illuminating than an Answer

2 Preliminaries

In this section, we will define first-order logic fragments employing some syntactic restrictions over the quantifier pattern and the vocabulary. These fragments consider formulas in the prenex normal form. We will also provide some basic definitions of parameterized complexity.

We assume some basic knowledge of propositional and first-order logic [Ebbinghaus et al., 2013], and of computational complexity [Papadimitriou, 2003]. Our main reference for prefix-vocabulary classes is [Börger et al., 2001].

2.1 Prefix-Vocabulary Fragments

Recall that a *vocabulary* τ is a finite set of relation, function and constant symbols. Each symbol $\sigma \in \tau$ is associated with a natural number, its *arity*(σ). The *arity* of τ is the maximum arity of its symbols. A τ-*structure* $\mathfrak{A}$ is a tuple $(A, R_1^{\mathfrak{A}}, \ldots R_r^{\mathfrak{A}}, f_1^{\mathfrak{A}}, \ldots, f_s^{\mathfrak{A}}, c_1^{\mathfrak{A}}, \ldots c_t^{\mathfrak{A}})$ such that A is a non-empty set, called the *domain*, and each $R_i^{\mathfrak{A}}$ is a relation under $A^{\text{arity}(R_i)}$ interpreting the symbol $R_i \in \tau$, each $f_i^{\mathfrak{A}}$ is a function from $A^{\text{arity}(f_i)}$ to A interpreting the symbol $f_i \in \tau$, and each $c_i^{\mathfrak{A}}$ is an element of A interpreting the symbol c_i. We assume structures with finite domain, and, without loss of generality, we use a domain of naturals $\{1, \ldots, n\}$, denoted by $[n]$.

A τ-term is a variable x, a constant c, or an m-ary function symbol f applied to τ-terms $t_1, t_2, \ldots, t_m$, $f(t_1, t_2, \ldots, t_m)$. If R is an m-ary relation symbol, and $t_1, t_2, \ldots t_m$ are τ-terms, then $R(t_1, t_2, \ldots, t_m)$, and $t_1 = t_2$ are τ-*formulas*, which we call *atomic formulas*. If φ and ψ are τ-formulas, then $(\varphi \wedge \psi)$, $(\varphi \vee \psi)$, $\neg \varphi$ are τ-formulas. If x is a variable, and φ is a τ-formula, then $\forall x \varphi$ and $\exists x \varphi$ are τ-formulas. A *sentence* is a formula without free variables. A formula is in *prenex normal form* if it is of the form $Q_1 x_1 \ldots Q_\ell x_\ell \psi$, such that ψ is a quantifier-free formula and $Q_1 \ldots Q_\ell \in \{\exists, \forall\}^*$ is the *prefix*. We define the *quantifier rank* $\mathrm{qr}(\varphi)$ as the maximum number of nested quantifiers occurring in φ. If φ is an atomic formula, then $\mathrm{qr}(\varphi) = 0$. If $\varphi := \neg \varphi'$, then $\mathrm{qr}(\varphi) = \mathrm{qr}(\varphi')$. If $\varphi := (\psi \square \theta)$, where $\square \in \{\wedge, \vee\}$, then $\mathrm{qr}(\varphi) = \max\{\mathrm{qr}(\psi), \mathrm{qr}(\theta)\}$. If $\varphi := Qx\psi$, where $Q \in \{\exists, \forall\}$, then $\mathrm{qr}(\varphi) = \mathrm{qr}(\psi) + 1$.

We define five structural parameters for a first-order formula. For a fixed formula φ, we define τ_φ as the set of symbols occurring in the formula φ. Then we denote *the number of relation symbols* in τ_φ by #r(φ), *the number of function symbols* in τ_φ by #f(φ), *the maximum arity of a symbol* in τ_φ by ar(φ), *the number of terms* occurring in φ by $|T|$, where T is the set of terms in φ, and *the maximum size of a term* in

φ by $|\varphi_{\text{term}}|$. The last two parameters will be considered in Section 5.

Let $\mathfrak{A}$ be a τ-structure and $a_1, a_2, \ldots, a_m$ be elements of the domain. If $\varphi(x_1, x_2, \ldots x_m)$ is a τ-formula with free variables $x_1, x_2, \ldots x_m$, then we write $\mathfrak{A} \models \varphi(a_1, a_2, \ldots a_m)$ to denote that $\mathfrak{A}$ satisfies φ if $x_1, x_2, \ldots x_m$ are interpreted by $a_1, a_2, \ldots a_m$, respectively. If φ is a sentence, then we write $\mathfrak{A} \models \varphi$ to denote that $\mathfrak{A}$ satisfies φ, or that $\mathfrak{A}$ is a model of φ.

The *satisfiability problem* consists of deciding, given a formula φ, if there exists a model $\mathfrak{A}$ for φ or not.

From the previous descriptions, a prefix is a string in the alphabet $\{\exists, \forall\}$. Then a *prefix set* is a set of prefixes. Moreover, a prefix set is *closed* if it contains all subsequences of its prefixes. A prefix set is called *standard* if either it is the set of all prefixes or or a regular expression that corresponds to a string w on $\{\exists, \forall, \exists^*, \forall^*\}$. If it contains all prefixes, we denote it by *all*.

An *arity sequence* is a function p from $\mathbb{N}$ to $\mathbb{N} \cup \omega$, the union of the set of naturals first infinite ordinal. An arity sequence p is *standard* if it satisfies the following condition: $p(n) = \omega$ whenever the sum $p(n) + p(n+1) + \ldots$ is infinite. Every standard sequence can be given a succinct notation. If $p(i) = \omega$ for $i \in \mathbb{N}$, we denote the sequence by "all". For any other, we denoted by the sequence $(p(1), p(2), \ldots, p(m))$.

Definition 1 (Prefix-Vocabulary Classes [Börger et al., 2001]**).** A prefix-vocabulary fragment $[\Pi, \bar{p}, \bar{f}]$ is a set of first-order formulas in the prenex normal form, without equality, where Π is a standard prefix set, $\bar{p}$ is a relation arity sequence $(p_1, p_2, \ldots)$, and $\bar{f}$ is a function arity sequence $(f_1, f_2, \ldots)$ where $p_a, f_a \in \mathbb{N} \cup \{\omega\}$ are the number of relations and functions of arity a, respectively. Occasionally, we use **all** to denote an arbitrary sequence of arities, or an arbitrary prefix. We denote a zeros sequence $(0, 0, \ldots)$ by (0). In case $\bar{f} = (0)$, we may write $[\Pi, \bar{p}]$ instead of $[\Pi, \bar{p}, (0)]$. The prefix-vocabulary fragment $[\Pi, \bar{p}, \bar{f}]_=$ is defined in the same way, but allowing formulas with the equality symbol $=$.

For example, the class denoted by $[\exists^*, \text{all}, \text{all}]_=$ is the set of all prenex sentences with an arbitrary vocabulary and $=$ with prefix of the form

$$\exists x_1 \ldots \exists x_m.$$

Definition 2. A prefix-vocabulary class $[\Pi, p, f]$ is *standard* if Π, p and f are standard.

All classes handled in this paper are decidable standard classes. The central problem for us is the *satisfiability problem*. It consists of deciding, given a formula φ, if there exists a model $\mathfrak{A}$ for φ or not. In the general case, the first-order satisfiability SAT(FO) is undecidable [Turing, 1936; Church, 1936b,a], while for many fragments the problem is decidable. In Section 3, we are going to describe the decidable classes studied in terms of the Parameterized Complexity Theory.

2.2 Parameterized Complexity

A *parameterized problem* is a pair (Q, κ) where $Q \subseteq \Sigma^*$, for some finite alphabet Σ, is a decision problem[1] and κ is a polynomial-time computable function from Σ^* to natural numbers $\mathbb{N}$, called the *parameterization*. For an *instance* $x \in \Sigma^*$ of Q, $\kappa(x) = k$ is the *parameter* of x.

A *slice* of a parameterized problem (Q, κ) is the decision problem $(Q, \kappa)_\ell := \{x \in Q \mid \kappa(x) = \ell \in \mathbb{N}\}$. For a complete picture of parameterized complexity theory, we refer to the textbook [Flum & Grohe, 2006].

A canonical example of a parameterized problem is the parameterized satisfiability problem for propositional logic (*p*-SAT)[2] where a propositional formula φ is encoded over some finite alphabet Σ and $\kappa(\varphi)$ equals to the number of propositional variables of φ.

p-SAT	
Instance:	A propositional formula α.
Parameter:	Number of variables of α.
Problem:	Decide whether α is satisfiable.

We can also look at the satisfiability problem from the perspective of different parameterizations like **the number of clauses** or **structural parameters** from different representations of the clausal form of propositional formulas [Szeider, 2004].

A central concept of the parameterized analysis is a relaxed notion of tractability.

We say that a problem (Q, κ) is *fixed-parameter tractable* (fpt) if there is an algorithm that decides $x \in Q$ in time bounded by $f(\kappa(x)) \cdot |x|^{O(1)}$ for some computable function f. The class of all fixed-parameter tractable problems is denoted by **FPT**.

The propositional satisfiability is an NP-complete problem. However, it is fixed-parameter tractable with respect to the parameterization given by the number of variables, and considering an exhaustive search iterating overall 2^k propositional truth values, and for each iteration, evaluate it in

---

[1] As is common in complexity theory, a decision problem is described as a language over finite alphabets. We always assume Σ to be nonempty.

[2] The satisfiability problem with a restricted number of variables is also denoted by SAT(n) in the literature.

the input formula. This procedure runs in $O(2^k \cdot n)$ time where k is the number of propositional variables, and n is the size of the propositional formula.

Given the parameterized problems (Q, κ) and (Q', κ') in the alphabets Σ and Σ', respectively, an *fpt-reduction from* (Q, κ) *to* (Q', κ') is a mapping $R : \Sigma^* \to (\Sigma')^*$ such that: (i) For all $x \in \Sigma^*$ we have $(x \in Q \Leftrightarrow R(x) \in Q')$. (ii) R is computable by an *fpt-algorithm* (with respect to κ). That is, there is a computable function f such that $R(x)$ is computable in time $f(\kappa(x)) \cdot |x|^c$ for some constant c. (iii) There is a computable function $g : \mathbb{N} \to \mathbb{N}$ such that $\kappa'(R(x)) \leq g(\kappa(x))$ for all $x \in \Sigma^*$.

Let C be a parameterized class. A parameterized problem (Q, κ) is *C-hard under fpt-reductions* if every problem in C is fpt-reductible to (Q, κ). A parameterized problem (Q, κ) is *C-complete under fpt-reductions* if (Q, κ) is C-hard and $(Q, \kappa) \in C$.

It can be shown that FPT is closed under fpt-reduction. Let $(Q, \kappa), (Q', \kappa')$ be parameterized problems and (Q', κ') in FPT. If there is an fpt-reduction from (Q, κ) to (Q', κ'), then (Q, κ) is in FPT too.

Lemma 1. *[Flum & Grohe, 2006] FPT is closed under fpt-reduction.*

The class XP (for *slicewise polynomial*) is the parameterized analog of the exponential time class. A parameterized problem (Q, κ) is in XP, if there is an algorithm that decides if $x \in Q$ in at most $f(\kappa(x)) \cdot |x|^{g(\kappa(x))}$ steps, for some computable functions $f, g : \mathbb{N} \to \mathbb{N}$.

As a consequence of the previous definition, all fixed-parameter tractable problems are in XP, and then FPT $\subseteq$ XP. Conversely, it is not clear that if a parameterized problem has all of its slices computed in polynomial time, the parameterized problem is in FPT. [3]

The class para-NP is characterized as the class of all parameterized problems solvable by a non-deterministic fpt-algorithm. Then, we say that a parameterized problem (Q, κ) is in para-NP if there is a non-deterministic algorithm that decides if $x \in (Q, \kappa)$ in at most $f(\kappa(x)) \cdot |x|^{O(1)}$ steps, such that f is a computable function.

The following theorem says that, when we find a finite set of NP-hard slices of a parameterized problem (Q, κ), then (Q, κ) is para-NP-hard. In Corollary 1 of Section 4, we apply the following result to conclude that the satisfiability problem for the monadic fragment of FO parameterized by the quantifier rank is para-NP-hard.

---

[3]Consider, for example, the problem of the k independent set, determining whether there is a set of k vertices in a graph for which there are no edges between them. It is an NP-complete problem, and for the parameterized complexity, it is W[1]-complete. The independent set problem has all its slices in PTIME, but it is unlikely to be in FPT. This fact would imply a collapse between FPT and W[1], which are believed to be distinct classes.

Theorem 3. *[Flum & Grohe, 2006] Let (Q, κ) be a parameterized problem, and non-trivial i.e. $\emptyset \subsetneq Q \subsetneq \Sigma^*$. Then (Q, κ) is paraNP-hard under fpt-reductions if, and only if, a union of finitely many slices of (Q, κ) is NP-hard i.e. there are $\ell, m_1, \ldots, m_\ell$ such that*

$$(Q, \kappa)_{m_1} \cup (Q, \kappa)_{m_2} \cup \cdots \cup (Q, \kappa)_{m_\ell}$$

is NP-hard under polynomial-time reductions.

3 Decidable Classes and Complexity Issues

Let us recall that, after the negative solution for *das Entsheidungsproblem* in 1936, the decision problem for FO changed completely to a *classification problem* of the fragments which are decidable or undecidable for the satisfiability, i.e. for $X \subseteq FO$, if SAT(X) is decidable, or undecidable. Before 1936, some logicians showed a mechanical procedure to decide the satisfiability of some fragments of FO. In [Börger et al., 2001], these classes are denoted as *classical classes*, and we display in Table 3.

Table 3: Classical Prefix-Vocabulary classes [Börger et al., 2001].

Prefix-Vocabulary Class	Reference
[all, (ω)]$_{(=)}$	[Löwenheim, 1915]
[∃*∀*, all]	[Ramsey, 1987]
[∃*∀∃*, all]	[Ackermann, 1928]
[∃*∀²∃*, all]	[Gödel, 1932; Kalmár, 1933; Schütte, 1934]

The Definition 1 give us a simple way to explore prenex classes of formulas constructively. A theorem by Gurevich [1966, 1969, 1976] allows a finite classification of maximal decidable standard classes for satisfiability[4]. These classes are called *maximal decidable classes* (see Table 2). The maximal decidable classes (1-6) satisfies the finite model property, i.e., all its satisfiable formulas have a finite model. The classes [all, $(\omega), (1)$]$_=$ and [∃*∀∃*, (all), (1)]$_=$ are maximal decidable with infinity axioms, satisfiable formulas without finite models.

All results described in [Bustamante et al., 2018] and reviewed in Section 4 rely on the finite model property, and then we focus on maximal standard classes concerning this property (see Table 4).

---
[4]We point out Section 2.3 of [Börger et al., 2001].

Table 4: Prefix-Vocabulary classes maximal for the finite model property [Börger et al., 2001].

Prefix-Vocabulary Class	Reference
(1) $[\text{all}, (\omega)]_=$	[Löwenheim, 1915]
(2) $[\exists^*\forall^*, \text{all}]_=$	[Bernays & Schönfinkel, 1928; Ramsey, 1987]
(3) $[\exists^*\forall\exists^*, \text{all}]_=$	[Ackermann, 1928]
(4) $[\exists^*\forall^2\exists^*, \text{all}]$	[Gödel, 1932; Schütte, 1934]
(5) $[\text{all}, (\omega), (\omega)]$	[Löb, 1967; Gurevich, 1969]
(6) $[\exists^*\forall\exists^*, \text{all}, \text{all}]$	[Gurevich, 1973]
(7) $[\exists^*, \text{all}, \text{all}]_=$	[Gurevich, 1976]
(8) $[\forall^*, (\omega), (1)]_=$	[Ash, 1975]
(9) $[\exists^*\forall, \text{all}, (1)]_=$	[Grädel, 1989]

The strategy for decidability for most of these classes uses the *finite model property*. For a class X with the finite model property, one can think of an algorithm that iterates over the structure size. For each possible structure, $\mathfrak{A}$ with that fixed size, it verifies whether $\mathfrak{A} \models \varphi$ and, simultaneously, verifies if $\neg\varphi$ is a valid sentence. Moreover, it is possible to obtain an upper bound on the size of the structure. The following lemma specifies the size of the model for formulas in the classical classes, and we will use them in the results of Section 4.

Lemma 2. *[Gurevich, 1976; Börger et al., 2001; Schütte, 1934]*

(i) Let ψ be a satisfiable sentence in $[\text{all}, \omega]$. Then ψ has a model with at most 2^m elements where ψ has m monadic predicates.

(ii) Let ψ be a satisfiable sentence in $[\text{all}, \omega]_=$. Then ψ has a model with at most $q \cdot 2^m$ elements where ψ has quantifier rank q and m monadic predicates.

(iii) Let $\psi := \exists x_1 \ldots \exists x_p \forall y_1 \ldots \forall y_m \varphi$ be a satisfiable sentence in $[\exists^*\forall^*, \text{all}]_=$. Then ψ has a model with at most $\max(1, p)$ elements.

(iv) Let $\psi := \exists x_1 \ldots \exists x_p \forall y_1 \forall y_2 \exists z_1 \ldots \exists z_m \varphi$ be a satisfiable sentence in prenex normal form containing t predicates of maximal arity h. Then ψ has a model with cardinality at most

$$4^{10tm^2 2^h (p+1)^{h+4}} + p.$$

(v) Let $\psi := \exists x_1 \ldots \exists x_p \varphi$ be a satisfiable sentence in $[\exists^*, all, all]_=$. Then ψ has a model with at most m elements where m is the number of terms occurring in φ, the value $|T_\varphi|$.

In some cases, an upper bound on the running time of the satisfiability problem can be found. Using nondeterminism, we can guess a structure with a size less than or equal to the size provided by the finite model property, and then we evaluate the input formula on this structure. For example, the satisfiability problem for $[all, (\omega)]$ is in $\text{NTIME}(2^{O(n)})$, where n is the size of the formula and, for the class $[\exists^*\forall^2\exists^*, all]$, the same problem is in $\text{NTIME}(2^{O(n/\log n)})$. The complexity of the satisfiability for most of these classes is addressed in [Lewis, 1980; Fürer, 1981; Grädel, 1989].

In [Börger et al., 2001, Section 6.4], some subclasses are denoted *classes of modest* due to the complexity of the satisfiability problem in P, NP, Co-NP, Σ_2^p, Π_2^p, and PSPACE. Those classes within P are trivialy in FPT by any parameterization and we exclude from our analysis. We also remove from this parameterized analysis those classes called *essentially finite classes*. A essentially finite class $[\Pi, s](=)$ if Π refers to a finite set of prefixes and s to a finite vocabulary in the following meaning: (i) Π uses only $\exists$ and $\forall$ without any occurrence $\exists^*, \forall*$ (a finite set of prefixes); and (ii) $s = (s_1, \ldots s_k)$ and $s_i \neq \omega$ (a finite vocabulary).

In Table 5, we summarize the description these modest complexity classes with their respective complexity result from [Börger et al., 2001].

We give an example of how this classification works for the monadic classes in Figure 1. The class $[all, (\omega)]$ and $[all, (\omega)]_=$ called the *Löwenhein class* and the *Löwenhein class with equality*, or, alternatively, relational monadic fragments. All subclasses below these are considered as classes of modest complexity. $[all, (\omega), (\omega)]$ the *full monadic class*, and $[all, (\omega), (1)]_=$ the *Rabin's class* are maximal with respect to decidability.

Considering the formal definition of the prefix-vocabulary class, we then question the parameterized complexity behavior of the satisfiability problem. For those classes, with high computational complexity, we are going to explore how the structural parameters, like the quantifier rank and the number of symbols, impact on their high complexity.

4 Fixed-parameter Tractability of Relational Classes

As we are going to consider the satisfiability problem in terms of different parameters, we need to define the parameterized version of the satisfiability problem.

Table 5: Prefix-vocabulary classes of modest complexity [Börger et al., 2001].

Prefix-Vocabulary Class	Complexity classification
$[\exists\forall^*, \text{all}]_=$ $[\exists^*\forall^u, \text{all}]_=$ for $u \in \mathbb{N}$ $[\exists^p\forall^2\exists^*, \bar{s}]$ for $p \in \mathbb{N}$ and $\bar{s}$ finite $[\exists^p\forall\exists^*, \bar{s}]_=$ for $p \in \mathbb{N}$ and $\bar{s}$ finite $[\Pi_t, (m)]_=$ $t, m \in \mathbb{N}$, and Π_t containing at most t universal quantifiers	NP
$[\exists^*, (0)]_=$ $[\exists^*, (1)]$ $[\exists, (\omega)]$ $[\forall, (\omega)]$	NP-complete
$[\exists^p\forall^*, \bar{s}]_=$ for $p \in \mathbb{N}$ and $\bar{s}$ finite $[\Sigma_t, (m)]_=$ for $t, m \in \mathbb{N}$, and Σ_t containing at most t existential quantifiers	Co-NP
$[\exists^2\forall^*, (0)]_=$ $[\exists^2\forall^*, (1)]$ $[\forall^*\exists, (0)]_=$ $[\forall^*\exists, (1)]$ $[\forall\exists\forall^*, (0)]_=$ $[\forall\exists\forall^*, (1)]$	Co-NP-complete
$[\exists^*\forall^*, (0)]_=$ $[\exists^*\forall^*, (1)]$ $[\exists^2\forall^*, (\omega)]$	Σ_2^p-complete
$[\forall^*\exists^*, (0)]_=$ $[\forall^*\exists^*, (\omega)]$	Π_2^p-complete
$[\exists^*\forall\exists, (0,1)]$ $[\forall\exists, (\omega)]$	PSPACE-complete

A Question is More Illuminating than an Answer

Figure 1: The inclusion relation for monadic classes with modest complexity on Löwenheim's classes, and the maximal classes $[\text{all}, (\omega), (\omega)]$, $[\text{all}, (\omega), (1)]_=$

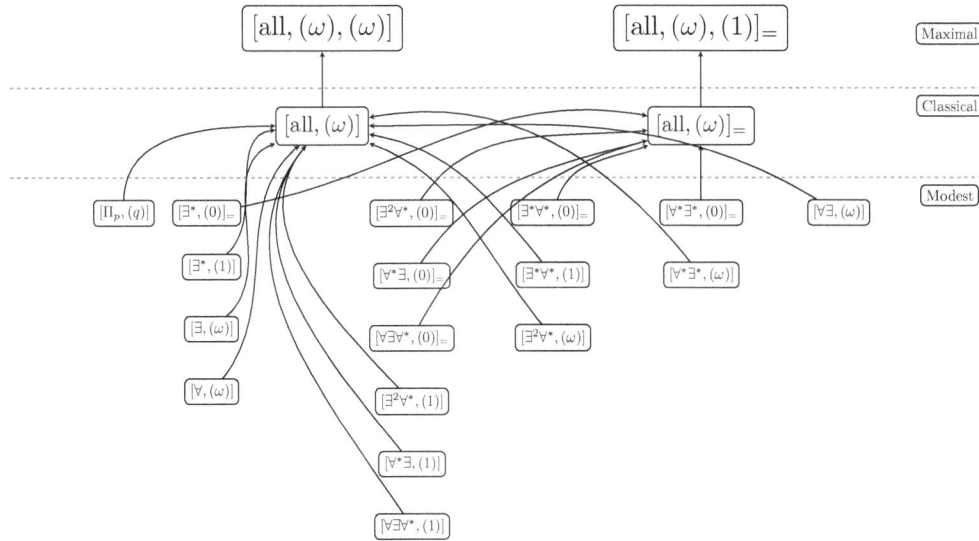

Definition 4 (The Parameterized Satisfiability Problem). Let X be a fragment of first-order logic, and κ be some parameterization. We define $p\text{-}\kappa\text{-SAT}(X)$ in the following way:

$p\text{-}\kappa\text{-SAT}(X)$	
Instance:	A first-order formula $\varphi \in X$.
Parameter:	κ, some parameterization.
Problem:	Decide whether φ is a satisfiable sentence.

If we consider a list of parameters L, we denote $p\text{-}[L]\text{-SAT}(X)$ as the parameterized satisfiability problem with parameterization $\kappa(\varphi) := \sum_{\iota \in L} \iota(\varphi)$ such that ι is a parameterization function.

The parameters considered here are:

- $\text{qr}(\varphi)$, the *quantifier rank*,
- $\#\text{r}(\varphi)$, the *number of relation symbols* in τ_φ,
- $\#\text{f}(\varphi)$, the *number of function symbols* in τ_φ,
- $\text{ar}(\varphi)$, the *arity* of τ_φ,

- $|T_\varphi|$, the *number of terms* occurring in φ where T_φ is the set of terms in φ, and

- $|\varphi_{\text{term}}|$, the maximum size of a term in φ.

A combination of these parameters leads to different parameterized problems. For example, p-qr-SAT(X) is the parameterized satisfiability for the class X when considering the quantifier rank as the parameter. Similarly, for p-#r-SAT(X) when considering the number of relation symbols as the parameter, and p-[qr, #r]-SAT(X) when considering both parameters.

Our strategy to prove that p-[L]-SAT(X) is fixed-parameter tractable, for some prefix-vocabulary class X and some list of parameters L, is to present an fpt-reduction to the propositional satisfiability problem p-SAT. As we know, p-SAT is in FPT, and FPT is closed under fpt-reductions, then we obtain that p-[L]-SAT(X) is in FPT too[1]. The essential tools for these results are the finite model property and the conversion to propositional formulas that we outline below.

In the conversion of a first-order formula into a propositional one, the finite model property provides a bound on the number of propositional variables since they are used to encode the structure. In the next subsection, we present our first result, and it will act as a prototypical argument for the other proofs.

4.1 The Löwenheim Class and the Löwenheim Class with Equality

Here we establish our first results of fixed-parameter tractability also described in [Bustamante et al., 2018]. For the monadic fragments [all, (ω)] and [all, (ω)]$_=$, we can show that the parameterized satisfiability problem is in FPT when parameterized by the quantifier rank and the number of relation symbols.

Theorem 5. *The satisfiability problem p-[qr,#r]-SAT([all, (ω)]) is in FPT.*

Proof. We give an fpt-reduction to p-SAT. Let $\varphi \in$ [all, (ω)] be a satisfiable formula with r monadic relation symbols and quantifier rank q. By (Lemma 2.(i) from Section 3), there is a model with at most 2^r elements. As we transform φ into a propositional formula φ^*, we represent each relation by 2^r propositional variables. More precisely, for each relation R_i with $1 \leq i \leq r$, and for each element $j \in [2^r]$, we use the variable p_{ij}

---
[1] Another way to see the fixed-parameter tractability for these problems can be obtained by cycling over the structures up to some domain size limits imposed by the finite model property.

to represent the truth value of $R_i(j)$. The translation works as follows. Inductivelly, we apply the conversion of quantifiers into big connectives ranging over the domain. We formally define this conversion as:

$$\varphi^* = \begin{cases} p_{ij} & \text{If } \varphi := R_i(j) \text{ is an atomic formula;} \\ \neg(\psi^*) & \text{If } \varphi := \neg\psi; \\ \psi^* \circ \vartheta^* & \text{If } \varphi := \psi \circ \vartheta \text{ for } \circ \in \{\wedge, \vee\}; \\ \bigvee_{j \in [2^r]} (\psi[x/j])^* & \text{If } \varphi := \exists x \psi; \\ \bigwedge_{j \in [2^r]} (\psi[y/j])^* & \text{If } \varphi := \forall y \psi. \end{cases}$$

It is easy to see that φ has a model if and only if φ^* is a satisfiable formula. Each inductive step constructs a formula of size $O(2^r \cdot |\psi|)$, and the whole process takes $O((2^r)^q \cdot n)$ where n is the size and q is the quantifier rank of φ. As the number of proposition variables is bounded by $r \cdot 2^r$, this leads to the desired fpt-reduction. □

The satisfiability problem for the Löwenheim class with equality [all, $(\omega)]_=$ is also in FPT when considering the quantifier rank and the number of monadic relations.

Theorem 6. *The satisfiability problem p-[qr,#r]-SAT([all, $(\omega)]_=$) is in FPT.*

Proof. Using the same idea of Theorem 5, and the finite model property from (Lemma 2.(ii) in Setion 3), we can describe an fpt-reduction from p-[qr,#r]-SAT([all, $(\omega)]_=$) to p-SAT.

Take a satisfiable formula $\varphi \in $ [all, $(\omega)]_=$ with at most $r = \#r(\varphi)$ unary symbols and quantifier rank $q = \text{qr}(\varphi)$, there is a model with at most $q \cdot 2^r$ elements by (Lemma 2.(ii) from Section 3). The number of steps on the conversion is bounded by $O((q \cdot 2^r)^q \cdot n)$ such that n is the inital size. Each atomic formula, including those with equality symbol, is converted into a propositional variable or a predicate constant, respectively, and this number is a function of the size of the domain and the size of the vocabulary, hence a function of q and r. For the equality symbol, for each occurrence of $i = j$ after the translation, we substitute for "true", if $i = j$, or "false" in the other case. Hence, this process can be considered fixed-parameter tractable. □

However, when we choose the quantifier rank as the parameter, it is unlikely to obtain an fpt-algorithm for the satisfiability of the Löwenheim's class.

Proposition 7. *Unless $P = NP$, p-qr-SAT([all, (ω)]) is not in XP.*

Proof. Assume, by contradiction, that p-qr-SAT$([\text{all},(\omega)])$ is in XP. Then there is an algorithm that solves the problem in time $f(q) \cdot n^{g(q)}$, where n is the size of the formula, q is the quantifier rank of the input formula, and f and g are computable functions. Hence, for the first slice of the problem, $[\exists,(\omega)] \cup [\forall,(\omega)]$ (see Table 5), there is an algorithm that runs in $f(1) \cdot n^{g(1)}$, polynomial time. This is a contradiction with the fact that these problems are NP-complete and with the reasonable assumption that $P \neq NP$. □

As a consequence of Theorem 3 and that $[\exists,(\omega)]$ and $[\forall,(\omega)]$ are NP-complete problems, we have that p-qr-SAT$[\text{all},(\omega)]$ is para-NP-hard under fpt-reductions.

Corollary 1. *p-qr-SAT$([\text{all},(\omega)])$ is paraNP-hard.*

4.2 The Bernays-Schönfinkel-Ramsey Class

Taking the same idea from Theorem 5 for the monadic class, if we choose a parameter that bounds the number of propositional variables, and a conversion procedure that can be conducted in FPT, we can provide an fpt-reduction for the parameterized satisfiability of prefix-vocabulary classes to p-SAT. This is the case of the Bernays-Schönfinkel-Ramsey class $[\exists^*\forall^*, \text{all}]_=$ when parameterized by the quantifier rank, number of relations, and arity of τ_φ.

Theorem 8. *p-[qr, #r, ar]-SAT$([\exists^*\forall^*, \text{all}])$ and p-[qr, #r, ar]-SAT$([\exists^*\forall^*, \text{all}]_=)$ are in FPT.*

Proof. Let $\varphi := \exists x_1 \ldots \exists x_k \forall y_1 \ldots \forall y_\ell \psi$ be a satisfiable formula in $[\exists^*\forall^*, \text{all}]$. By (Lemma 2.(iii) from Section 3), φ has a model of size at most $k \leq \text{qr}(\varphi)$. Then, we need at most $\#\text{r}(\varphi) \cdot k^{\text{ar}(\varphi)}$ propositional variables to represent the whole structure data. Applying the conversion described in Theorem 5 considering the finite domain and relations with arity greater than 1, we will produce a satisfiable propositional formula with the number of variables bounded by $g(\text{qr}(\varphi), \#\text{r}(\varphi), \text{ar}(\varphi))$, for some computable function g. To handle atomic formulas with relation symbols with arity greater than 1, we need to consider a propositional variable for each tuple of elements and each symbol. Clearly, this is a fixed-parameter reduction, and the problem is in FPT.

For Ramsey's class $[\exists^*\forall^*, \text{all}]_=$, we apply the same reduction used here. □

4.3 The Gödel-Kalmár-Shütte and Akermann Classes

For the Gödel-Kalmár-Shütte and Akermann classes, by (Lemma 2.(iv) from Section 3), the finite model property provides a model with size related to the parameters of the class definition (the number of existential quantifiers, the number of relations, the maximum arity). If we consider the quantifier rank, the number of relation symbols, and the arity of the vocabulary as the parameters, we can show it is in FPT.

Theorem 9. p-[qr, #r, ar]-SAT($[\exists^*\forall^2\exists^*, \text{all}]$) *is in FPT.*

Proof. Let $\varphi := \exists x_1 \ldots \exists x_k \forall y_1 \forall y_2 \exists z_1 \ldots \exists z_\ell \psi$ be a first-order formula with r relation symbols of maximum arity $\text{ar}(\varphi)$. By (Lemma 2.(iv) from Section 3), there is a model of size bounded by $s := 4^{10 \cdot r \cdot \ell^2 \cdot 2^{\text{ar}(\varphi)} \cdot (k+1)^{\text{ar}(\varphi)+4}} + k$ that satisfies φ. There is a bound for a satisfiable structure of φ. Considering the conversion to propositional formula, the number of propositional variables will be bounded by $r \cdot s^{\text{ar}(\varphi)}$. By structural induction on φ, apply the conversion of existential quantifiers into big disjunctions, and universal quantifiers into big conjunctions. Then it introduces one propositional variable to each possible assignment of tuples and relation symbols. This conversion is clearly a function of $s, r, \text{ar}(\varphi)$ and n, the size of φ. This lead to an fpt-reduction to p-SAT. □

The same argument can be applied to the Ackermann's class $[\exists^*\forall\exists^*, \text{all}]$.

Corollary 2. p-[qr, #r, ar]-SAT($[\exists^*\forall\exists^*, \text{all}]$) *is in FPT.*

The results presented in this section are summarized in Table 6.

Table 6: The parameterized complexity of the classical solvable cases.

Problem	Result
p-[qr, #r]-SAT($[\text{all}, (\omega)]$)	FPT (Theorem 5)
p-qr-SAT($[\text{all}, (\omega)]$)	paraNP-hard (Corollary 1)
p-[qr, #r]-SAT($[\text{all}, (\omega)]_=$)	FPT (Theorem 6)
p-[qr, #r, ar]-SAT($[\exists^*\forall^*, \text{all}]_{(=)}$)	FPT (Theorem 8)
p-[qr, #r, ar]-SAT($[\exists^*\forall\exists^*, \text{all}]$)	FPT (Corollary 2)
p-[qr, #r, ar]-SAT($[\exists^*\forall^2\exists^*, \text{all}]$)	FPT (Theorem 9)

4.4 Modest Complexity Prefix-Vocabulary Classes

In this subsection, we analyze the parameterized complexity of prefix-vocabulary classes with modest complexity. We summarize our results on Table 7. For these classes, it is possible to point out a parameter that puts the problem in FPT. First, for some of these classes, the fixed-parameter tractability follows directly from the inclusion in the relational monadic class. Then, we begin with a corollary of Theorem 5.

Corollary 3.

(i) p-#r-SAT(X) is in FPT for $X \in \{[\exists,(\omega)], [\forall,(\omega)], [\forall\exists,(\omega)]\}$.

(ii) p-[qr, #r]-SAT(X) is in FPT for $X \in \{[\exists^2\forall^*,(\omega)], [\forall^*\exists^*,(\omega)]\}$.

Proof. Let $\varphi \in X$ and $r = \#r(\varphi)$. We already know that the finite model property gives us, for a satisfiable formula, a model of size at most 2^r. Then the same conversion procedure used in Theorem 5 leads to an fpt-reduction to p-SAT. In these cases, the quantifier rank is a constant.

The claim *(ii)* follows directly from Theorem 5. □

Again, we can apply (Lemma 2.(i) from Section 3) to give an fpt-reduction for the satisfiability problem of some classes with modest complexity.

Theorem 10. *p-qr-SAT(X) is in FPT for $X \in \{[\forall^*\exists,(1)], [\forall\exists\forall^*,(1)], [\exists^*\forall^*,(1)]\}$.*

Proof. Consider the satisfiability problem for $[\forall^*\exists,(1)]$. By (Lemma 2.(i) from Section 3), a satisfiable formula $\varphi := \forall x_1 \ldots \forall x_k \exists y \psi$ with one monadic relation and quantifier rank $q = \mathrm{qr}(\varphi)$ has a model with size at most 2. Applying the same conversion of Theorem 5, the number of steps will be bounded by $2^q \cdot n$ where n is the size of φ. This will lead to an fpt-reduction to a propositional formula with two propositional variables.

The same reduction works for $[\forall\exists\forall^*,(1)]$ and $[\exists^*\forall^*,(1)]$. □

In the sequence, we can use (Lemma 2.(ii) from Section 3) to obtain a reduction from the satisfiability of some classes with modest complexity to p-SAT when we use the quantifier rank as a parameter.

Theorem 11. *p-qr-SAT(X) is in FPT for $X \in \{ [\Pi_t,(m)]_=, [\Sigma_t,(m)]_=, [\forall^*\exists,(0)]_=, [\forall\exists\forall^*,(0)]_=, [\forall^*\exists^*,(0)]_=\}$.*

Proof. Consider the satisfiability problem for the class $[\Pi_t, (m)]_=$, the class of formulas in the prenex normal form with at most t universal quantifiers, m unary relations and $=$. Let φ be a satisfiable formula in $[\Pi_t, (m)]_=$ with at most m monadic relations and quantifier rank q. Then, by (Lemma 2.*(ii)* from Section 3), φ has a model with size at most $q \cdot 2^m$. Each monadic relation can be represented by $q \cdot 2^m$ propositional variables. So, in order to transform φ into a propositional formula, $q \cdot m \cdot 2^m$ propositional variables will be necessary to represent all relations. Applying the same conversion of Theorem 6, it will lead to an fpt-reduction to p-SAT.

The same reduction holds for $[\Pi_t, (q)]_=$ with at most t existential quantifiers. For $[\forall^*\exists, (0)]_=$, $[\forall\exists\forall^*, (0)]_=$, and $[\forall^*\exists^*, (0)]_=$, the size of the structure is bounded by $qr(\varphi)$ and the reduction follows in the same way. □

Formulas with a leading block of existential quantifiers can be handled with the finite model property by (Lemma 2. *(iii)* from Section 3).

Theorem 12. p-[qr, #r, ar]-SAT(X) is in FPT for $X \in \{[\exists\forall^*, all]_=, [\exists^*\forall^u, all]\}$.

Proof. Consider the satisfiability problem for $[\exists^*\forall^u, all]$. Let φ be a satisfiable formula in $[\exists^*\forall^u, all]$ with fixed natural u. So, $\varphi := \exists x_1 \ldots \exists x_k \forall y_1 \ldots \forall y_u \psi$. By (Lemma 2.*(iii)*) from Section 3), φ has a model with size at most $k \leq$ qr(φ). Let τ_φ the vocabulary of φ with maximum arity ar(φ) and size #r(φ). Applying the same conversion presented in Theorem 5, it will return a propositional formula with at most #r$(\varphi) \cdot$ qr$(\varphi)^{ar(\varphi)}$ propositional variables, and the whole process can be carried out by an fpt-algorithm.

For the class $[\exists\forall^*, all]_=$, the finite model property will provide a model of size 1, and each universally quantified variable could be handled as a dummy variable. □

Theorem 13. p-qr-SAT(X) is in FPT for $X \in \{[\exists^*, (0)]_=, [\exists^*, (1)], [\exists^p\forall^*, \bar{s}]_=, [\exists^2\forall^*, (0)]_=, [\exists^2\forall^*, (1)], [\exists^*\forall^*, (0)]_=\}$.

Proof. Consider the satisfiability problem for $[\exists^*\forall^*, (0)]_=$. By (Lemma 2.*(iii)* from Section 3), for a satisfiable formula $\varphi := \exists x_1 \ldots \exists x_k \forall y_1 \ldots \forall y_l \psi$ there is a model with size $k \leq$ qr(φ). Applying the same translation of Theorem 5, and considering that only the equality symbol is in τ_φ, the number of propositional variables obtained in the reduction is bounded by a function of qr(φ). This will lead to an fpt-reduction to p-SAT. □

Theorem 14. p-qr-SAT(X) is in FPT for $X \in \{ [\exists^p\forall^2\exists^*, \bar{s}], [\exists^p\forall\exists^*, \bar{s}]_=, [\exists^*\forall\exists, (0,1)]\}$.

Proof. Let $\varphi := \exists x_1 \ldots \exists x_p \forall y_1 \forall y_2 \exists z_1 \ldots \exists z_k \psi$ be a satisfiable formula in $[\exists^p\forall^2\exists^*, s]$ with a fixed vocabulary with r relation symbols of arity ar(φ).

By (Lemma 2.(iv) from Section 3), there is a model of size bounded by $\ell := 4^{10r \cdot k^2 2^{\mathrm{ar}(\varphi)} (p+1)^{\mathrm{ar}(\varphi)+4}} + p$ that satisfies φ. As $p, r, \mathrm{ar}(\varphi)$ are constants, the size of this model is a function of k.

Then we can describe each relation with at most $\ell^{\mathrm{ar}(\varphi)}$ propositional variables. All relations are described by a binary string with length $r \cdot \ell^{\mathrm{ar}(\varphi)}$. By structural induction on φ, apply the conversion procedure of Theorem 5. Introduce one propositional variable to each possible assignment of a tuple to a relation symbol, which is a function of k. This conversion is a function of k and n, the size of φ. This process leads to an fpt reduction to p-SAT. □

The results for modest complexity classes are summarized in Table 7.

5 Fixed-parameter Tractability of Classes with Function Symbols

Now we consider the functional classes [all, $(\omega), (\omega)$] and [$\exists^*$, all, all]$_=$. If we look at those classes that are maximal concerning the finite model property (Table 4), it will remain the analysis of those with function symbols. We can apply the same argument of the previous sections for the classes that have an upper bound on the size of the structure.

To handle prefix-vocabulary classes with function symbols, we can replace the function symbols to reduce it to a relational class with finite model property or find a finite model based on possible ground terms constructed by the formula. We can then prove the fixed-parameter tractability of these functional classes when parameterized by the parameters that we already considered (quantifier rank, number of relations, and the arity of the vocabulary) with the number of functions the number of terms, and the maximum size of a term.

5.1 The Löb-Gurevich class [all, $(\omega), (\omega)$]

For the class [all, $(\omega), (\omega)$], we can show that the parameterized satisfiability problem is in FPT when parameterized by the quantifier rank, the number of monadic relation symbols, the number of unary function symbols, and the maximum size of the terms. We need to modify a proof of Proposition 5 from [Grädel, 1989], removing the unary function symbols. The lemma shows that every formula φ in [all, $(\omega), (\omega)$] of length n can be converted into a formula $\psi \in$ [all, $(n), (0)$] satisfiable over the same domains as φ.

Table 7: Prefix-vocabulary classes of modest complexity in which their satisfiability problem is in FPT with respect to some parameter.

Prefix-vocabulary Class	Complex. SAT	Complex. p-κ-SAT Param. κ	Result
$[\exists\forall^*, all]_=$		(qr+vs+ar)	FPT (Theo. 12)
$[\exists^*\forall^u, all]_=$ for $u \in \mathbb{N}$		(qr+vs+ar)	FPT (Theo. 12)
$[\exists^p\forall^2\exists^*, \bar{s}]$ for $p \in \mathbb{N}$ and $\bar{s}$ finite	NP	qr	FPT (Theo. 14)
$[\exists^p\forall\exists^*, \bar{s}]_=$ for $p \in \mathbb{N}$ and $\bar{s}$ finite		qr	FPT (Theo. 14)
$[\Pi_t, (m)]_= t, m \in \mathbb{N}$, Π_t containing at most t universal quantifiers		qr	FPT (Theo. 11)
$[\exists^*, (0)]_=$		qr	FPT (Theo. 13)
$[\exists^*, (1)]$	NP-complete	qr	FPT (Theo. 13)
$[\exists, (\omega)]$		vs	FPT (Cor. 3)
$[\forall, (\omega)]$		vs	FPT (Cor. 3)
$[\exists^p\forall^*, \bar{s}]$ for $p \in \mathbb{N}$ and $\bar{s}$ finite $|\tau|$	Co-NP	qr	FPT (Theo. 13)
$[\Sigma_t, (m)]_=$ for $t, m \in \mathbb{N}$, and Σ_t containing at most t existential quantifiers		qr	FPT (Theo. 11)
$[\exists^2\forall^*, (0)]_=$		qr	FPT (Theo. 13)
$[\exists^2\forall^*, (1)]$		qr	FPT (Theo. 13)
$[\forall^*\exists, (0)]_=$	Co-NP-complete	qr	FPT (Theo. 11)
$[\forall^*\exists, (1)]$		qr	FPT (Theo. 10)
$[\forall\exists\forall^*, (0)]_=$		qr	FPT (Theo. 11)
$[\forall\exists\forall^*, (1)]$		qr	FPT (Theo. 10)
$[\exists^*\forall^*, (0)]_=$		qr	FPT (Theo. 13)
$[\exists^*\forall^*, (1)]$	Σ_p^2-complete	qr	FPT (Theo. 10)
$[\exists^2\forall^*, (\omega)]$		(qr+vs)	FPT (Cor. 3)
$[\forall^*\exists, (0)]_=$	Π_p^2-complete	qr	FPT (Theo. 11)
$[\forall^*\exists, (\omega)]$		(qr+vs)	FPT (Cor. 3)
$[\exists^*\forall\exists, (0,1)]$	PSPACE-complete	qr	FPT (Theo. 14)
$[\forall\exists, (\omega)]$		vs	FPT (Cor. 3)

Lemma 3. *[Grädel, 1989] Let φ be a formula in the prenex normal form with size n, quantifier rank q, r monadic relations, f unary function symbols, and with terms of the form $f^i x_j$ such that $i < t$ for some constant $t < n$. Then, there is a formula ψ without functions satisfiable in the same domain of φ.*

Proof. Consider a first-order sentence φ as in the claim. For each $R_i, f_j \in \tau_\varphi$, we introduce a new monadic relation $Q_{\overline{ij}}$ for a term in the form $R_i f_j t$

for some arbitrary term t. Then φ is satisfiable over the same domains as

$$\varphi[R_i f_j t / Q_{\overline{ij}} t] \wedge \forall x (R_i f_j x \leftrightarrow Q_{\overline{ij}} x).$$

To each possible pair of a relation and function symbol, apply the previous process.

Occasionally, we reach $Q_{\overline{a}} f_i t$ for some index $\overline{a}$. Thus, a new monadic relation $Q_{\overline{a}i}$ should be added. Taking a nested sequence of functions with size at most t, this will lead to a maximum of $O(r \cdot s^t)$ new monadic relation symbols.

Then we arrive at $\alpha \wedge \forall x \beta$ where α is a formula without function symbols and β a conjunction of formulas of the form $R_i f_j x \leftrightarrow Q_{\overline{ij}} x$ and $Q_{\overline{a}} f_j x \leftrightarrow Q_{\overline{a}j} x$. Let $f_1, \ldots, f_s$ be function symbols in β. Then, $\forall x \beta$ is the Skolem normal form of $\forall \exists y_1 \ldots \exists y_s \beta[f_i x / y_i]$ which is a relational formula, and

$$\psi := \alpha \wedge \forall x \exists y_1 \ldots \exists y_s \beta[f_i x / y_i]$$

is satisfiable over the same domains as φ. So ψ have at most $O(r + r \cdot s^t)$ monadic relation symbols with the quantifier rank bounded by $q + s + 1$. Thus ψ is in $[\text{all}, (r + r \cdot s^t), (0)]$. □

Using the previous lemma, we can achieve the fixed-parameter tractability.

Theorem 15. *p-[qr, #r, #f, $|\varphi_{\text{term}}|$]-SAT([all, $(\omega), (\omega)$] is in FPT.*

Proof. We give an fpt-reduction to p-SAT. Let $\varphi \in [\text{all}, (\omega), (\omega)]$ be a satisfiable formula with size n, quantifier rank q, r monadic relation symbols, s function symbols, and terms with size at most t. Using the Lemma 3, we obtain a first-order formula $\psi \in [\text{all}, (r + r \cdot s^t), (0)]$. By (Lemma 2-(i) from Section 3), there is a model with at most $2^{(r+r \cdot s^t)}$ elements with quantifier rank bounded by $q + s + 1$. The next step follows similarly to the relational case. As we will transform ψ into a propositional formula ψ^*, we will represent each relation by $2^{(r+r \cdot s^t)}$ propositional variables. More precisely, for each relation R_i with $1 \leq i \leq (r + r \cdot s^t)$, and for each element $j \in [2^{(r+r \cdot s^t)}]$, we use the variable p_{ij} to represent the truth value of $R_i(j)$. The translation to propositional sentence follows in the same way as in Theorem 5. Once again, it is easy to see that φ has a model if and only if ψ^* is a satisfiable formula. Each inductive step constructs a formula of size $2^{(r+r \cdot s^t)} \cdot |\varphi|$, and the whole process takes $O((2^{(r+r \cdot s^t)})^{q+s+1} \cdot n)$ where n is the size of φ. As the number of variables is bounded by $(r + r \cdot s^t) \cdot 2^{(r+r \cdot (s)^t)}$, this leads to the desired fpt-reduction. □

5.2 The Existential class $[\exists^*, \text{all}, \text{all}]_=$

The existential fragment with equality is one of the decidable cases that are maximal concerning the finite model property, and its satisfiability is NP-complete [Börger et al., 2001, pg. 304]. The finite model property, then, is obtained through the size of the set of terms occurring in a given existential formula (Lemma 2.(v) from Section 3). In this case, we add the number of terms in the parameterization function. For all terms in the form $s = fs_1 \ldots s_k \in T_\varphi$, we also consider $s_1 \ldots s_k$ and their sub-terms in T_φ.

Theorem 16. $p\text{-}[\#r, \#f, ar, |T_\varphi|]\text{-SAT}([\exists^*, \text{all}, \text{all}]_=)$ is in FPT.

Proof. Let φ be in the form $\exists x_1 \ldots \exists x_q \psi$, and let $T_\varphi = \{t_1 \ldots, t_m\}$ be the set of terms occurring in φ with $|T_\varphi| = m$. Also consider that the number of relation and function symbols are r and ℓ, respectively, and that the arity of τ_φ is a.

For each $t \in T_\varphi$, let x_t be a fresh variable, except for the case where t is a variable, say $t = y$, where we set $x_y = y$. For each k-ary function symbol f, let Q_f be a new $k+1$-ary relation symbol. For each term $t \in T_\varphi$ of the form $t = ft_1, \ldots, t_k$, we define the formula

$$\gamma_t = Q_f(x_{t_1}, \ldots, x_{t_k}, x_t)$$

We define ψ' from ψ by replacing each atomic formula $R(t_1, \ldots, t_k)$ with $R(x_{t_1}, \ldots, x_{t_k})$. Now consider the following formula

$$\theta = \exists x_{t_1} \ldots \exists x_{t_m} \left(\psi' \wedge \bigwedge_{t \in T_\varphi} \gamma_t \right)$$

The idea behind these transformations is to eliminate the function symbols. In their places, we use relations that are interpreted as the graph of the corresponding functions. If we take any model of φ we can get a model of θ with the same domain if we interpret Q_f as the graph of f. The converse is not necessarily true, since this transformation does not guarantee that the interpretation of Q_f in a model that satisfies θ is a function.

We have to add the condition that Q_f is the graph of a function. For each k-ary function symbol f, consider the formula

$$\alpha_f = \forall x_1, \ldots, x_k \exists^{=1} y (Q_f(x_1, \ldots, x_k, y)).$$

Now let φ' be the formula

$$\varphi' = \left(\bigwedge_{f \in \tau} \alpha_f\right) \wedge \theta.$$

We then apply the conversion to propositional logic described in Theorem 5. By (Lemma 2.(v) from Section 3), any satisfiable formula in $[\exists^*, \text{all}, \text{all}]_=$ has a model of size at most m, where m is the number of terms occurring in φ. We estimate the size of $(\varphi')^*$ and the number of propositional variables produced in this conversion. We consider α_f and θ separately.

For the transformation of $\left(\bigwedge_{f \in \tau} \alpha_f\right)$ into a propositional formula, each α_f contributes with m^{k+2} propositional variables ($\exists^{=1}$ counts as two quantifiers). So, considering all conjuctions we have at most $m^{a+2} \times \ell$ propositional variables where a is the arity of the vocabulary. The size of this part is bounded by $O(m^{a+2} \times \ell)$.

For θ, the conversion will produce at most $(r + m) \times m^a$ propositional variables, and the final size is $O((|\psi|+m) \times m^m)$. The propositional variables obtained from θ includes those from $\left(\bigwedge_{f \in \tau} \alpha_f\right)$, and the final size of the $(\varphi')^*$ bounds the running time of the whole process.

At the end of this process, we will achieve an fpt-reduction to p-SAT. □

6 Concluding Remarks

We extend the fixed-parameter results developed in [Bustamante et al., 2018] to classes with symbol functions. Let us recall that, for some prefix-vocabulary classes, satisfiability was shown to be in FPT. We added different parameters from Definition 1 associated with the usage of symbol functions. Combining the choice of parameterization and finite model property, we could construct fixed-parameter reductions to the propositional satisfiability parameterized by the number of propositional variables.

We reviewed, in Section 4, that, for all relational classes (maximal and modest complexity), the satisfiability problem parameterized by the size of the vocabulary, the quantifier rank, and the maximum arity is fixed-parameter tractable. For example, we observed that the satisfiability problem of Lowenheim's class $[\text{all}, \omega]$ is within FPT when parameterized by the number of monadic relations and the quantifier rank (Theorem 5). However, when just the quantifier rank is considered, the problem is not in XP, unless P$\neq$NP (Proposition 7).

We expanded the idea of fixed-parameter tractability from relational classes to functional ones. In Section 5, we achieved two fixed-parameter results for the satisfiability of $[\text{all}, (\omega), (\omega)]$ and $[\exists^*, \text{all}, \text{all}]_=$.

Now, we outline some directions to improve the understanding of the parameterized complexity of the prefix-vocabulary classes.

1. **Reducing the parameterization.** For $[\exists^*\forall^*, \text{all}]_=$, $[\exists^*\forall\exists^*, \text{all}]$, and $[\exists^*\forall^2\exists^*, \text{all}]$, the fixed-parameter tractability is obtained when we consider $\text{qr}(\varphi) + \#\text{r}(\varphi) + \text{ar}(\varphi)$ as our parameterization function in Theorems 8 and 9. One question remains: What is the parameterized complexity of the satisfiability for these classes when some terms ($\text{qr}(\varphi)$, $\#\text{r}(\varphi)$, and $\text{ar}(\varphi)$) are not considered in the parameterization function? The same question arises in the context of functional classes developed in Section 5.

2. **Functional classes.** Functional classes were not thoroughly investigated. As future work, we indicate a further investigation of the parameterized complexity of the remaining maximal classes concerning finite model theory (Table 8) and some functional classes with modest complexity (Table 9).

Table 8: Remaining prefix-vocabulary classes maximal for the finite model property.

Prefix-Vocabulary Class	Reference
$[\exists^*\forall\exists^*, \text{all}, \text{all}]$	[Gurevich, 1973]
$[\forall^*, (\omega), (1)]_=$	[Ash, 1975]
$[\exists^*\forall, \text{all}, (1)]_=$	[Grädel, 1989]

Table 9: Modest complexity prefix-vocabulary classes with functions.

Prefix-Vocabulary Class	Complexity classification
$[\exists^*\forall^m, (0), (1)]_=$	NP-complete
$[\exists^*\forall\exists^*, (0), (1)]_=$	NP-complete
$[\exists^k\forall^*, (0), (1)]_=$	Co-NP-complete
$[\exists^*\forall^*, (0), (1)]_=$	Σ_2^p-complete

References

Achilleos, A., Lampis, M. & Mitsou, V. (2012), 'Parameterized modal satisfiability', *Algorithmica* **64**(1), 38–55.

Ackermann, W. (1928), 'Zum hilbertschen aufbau der reellen zahlen', *Mathematische Annalen* **99**(1), 118–133.

Ash, C. J. (1975), 'Sentences with finite models', *Mathematical Logic Quarterly* **21**(1), 401–404.

Bernays, P. & Schönfinkel, M. (1928), 'Zum entscheidungsproblem der mathematischen logik', *Mathematische Annalen* **99**(1), 342–372.

Börger, E., Grädel, E. & Gurevich, Y. (2001), *The classical decision problem*, Springer-Verlag Berlin Heidelberg, Berlin Heidelberg.

Bustamante, L. H. (2019), Parameterized complexity investigations on the first-order satisfiability and mathing problems, PhD thesis, Federal University of Ceará.

Bustamante, L. H., Martins, A. T. & Ferreira, F. M. (2018), The parameterized complexity of some prefix-vocabulary fragments of first-order logic, *in* L. S. Moss, R. de Queiroz & M. Martinez, eds, 'Logic, Language, Information, and Computation', Springer Berlin Heidelberg, Berlin, Heidelberg, pp. 163–178.

Bustamante, L. H., Martins, A. T. & Ferreira, F. M. (2019), 'Fixed-parameter tractability of some classes of formulas in fo with functions', 19th Brazilian Logic Conference (EBL2019).

Church, A. (1936a), 'A note on the entscheidungsproblem', *The journal of symbolic logic* **1**(1), 40–41.

Church, A. (1936b), 'An unsolvable problem of elementary number theory', *American journal of mathematics* **58**(2), 345–363.

Courcelle, B. (1990), 'The monadic second-order logic of graphs. i. recognizable sets of finite graphs', *Information and Computation* **85**(1), 12–75.

de Haan, R. & Szeider, S. (2016), Parameterized complexity results for symbolic model checking of temporal logics, *in* 'Proceedings of the Fifteenth International Conference on Principles of Knowledge Representation and Reasoning', AAAI Press, Cape Town, South Africa, pp. 453–462.

Denton, J. S. (1963), Applications of the Herbrand theorem, PhD thesis, Harvard University.

Downey, R. G. & Fellows, M. R. (1992a), Fixed-parameter intractability, in '[1992] Proceedings of the Seventh Annual Structure in Complexity Theory Conference', IEEE, Boston, pp. 36–49.

Downey, R. G. & Fellows, M. R. (1992b), 'Fixed-parameter tractability and completeness', Congressus Numerantium pp. 161–161.

Downey, R. G. & Fellows, M. R. (2012), Parameterized complexity, Springer Science & Business Media, New York.

Ebbinghaus, H.-D., Flum, J. & Thomas, W. (2013), Mathematical logic, Springer Science & Business Media, Berlin ; New York.

Flum, J. & Grohe, M. (2006), Parameterized complexity theory, Springer Science & Business Media, Berlin, Heidelberg.

Fürer, M. (1981), 'Alternation and the Ackermann case of the decision problem', LâĂŹEnseignement Math 27, 137–162.

Genenz, J. (1965), Untersuchungen zum Entscheidungsproblem im Prädikatenkalkül der ersten Stufe, PhD thesis, Logik und Grundlagenforschung, Universität Münster,.

Gödel, K. (1932), 'Zum intuitionistischen aussagenkalkül', Anzeiger der Akademie der Wissenschaften in Wien 69, 65–66.

Gödel, K. (1933), 'Zum entscheidungsproblem des logischen funktionenkalküls', Monatshefte für Mathematik 40(1), 433–443.

Goldfarb, W. D. (1984), 'The unsolvability of the Gödel class with identity', The Journal of Symbolic Logic 49(4), 1237–1252.

Gottlob, G., Scarcello, F. & Sideri, M. (2002), 'Fixed-parameter complexity in ai and nonmonotonic reasoning', Artificial Intelligence 138(1-2), 55–86.

Grädel, E. (1989), Complexity of formula classes in first order logic with functions, in 'International Conference on Fundamentals of Computation Theory', Springer, Springer Berlin Heidelberg, Berlin, Heidelberg, pp. 224–233.

Gurevich, Y. (1966), 'On the algorithmc decision of the satisfiability of predicate logic formulas', Algebra i Logika 5, 25–55.

Gurevich, Y. (1969), 'The decision problem for the logic of predicates and of operations', *Algebra and Logic* **8**(3), 160–174.

Gurevich, Y. (1973), 'Formulas with one ∀', *Selected Questions in Algebra and Logic; in memory of A. Mal cev* pp. 97–110.

Gurevich, Y. (1976), 'The decision problem for standard classes', *The Journal of Symbolic Logic* **41**(2), 460–464.

Hilbert, D. & Ackerman, W. (1928), *Theoretische Logik*, Springer, Berlin Heidelberg.

Kahr, A. (1962), Improved reductions of the entscheidungsproblem to subclasses of ∀∃∀ formulas, *in* 'Proc. Symp. on Mathematical Theory of Automata', Brooklyn Polytechnic Institute, Brooklyn, New York, pp. 57–70.

Kalmár, L. (1933), 'Über die erfüllbarkeit derjenigen zählausdrücke, welche in der normalform zwei benachbarte allzeichen enthalten', *Mathematische Annalen* **108**(1), 466–484.

Kalmár, L. & Surányi, J. (1947), 'On the reduction of the decision problem. second paper. godel prefix, a single binary predicate', *Journal of Symbolic Logic* **12**(3), 65–73.

Kalmár, L. & Surányi, J. (1950), 'On the reduction of the decision problem. third paper. pepis prefix, a single binary predicate', *The Journal of Symbolic Logic* **15**(3), 161–173.

Kostyrko, V. (1964), 'The reduction class ∀∃^{n}∀', *Algebra i Logika Sem* **3**, 45–65.

Lewis, H. R. (1980), 'Complexity results for classes of quantificational formulas', *Journal of Computer and System Sciences* **21**(3), 317–353.

Löb, M. (1967), Decidability of the monadic predicate calculus with unary function symbols, *in* 'Journal of Symbolic Logic', Vol. 32, p. 563.

Löwenheim, L. (1915), 'Über möglichkeiten im relativkalkül', *Mathematische Annalen* **76**(4), 447–470.

Lück, M., Meier, A. & Schindler, I. (2017), 'Parametrised complexity of satisfiability in temporal logic', *ACM Transactions on Computational Logic (TOCL)* **18**(1), 1.

Meier, A., Ordyniak, S., Ramanujan, M. & Schindler, I. (2019), 'Backdoors for linear temporal logic', *Algorithmica* **81**(2), 476–496.

Papadimitriou, C. H. (2003), *Computational complexity*, John Wiley and Sons Ltd., USA.

Pfandler, A., Rümmele, S., Wallner, J. P. & Woltran, S. (2015), On the parameterized complexity of belief revision, *in* 'Proceedings of the 24th International Conference on Artificial Intelligence', IJCAI'15, AAAI Press, Buenos Aires, Argentina, pp. 3149–3155.

Rabin, M. O. (1969), 'Decidability of second-order theories and automata on infinite trees', *Transactions of the american Mathematical Society* **141**, 1–35.

Ramsey, F. P. (1987), On a problem of formal logic, *in* 'Classic Papers in Combinatorics', Birkhäuser Boston, Boston, MA, pp. 1–24.

Schütte, K. (1934), 'Untersuchungen zum entscheidungsproblem der mathematischen logik', *Mathematische Annalen* **109**(1), 572–603.

Seese, D. (1996), 'Linear time computable problems and first-order descriptions', *Math. Struct. in Comp. Science* **6**, 505–526.

Shelah, S. (1977), 'Decidability of a portion of the predicate calculus', *Israel Journal of Mathematics* **28**(1-2), 32–44.

Skolem, T. A. (1920), *Logisch-kombinatorische Untersuchungen über die Erfüllbarkeit oder Beweisbarkeit mathematischer Sätze, nebst einem Theoreme über dichte Mengen, von Th. Skolem*, J. Dybwad, as reprinted in Skolem 1970.

Surányi, J. (1959), 'Reduktionstheorie des entscheidungsproblems im prädikatenkalkül der ersten stufe', *Verlag der Ungarischen Akademie der Wissenschaften*.

Szeider, S. (2004), On fixed-parameter tractable parameterizations of sat, *in* 'Theory and Applications of Satisfiability Testing', Vol. 2919, Springer Science & Business Media, Springer Berlin Heidelberg, Berlin, Heidelberg, pp. 188–202.

Turing, A. M. (1936), 'On computable numbers, with an application to the entscheidungsproblem', *J. of Math* **58**(345-363), 5.

An Objectual Semantics for First-Order LFI1 with an Application to Free Logics[‡]

Walter Carnielli[*]   Henrique Antunes[†]

[*] Department of Philosophy
University of Campinas
walterac@unicamp.br

[†] Department of Philosophy
Federal University of Minas Gerais
antunes.henrique@outlook.com

This paper is dedicated to Professor Paulo A.S. Veloso, friend and co-author, for his brilliant career in logic, philosophy, and computer science, and for his breadth of intellectual interest.

Abstract

In this paper we present an objectual semantics for the first-order version of **LFI1**, along with a corresponding natural deduction system, $\mathcal{N}$. After proving the completeness of $\mathcal{N}$, we show how **LFI1** can be adapted to yield a free logic of formal inconsistency, which we dub **FLFI1**. **FLFI1** results from some slight modifications on $\mathcal{N}$, and the new system is then proved to be sound and complete with respect to a non-inclusive dual-domain semantics. Finally, we illustrate how combining **FLFI1**'s consistency operator, ◦, with its existence predicate, E, may used to express certain ontological theses concerning the relationships between the notions of consistency and existence.

Keywords: Logics of Formal Inconsistency, Objectual Semantics, Free Logics

---

[‡]The first author acknowledges support from the National Council for Scientific and Technological Development (CNPq), Brazil, under research grant 307376/2018-4. The second author is a post-doctoral researcher at the Department of Philosophy of the Federal University of Minas Gerais.

Introduction

Among the vast number of paraconsistent systems that comprise the *logics of formal inconsistency* (**LFI**s), **LFI1** is perhaps one of the best-known, alongside such logics as da Costa's C_n systems [da Costa, 1974] and **mbC**[1]. **LFI1** was presented for the first time by Carnielli et al. [2000], where the authors proposed it as logical framework to be used for modeling evolutionary databases (see also [de Amo et al., 2002]). In [Carnielli et al., 2000], **LFI1** is presented as a 3-valued logic whose language includes an inconsistency operator, •, and it is proven there to be definitionally equivalent to the logic J_3 of D'Ottaviano & da Costa [1970, 1985]. Moreover, the authors also establish that (the propositional fragment of) **LFI1** is *maximal* w.r.t. to classical logic, meaning that whenever **LFI1** is extended with a valid schema of classical logic (that is not also valid in **LFI1**) this causes the two logics to coincide.

Since the publication of [Carnielli et al., 2000], **LFI1** has reappeared a few times in the literature (in, for example, [Carnielli & Coniglio, 2016; Omori & Waragai, 2011; Sano & Omori, 2013]). However, to the best of the authors' knowledge, all of those works have invariably interpreted the quantifiers of **LFI1** in a substitutional fashion; that is, in terms of the truth of *substitution instances* of quantified formulas:

$\forall x A$ is true in a structure iff $A(\overline{a}/x)$ is true for every element a of D;

$\exists x A$ is true in a structure iff $A(\overline{a}/x)$ is true for some element a of D

(where the relevant language is required to contain at least one individual constant $\overline{a}$ for each element a of the domain, and for each $a \in D$, $A(\overline{a}/x)$ is a substitution instance of A).

Although there is nothing technically objectionable about this way of handling the semantics of the quantifiers of **LFI1**, there are some reasons that might favor the adoption of an objectual interpretaion of $\forall$ and $\exists$, instead. According to the objectual interpretation, the semantics value of a quantified formula, say, $\forall x A$, is not stated in terms of the truth of its substitution instances, as in the case of a substituional semantics, but rather in terms of the *satisfaction* of its immediate syntatic constituent, A – which is usually not a sentence and so does not classify as the sort of thing we may appropriately call either true of false:

s satisfies $\forall x A$ in D iff $s(a/x)$ satisfies A, for every element a of D;

---
[1] See [Carnielli et al., 2007] and [Carnielli & Coniglio, 2016] for thorough investigations on the logics of formal inconsistency.

s satisfies $\exists x A$ in D iff $s(a/x)$ satisfies A, for some element a of D

(where s is an *assignment* of elements of the domain, D, to the first-order variables, and $s(a/x)$ is the assignment that is just like s except for assigning the element a of D to x^2).

From a purely a technical perspective, opting for an objectual interpretation of the quantifiers makes very little difference over choosing a substitutional interpretation. As a matter of fact, a substitutional semantics is in general easier to work with, for it does not require one to prove several of the technical lemmas about the satisfaction relation that are otherwise required in an objectual semantics[3]. But the extra work involved when dealing with objectual quantifiers also brings with it some small benefits, such as, for example, the ability to give a uniform semantical treatment of open and closed formulas alike through the notion of satisfaction – which in turn allows one to take open formulas for what they are, rather than for their universal closures[4,5].

However, the real differences between the two interpretations are most evident when one takes into account the particular ontological role that recent philosophical tradition has invested objectual quantifiers with. More specifically, Quine [1948, 1969] has famously argued for his *criterion of ontological commitment*, according to which in order to determine the objects a certain regimented theory assumes to exist, one should look for its quantified consequences of the form $\exists x A$, where $\exists$ is invariably interpreted as an objectual quantifier. Although Quine intended his criterion to apply to theories formalized on the basis of classical logic, there is no specific reason why it could not be applied to other logics as well. In particular, there is no reason why it could not be applied to **LFI1**, *as long as its quantifiers are given an objectual interpretation*.

---

[2]In some construals, including the one originally presented by Tarski [1983], s may be alternatively characterized as an infinite sequence of elements of D.

[3]These results include, for example, one or another version of the *Substitution Lemma* or results concerning the semantic equivalence of formulas that differ only in some of their bound variables (see sections 2.1 and 2.2).

[4]As a result, while adopting an objectual semantics for a first-order logic one is free to come up with a deductive system that allows for derivations in which open formulas occur and that generates a syntactic consequence relation, $\vdash$, that *completely* matches its semantic counterpart, $\vDash$ (assuming that the system in question is indeed complete). On the other hand, in a substitutional semantics, where only closed formulas are allowed as the relata of $\vDash$, whenever a completeness result with respect to a deductive system of this sort is stated as '$\Gamma \vdash A$ if and only if $\Gamma \vDash A$', this must be interpreted as expressing that $\Gamma \vdash A$ holds if and only if $\forall(\Gamma) \vDash \forall(A)$, where $\forall(\Gamma)$ and $\forall(A)$ are respectively the universal closures of the elements of Γ and the universal closure of A.

[5]For a discussion of the distinction between substitutional and objectual quantifiers, see [Hand, 2007].

In order to fill in this gap, in this paper we provide a pure objectual semantics for the first-order version of **LFI1**, taking its language to include a *consistency operator* ∘ (instead of •) and the identity predicate, =. The language will also be characterized in such a way as to allow for the presence of function letters. **LFI1** will be presented here as natural deduction system, $\mathcal{N}$, and while proving the completeness of $\mathcal{N}$ we shall attempt to indicate which of the steps of the proof would be otherwise unnecessary if we were dealing with a substitutional semantics instead. Moreover, we shall demonstrate some of the practical advantages of having an objectual semantics for **LFI1** ready at our disposal by showing how it can be modified to yield a corresponding free logic of formal inconsistency, which we call **FLFI1**.

The paper is structured as follows: in Section 1 we present the logic **LFI1** and provide it with both an objectual semantics and a corresponding natural deduction system, $\mathcal{N}$, which is then proved, in Section 2, to be sound and complete. In Section 3 we present the logic **FLFI1** and prove it to be sound and complete with respect to a non-inclusive dual-domain semantics where the variables are interpreted as in an E^+-logic. We also indicate there how having ∘ and the existence predicate, **E**, among the logical primitives of **FLFI1** allows us to express some interesting ontological theses concerning the relationships between the notions of consistency and existence.

1 The Logic LFI1

We start by characterizing some preliminary syntactic notions, and move on to define the fundamental concepts of a Tarskian objectual semantics, such as *structure, denotation, satisfaction, truth*, etc. The semantics we shall adopt here is a so-called $\mathscr{P}(\{0,1\})$-*semantics*, where formulas are assigned one or more values from the set $\{0,1\}$. This idea will be implemented by interpreting each formula, A, as having one of the sets, $\{1\}, \{0\}, \{1,0\}$, as its semantic value (*the set of truth-values of A*). If a formula A receives the value $\{1\}$, $\{0\}$, or $\{1,0\}$ (in given structure and w.r.t. to a given assignment), we shall sometimes express this by saying that A is *true, false*, or *true and false*, respectively. In spite of our use of the words 'true' and 'false' (here and elsewhere), however, it should be emphasized that this is by no means supposed to be understood as expressing the commitment of either author to *dialetheism* – i.e., the philosophical thesis that there are some true contradictions. In most places, these words will merely serve stylistic purposes, whereas in a few others, as in Section

3, they will have to be taken at face value. We leave to the reader to the task of disambiguating between those two usages. Finally, it should be mentioned that although it is customary to differentiate the propositional fragment and the first-order version of **LFI1** by calling them respectively **LFI1** and **LFI1***, we shall use the former throughout to refer to the first-order version, given that we shall not regard **LFI1**'s propositional fragment as a separate system in this paper.

1.1 Syntax and Semantics

The logical vocabulary of **LFI1** will be taken here as comprising the connectives $\neg, \wedge, \vee, \rightarrow$, and the *consistency operator*, $\circ$; the quantifiers $\forall$ and $\exists$; denumerably many first-order variables; and a few punctuation symbols. An **LFI1**-first-order language will be completely determined by specifying its first-order signature:

Definition 1. A first-order *signature* $\mathcal{S}$ is a triple $\langle \mathcal{C}, \mathcal{F}, \mathcal{R} \rangle$ such that:

(1) $\mathcal{C}$ is a set;
(2) $\mathcal{F} = \langle \mathcal{F}_n \rangle_{n \geq 1}$ is a family of disjoint sets;
(3) $\mathcal{P} = \langle \mathcal{P}_n \rangle_{n \geq 1}$ is a family of disjoint sets such that $= \in \mathcal{P}_2$;
(4) $\mathcal{C}, \bigcup_{n \geq 1} \mathcal{F}_n, \bigcup_{n \geq 1} \mathcal{P}_n, \mathcal{V} \cup \{(,),,,\neg,\circ,\wedge,\vee,\rightarrow,\forall,\exists\}$ are pairwise disjoint.

The elements of $\mathcal{V} = \{v_n : n \in \mathbb{N}\}$ are the *individual variables*, which will remain constant across every first-order signature. The symbols x, y, z, and w will be used here as metavariables ranging over $\mathcal{V}$. Given $\mathcal{S} = \langle \mathcal{C}, \mathcal{F}, \mathcal{P} \rangle$, the elements of $\mathcal{C}$ are the *individual constants* of $\mathcal{S}$; and, for each $n \geq 1$, $\mathcal{F}_n$ and $\mathcal{P}_n$ are respectively its set of n-ary *function letters* and of n-ary *predicate letters* of $\mathcal{S}$. Notice that the *identity predicate*, $=$, is included among the binary predicate letters of every first-order signature. The elements of the set:

$$[\mathcal{S}] = \mathcal{C} \cup \bigcup_{n \geq 1} \mathcal{F}_n \cup \bigcup_{n \geq 1} \mathcal{P}_n \cup \{\forall, \exists\}$$

will be called $\mathcal{S}$-*parameters*. These are the symbols whose interpretation may vary across different first-order structures (see Definition 2 below).

The set of *terms* and the set of *formulas* generated by $\mathcal{S}$ are defined in the usual way. Henceforth, we shall denote them by $Term(\mathcal{S})$ and $Form(\mathcal{S})$, respectively, reserving the symbols t and u, with or without subscripts, to be used as metavariables for terms, and $A, B, C, D, \ldots$ to be used as metavariables for formulas. Moreover, we shall assume the usual definitions of a bound/free occurrence of a variable of (in) a formula, and of a term free for a variable in a formula. The notation $A(t_1/t_2)$ (and also $u(t_1/t_2)$) will be used to denote the result of substituting all free occurrences of the term t_2 in the formula A (in the term u) by t_1[6]. Finally, the connective $\leftrightarrow$ will receive its usual definition in terms of $\rightarrow$ and $\wedge$: $A \leftrightarrow B =_{df} (A \rightarrow B) \wedge (B \rightarrow A)$.

Definition 2. Let $\mathcal{S} = \langle \mathcal{C}, \mathcal{F}, \mathcal{P} \rangle$ be a first-order signature. An $\mathcal{S}$-*structure* $\mathfrak{A}$ (or $\mathcal{S}$-**LFI1**-*structure*) is a function on $[\mathcal{S}]$ that satisfies the following conditions:

(1) $\mathfrak{A}(\forall) = \mathfrak{A}(\exists)$ is a non-empty set $|\mathfrak{A}|$ (the *domain* of $\mathfrak{A}$);

(2) For every $c \in \mathcal{C}$, the interpretation, $c^{\mathfrak{A}}$, of c is an element of $|\mathfrak{A}|$;

(3) For every $f \in \mathcal{F}_n$, the interpretation, $f^{\mathfrak{A}}$, of f is a function $g : |\mathfrak{A}|^n \longrightarrow |\mathfrak{A}|$;

(4) For every $P \in \mathcal{P}_n$, the interpretation, $P^{\mathfrak{A}}$, of P is a pair $\langle \mathcal{P}_+^{\mathfrak{A}}, \mathcal{P}_-^{\mathfrak{A}} \rangle$ such that $\mathcal{P}_+^{\mathfrak{A}} \cup \mathcal{P}_-^{\mathfrak{A}} = |\mathfrak{A}|^n$;

(5) $=_+^{\mathfrak{A}}$ is the identity relation on $|\mathfrak{A}|$ (i.e., $=_+^{\mathfrak{A}}$ is the set $\{\langle a, a \rangle : a \in |\mathfrak{A}|\}$).

In clause (4) of the above definition, $P_+^{\mathfrak{A}}$ and $P_-^{\mathfrak{A}}$ are called respectively the *extension* and the *anti-extension* of P, and designate the set of n-tuples (of elements of $|\mathfrak{A}|$) that satisfy P and the set of n-tuples that satisfy its negation. The requirement that $P_+^{\mathfrak{A}}$ and $P_-^{\mathfrak{A}}$ be exhaustive amounts to the fact that **LFI1** does not allow for truth value gaps (and so is not a paracomplete logic). There is, however, no similar requirement to the effect that they must be exclusive. As far as the definition goes, $P_+^{\mathfrak{A}}$ and $P_-^{\mathfrak{A}}$ may have a non-empty intersection, in which case P would have contradictory instances. Clause (5) guarantees that $=$ has at least some of the properties we expect of an identity predicate. Notice that even though $\langle a, b \rangle \in =_+^{\mathfrak{A}}$ if and only if $a = b$, and $\langle a, b \rangle \in =_-^{\mathfrak{A}}$, if $a \neq b$ (given that $(=_+^{\mathfrak{A}} \cup =_-^{\mathfrak{A}}) = |\mathfrak{A}|^2$), it may still happen that $\langle a, a \rangle \in =_-^{\mathfrak{A}}$, for some $a \in |\mathfrak{A}|$.

A function $s : \mathcal{V} \longrightarrow |\mathfrak{A}|$ from the set of individual variables, $\mathcal{V}$, to the domain of $\mathfrak{A}$ will be called an $\mathfrak{A}$-*assignment*. Given an assignment s and an element a of $|\mathfrak{A}|$, we shall use $s(a/x)$ to denote the $\mathfrak{A}$-assignment that is just like s, except that it assigns a to x (i.e, $s(a/x)(x) = a$). Notice, however, that if the original s is such that $s(x) = a$, then s and $s(a/x)$ will not differ at all. We shall sometimes say that two assignments *agree* on certain variables in order to express that they assign the same values to *those* variables, allowing for the possibility that they may differ on the values of the other variables.

---

[6] An occurrence of a term t is free in a formula A if none of the variables that occur in it lie within the scope of a quantifier that binds one of those variables.

Definition 3. Let $\mathfrak{A}$ be an $\mathcal{S}$-structure, s an $\mathfrak{A}$-assignment, and $t \in Term(\mathcal{S})$. The *s-denotation* of t in $\mathfrak{A}$ (denoted by $t^{\mathfrak{A}}[s]$) is inductively defined by:

(1) If t is an individual variable (i.e., $t \in \mathcal{V}$), then $t^{\mathfrak{A}}[s] = s(t)$;
(2) If t is an individual constant (i.e., $t \in \mathcal{C}$), then $t^{\mathfrak{A}}[s] = t^{\mathfrak{A}}$;
(3) If $t = f(t_1, \ldots, t_n)$, then $t^{\mathfrak{A}}[s] = f^{\mathfrak{A}}(t_1^{\mathfrak{A}}[s], \ldots, t_n^{\mathfrak{A}}[s])$.

Definition 4. Let $\mathfrak{A}$ be an $\mathcal{S}$-structure, s an $\mathfrak{A}$-assignment, and $A \in Form(\mathcal{S})$. The *set of truth values* of A under $\mathfrak{A}$ and s (denoted by $A^{\mathfrak{A}}[s]$) is inductively defined as follows:

(1) $1 \in P(t_1, \ldots, t_n)^{\mathfrak{A}}[s]$ iff $\langle t_1^{\mathfrak{A}}[s], \ldots, t_n^{\mathfrak{A}}[s] \rangle \in P_+^{\mathfrak{A}}$;
      $0 \in P(t_1, \ldots, t_n)^{\mathfrak{A}}[s]$ iff $\langle t_1^{\mathfrak{A}}[s], \ldots, t_n^{\mathfrak{A}}[s] \rangle \in P_-^{\mathfrak{A}}$
(2) $1 \in (\neg A)^{\mathfrak{A}}[s]$ iff $0 \in A^{\mathfrak{A}}[s]$;
      $0 \in (\neg A)^{\mathfrak{A}}[s]$ iff $1 \in A^{\mathfrak{A}}[s]$
(3) $1 \in (A \wedge B)^{\mathfrak{A}}[s]$ iff $1 \in A^{\mathfrak{A}}[s]$ and $1 \in B^{\mathfrak{A}}[s]$;
      $0 \in (A \wedge B)^{\mathfrak{A}}[s]$ iff $0 \in A^{\mathfrak{A}}[s]$ or $0 \in B^{\mathfrak{A}}[s]$
(4) $1 \in (A \vee B)^{\mathfrak{A}}[s]$ iff $1 \in A^{\mathfrak{A}}[s]$ or $1 \in B^{\mathfrak{A}}[s]$;
      $0 \in (A \vee B)^{\mathfrak{A}}[s]$ iff $0 \in A^{\mathfrak{A}}[s]$ and $0 \in B^{\mathfrak{A}}[s]$
(5) $1 \in (\circ A)^{\mathfrak{A}}[s]$ iff $1 \notin A^{\mathfrak{A}}[s]$ or $0 \notin A^{\mathfrak{A}}[s]$;
      $0 \in (\circ A)^{\mathfrak{A}}[s]$ iff $1 \in A^{\mathfrak{A}}[s]$ and $0 \in A^{\mathfrak{A}}[s]$
(6) $1 \in (A \to B)^{\mathfrak{A}}[s]$ iff $1 \notin A^{\mathfrak{A}}[s]$ or $1 \in B^{\mathfrak{A}}[s]$;
      $0 \in (A \to B)^{\mathfrak{A}}[s]$ iff $1 \in A^{\mathfrak{A}}[s]$ and $0 \in B^{\mathfrak{A}}[s]$
(7) $1 \in (\forall x A)^{\mathfrak{A}}[s]$ iff $1 \in A^{\mathfrak{A}}[s(a/x)]$, for every $a \in |\mathfrak{A}|$;
      $0 \in (\forall x A)^{\mathfrak{A}}[s]$ iff $0 \in A^{\mathfrak{A}}[s(a/x)]$, for some $a \in |\mathfrak{A}|$
(8) $1 \in (\exists x A)^{\mathfrak{A}}[s]$ iff $1 \in A^{\mathfrak{A}}[s(a/x)]$, for some $a \in |\mathfrak{A}|$;
      $0 \in (\exists x A)^{\mathfrak{A}}[s]$ iff $0 \in A^{\mathfrak{A}}[s(a/x)]$, for every $a \in |\mathfrak{A}|$

We shall say an $\mathfrak{A}$-assignment s *satisfies* $A \in Form(\mathcal{S})$ in $\mathfrak{A}$ (denoted by $\mathfrak{A}, s \models A$) if and only if $1 \in A^{\mathfrak{A}}[s]$, and A *true* in $\mathfrak{A}$ (denoted by $\mathfrak{A} \models A$) if and only if $\mathfrak{A}, s \models A$, for every $\mathfrak{A}$-assignment s. A is a *logical consequence* of $\Gamma \subseteq Form(\mathcal{S})$ (denoted by $\Gamma \models A$) if and only if for every structure $\mathfrak{A}$ and every $\mathfrak{A}$-assignment s, if $\mathfrak{A}, s \models B$, for every $B \in \Gamma$, then $\mathfrak{A}, s \models A$. Finally, A is *logically valid* (denoted by $\models A$) if and only if $\emptyset \models A$.

From the definitions above, it can then be easily proven that **LFI1** does not allow for truth-value gaps:

Proposition 5. Let $\mathfrak{A}$ be an $\mathcal{S}$-structure and s an $\mathfrak{A}$-assignment. For every $A \in Form(\mathcal{S})$, either $1 \in A^{\mathfrak{A}}[s]$ or $0 \in A^{\mathfrak{A}}[s]$.

1.2 A Natural Deduction System

We shall adopt here a natural deduction calculus, $\mathcal{N}$, for **LFI1**. The system is composed of the following rules:

$$\dfrac{A \quad B}{A \wedge B}\wedge I \qquad \dfrac{A \wedge B}{A}\wedge E \qquad \dfrac{A \wedge B}{B}\wedge E$$

$$\dfrac{A}{A \vee B}\vee I \qquad \dfrac{B}{A \vee B}\vee I \qquad \dfrac{A \vee B \quad \begin{array}{c}[A]\\\vdots\\C\end{array} \quad \begin{array}{c}[B]\\\vdots\\C\end{array}}{C}\vee E$$

$$\dfrac{\begin{array}{c}[A]\\\vdots\\B\end{array}}{A \to B}\to I \qquad \dfrac{A \to B \quad A}{B}\to E \qquad \dfrac{}{A \vee \neg A}PEM \qquad \dfrac{A}{\neg\neg A}DN$$

$$\dfrac{\neg(A \wedge B)}{\neg A \vee \neg B}\wedge DM \qquad \dfrac{\neg(A \vee B)}{\neg A \wedge \neg B}\vee DM \qquad \dfrac{\neg(A \to B)}{\neg A \wedge B}\to DM$$

$$\dfrac{\circ A \quad A \quad \neg A}{B}EXP^{\circ} \qquad \dfrac{\neg\circ A}{A}\neg\circ E \qquad \dfrac{\neg\circ A}{\neg A}\neg\circ E$$

$$\dfrac{A}{\forall x A}\forall I \qquad \dfrac{\forall x A}{A(t/x)}\forall E \qquad \dfrac{A(t/x)}{\exists x A}\exists I \qquad \dfrac{\exists x A \quad \begin{array}{c}[A]\\\vdots\\B\end{array}}{B}\exists E$$

$$\dfrac{\neg \forall x A}{\exists x \neg A}\forall DM \qquad \dfrac{\neg \exists x A}{\forall x \neg A}\exists DM \qquad \dfrac{}{t=t}=I \qquad \dfrac{t=u \quad A(t/u)}{A}=E$$

As usual, enclosing a formula A in square brackets is supposed to indicate that A is being regarded as a hypothesis – either discharged or undischarged; and the use of the double line in DN, $\wedge DM$, $\vee DM$, $\forall DM$, etc. indicates that these rules may be applied in either direction. The notion of a *derivation* in $\mathcal{N}$ can be inductively defined in terms of trees by adapting the definition in [van Dalen, 2013, p. 34]. For our purposes, it suffices to say that a derivation is a tree composed of (labelled) formulas whose *nodes* are either a hypothesis or the result of applying one of the rules above. The notation $\Gamma \vdash A$ will be used to express that there is a derivation $\mathcal{D}$ in $\mathcal{N}$ such that A is the last formula that occurs in $\mathcal{D}$ (its *conclusion*) and all of $\mathcal{D}$'s undischarged hypotheses belong to Γ ($\vdash A$ will be regarded as a shorthand for $\varnothing \vdash A$). We shall say that an occurrence of a formula A in a derivation *depends* on a hypothesis B if B is an undischarged hypothesis in the derivation of (that occurrence of) A. Finally, the rules for the quantifiers and rule $=E$ are subject to the following restrictions: in $\forall I$, x must not be free in any hypothesis on which A depends; in $\exists E$, x must be free neither in B nor in any hypothesis on which it depends, except A; in $\forall E$ and $\exists I$, t must be free for x in A; and in $=E$, t must be free for u in A – that is, for every variable x that occurs in u, t must be free for x in A.

The following properties of **LFI1** will be useful later on:

Proposition 6. Let $\mathcal{S}$ be a first-order signature, $t \in Term(\mathcal{S})$, $A, B \in Form(\mathcal{S})$, and $x \in \mathcal{V}$. Then:

(1) If $\Gamma, A \vdash B$, then $\Gamma \vdash A \to B$;
(2) If $\Gamma, A \vdash \neg A$, then $\Gamma \vdash \neg A$;
(3) If $\Gamma, \neg A \vdash A$, then $\Gamma \vdash A$;
(4) If x is not free in Γ and $\Gamma \vdash A$, then $\Gamma \vdash \forall x A$;
(5) If x is not free in $\Gamma \cup \{B\}$, $\Gamma \vdash \exists x A$, and $\Gamma, A \vdash B$, then $\Gamma \vdash B$.
(6) $\vdash \neg \forall x \neg A \leftrightarrow \exists x A$
(7) $\vdash \neg \exists x \neg A \leftrightarrow \forall x A$

Proof: The proofs of (1)-(7) pose no difficulties. For the sake of illustration, let us prove (6) and leave remaining items to the reader:

$$\cfrac{\cfrac{[\neg \forall x \neg A]^2}{\exists x \neg \neg A} \forall DM}{\cfrac{\exists x A}{\neg \forall x \neg A \to \exists x A}} \quad \cfrac{\cfrac{[\neg \neg A]^1}{A} DN}{\exists x A} \exists I \;\; \exists E_1 \quad [\exists x A]^4 \quad \cfrac{\cfrac{[A]^3}{\neg \neg A} DN}{\exists x \neg \neg A} \exists I$$

$$\cfrac{\to I_2}{\neg \forall x \neg A \leftrightarrow \exists x A} \wedge I$$

∎

Notice that we have used numerical subscripts to indicate at which step a hypothesis used in the derivation was effectively discharged. From now on, we shall continue to use this convention, and adopt the further conventions of enclosing undischarged hypotheses in brackets (with no superscripts).

One of the distinguishing features of the **LFI**s (or, at least, the ones that are based on the positive fragment of classical logic) is that they have enough resources to recover every classical inference, even though they are, strictly speaking, subclassical logics. Proof-theoretically, this fact is usually expressed by the *Derivability Adjustment Theorem* (DAT), which establishes that every classical derivation can be converted into a derivation in a corresponding **LFI** by (i) replacing every application of the rule (axiom) of explosion by its **LFI**-analogue (in our case EXP°), and by (ii) supplementing the original derivation with some consistency assumptions (i.e., premisses to the effect that certain formulas are consistent). Now, one interesting aspect of **LFI1** is that it allows us to push those consistency assumptions all the way down to the atomic level: by assuming that some of the predicates that occur in a given classical derivation are universally consistent, i.e., that $\forall x_1 \ldots \forall x_n \circ P(x_1, \ldots, x_n)$ holds, we can reconstruct that original classical derivation as a corresponding derivation in **LFI1**. This property is a consequence of the fact that **LFI1**'s consistency operator, $\circ$, propagates across every sentential connective and quantifier; that is, that the following schemas can all be derived in **LFI1**[7]:

$$\circ A \to \circ \neg A \qquad \circ A \to \circ\circ A \qquad \circ A \to \circ\neg\circ A$$
$$(\circ A \wedge \circ B) \to \circ(A \wedge B) \qquad (\circ A \wedge \circ B) \to \circ(A \vee B) \qquad \circ B \to \circ(A \to B)$$
$$\forall x \circ A \to \circ \forall x A \qquad \forall x \circ A \to \circ \exists x A$$

Theorem 7. (Derivability Adjustment Theorem (Atomic-Level Version)) Let $\mathcal{S} = \langle \mathcal{C}, \mathcal{F}, \mathcal{P} \rangle$ be a first-order signature, and let $\Gamma \cup \{A\} \subseteq Form(\mathcal{S})$ be such that $\circ$ occurs neither in (any element of) Γ nor in A. If there is a derivation of A from Γ in classical logic, then $\Gamma, \Delta \vdash A$, for some set $\Delta \subseteq \{\forall x_1 \ldots \forall x_n \circ P(x_1, \ldots, x_n) : P \text{ is predicate letter of } \mathcal{S}\}$.

Proof: Let us first fix a deductive system for classical logic: to simplify, and without any concerns about the obvious redundancies, let $\mathcal{N}_{\text{CL}}$ be the system formulated in terms of the same logical vocabulary as **LFI1**, except for the absence of $\circ$; and let the corresponding deductive system be composed of the same $\circ$-free rules of $\mathcal{N}$ together with the unrestricted version of the *rule of explosion*:

$$\frac{A \quad \neg A}{B} \, EXP$$

---

[7]The schemas $\circ A \to \circ\circ A$ and $\circ A \to \circ\neg\circ A$ follow trivially from, respectively, $\circ\circ A$ and $\circ\neg\circ A$.

Now, assume that there is a derivation $\mathcal{D}$ of A from Γ in $\mathcal{N}_{\mathrm{CL}}$, and suppose that $A_1, \ldots, A_n$ are all the formulas (if any) that occur in $\mathcal{D}$ as the first premise of an application of rule EXP. Let $P_1, \ldots, P_m$ be all the predicates that occur in A_i (where $P_i \in \mathcal{P}_{r_i}$, for every $1 \leq i \leq m$). By letting $\Delta_i = \{\forall x_1 \ldots \forall x_{r_i} \circ P_i(x_1, \ldots, x_{r_i}) : 1 \leq i \leq m\}$, it then follows, by induction on the complexity of A_i (and using the consistency-propagation properties listed above), that there is a derivation $\mathcal{D}_i$ in $\mathcal{N}$ of $\circ A_i$ from Δ_i[8]. Finally, let $\mathcal{D}'$ be the result of (i) replacing every application of EXP in $\mathcal{D}$ by a corresponding application of EXP° and (ii) placing $\mathcal{D}_i$ right above the additional premise:

$$\dfrac{\begin{array}{c}\mathcal{D}_i \\ \vdots \\ \circ A_i\end{array} \quad \begin{array}{c}\vdots \\ A_i\end{array} \quad \begin{array}{c}\vdots \\ \neg A_i\end{array}}{\begin{array}{c}B \\ \vdots\end{array}} \; EXP^{\circ}$$

By defining $\Delta = \bigcup_{1 \leq i \leq n} \Delta_i$, it follows that $\mathcal{D}'$ is a derivation of A from $\Gamma \cup \Delta$ in $\mathcal{N}$. Therefore, $\Gamma, \Delta \vdash A$.

∎

The theorem above represents a rather important feature of **LFI1**: it establishes that every classically valid reasoning that is not already valid in **LFI1** may be rendered so if we assume that only some of the predicates occurring in it admit of no contradictory instances; or, to frame it in a slightly different manner, that any classical theory can be losslessly replicated within the context of **LFI1** by simply assuming its predicates to be universally consistent. Notice that although the **LFI1**-analogue of a classical derivation $\mathcal{D}$ does not, in general, coincide with $\mathcal{D}$, it does not bring significant changes upon $\mathcal{D}$'s overall structure either, for, as the proof above clearly shows, it only adds a new branch at each place where EXP was originally applied. Later on, in Section 3, we shall prove a slight variation of the Derivability Adjustment Theorem, this time for a free logic version of **LFI1**[9].

---

[8]In general, if $P_1, \ldots, P_m$ are all the predicate symbols that occur in a $\circ$-free formula C (where $P_i \in \mathcal{P}_{r_i}$, for every $1 \leq i \leq m$), then:

$$\forall x_1 \ldots \forall x_{r_1} \circ P_1(x_1, \ldots, x_{r_1}), \ldots, \forall x_1 \ldots \forall x_{r_m} \circ P_m(x_1, \ldots, x_{r_m}) \vdash \circ C$$

[9]See [Antunes, 2019b] for a discussion of the philosophical significance of the Derivability Adjustment Theorem.

2 Soundness and Completeness

2.1 Soundness

The proof of the soundness of $\mathcal{N}$ requires two preliminary results. The first one tells us that if two assignments agree on all variables free in a formula A, then A will have the same truth values under either assignment. The second result is the **LFI1**-version of the *Substitution Theorem*: it establishes that substituting a term t for a variable x in a formula A and evaluating the result under an assignment s is equivalent to evaluating the original A under the assignment that is just like s, except that it assigns the value of t (under s) to x:

Lemma 8. Let $\mathfrak{A}$ be an $\mathcal{S}$-structure and $A \in Form(\mathcal{S})$. If s and s' are two $\mathfrak{A}$-assignments such that $s(x) = s'(x)$, for every variable x free in A, then $A^{\mathfrak{A}}[s] = A^{\mathfrak{A}}[s']$.

Proof: Notice first that $(*)$ $t^{\mathfrak{A}}[s] = t^{\mathfrak{A}}[s']$ – this can be proven by a straightforward induction on the complexity of $t \in Term(\mathcal{S})$. Given $(*)$, it then suffices to prove that $1 \in A^{\mathfrak{A}}[s]$ iff $1 \in A^{\mathfrak{A}}[s']$, and that $0 \in A^{\mathfrak{A}}[s]$ iff $0 \in A^{\mathfrak{A}}[s']$. The proof proceeds by induction on the complexity of A. If A is an atomic formula, then the result is an immediate consequence of $(*)$. If A is $\neg B$, $\circ B$, $B \wedge C$, $B \vee C$, or $B \to C$, then the result follows immediately from the induction hypothesis. Finally, if A is $\forall x B$ or $\exists x B$, notice that $s(a/x)$ and $s'(a/x)$ agree on all variables free in B, for every $a \in |\mathfrak{A}|$. By the induction hypothesis, $1 \in B^{\mathfrak{A}}[s(a/x)]$ iff $1 \in B^{\mathfrak{A}}[s'(a/x)]$, and $0 \in B^{\mathfrak{A}}[s(a/x)]$ iff $0 \in B^{\mathfrak{A}}[s'(a/x)]$, for every $a \in |\mathfrak{A}|$. Therefore, $1 \in A^{\mathfrak{A}}[s]$ iff $1 \in A^{\mathfrak{A}}[s']$, and $0 \in A^{\mathfrak{A}}[s]$ iff $0 \in A^{\mathfrak{A}}[s']$. ∎

Theorem 9. (Substitution Lemma) Let $\mathfrak{A}$ be an $\mathcal{S}$-structure and s an $\mathfrak{A}$-assignment. Let $t \in Term(\mathcal{S})$ and define s' to be the assignment $s(t^{\mathfrak{A}}[s]/x)$. If t is free for x in A, then $A^{\mathfrak{A}}[s'] = A(t/x)^{\mathfrak{A}}[s]$.

Proof: Notice first that $(*)$ $u^{\mathfrak{A}}[s'] = u(t/x)^{\mathfrak{A}}[s]$ – this can be proven by a straightforward induction on the complexity of $u \in Term(\mathcal{S})$. It suffices to show then that $1 \in A^{\mathfrak{A}}[s']$ iff $1 \in A(t/x)^{\mathfrak{A}}[s]$, and $0 \in A^{\mathfrak{A}}[s']$ iff $0 \in A(t/x)^{\mathfrak{A}}[s]$. The proof proceeds by induction on the complexity of A. If A is an atomic formula, then the result follows immediately from $(*)$. If A is $\neg B$, $\circ B$, $B \wedge C$, $B \vee C$, or $B \to C$, then the result follows immediately from the induction hypothesis. Finally, if A is $\forall y B$ or $\exists y B$, then there are two cases:

(1) x is not free in A: $A(t/x)$ is A itself and the result follows immediately from Lemma 8 above;

(2) x is free in A: Notice that since x is free in A, x and y are different variables. Thus, (i) $s(b/x)(a/y) = s(a/y)(b/x)$, for every $a, b \in |\mathfrak{A}|$. Furthermore, since t is free for x in A, y does not occur in t. Hence, s and $s(a/y)$ agree on all variables in t, for every $a \in |\mathfrak{A}|$. Thus, (ii) $t^{\mathfrak{A}}[s] = t^{\mathfrak{A}}[s(a/y)]$, for every $a \in |\mathfrak{A}|$. Now:

$$
\begin{array}{lll}
1 \in B^{\mathfrak{A}}[s'(a/y)] & \text{iff} \quad 1 \in B^{\mathfrak{A}}[s(t^{\mathfrak{A}}[s]/x)(a/y)] & \\
& \text{iff} \quad 1 \in B^{\mathfrak{A}}[s(a/y)(t^{\mathfrak{A}}[s]/x)] & \text{(i)} \\
& \text{iff} \quad 1 \in B^{\mathfrak{A}}[s(a/y)(t^{\mathfrak{A}}[s(a/y)]/x)] & \text{(ii)} \\
& \text{iff} \quad 1 \in B(t/x)^{\mathfrak{A}}[s(a/y)] & \text{IH}
\end{array}
$$

$$
\begin{array}{lll}
0 \in B^{\mathfrak{A}}[s'(a/y)] & \text{iff} \quad 0 \in B^{\mathfrak{A}}[s(t^{\mathfrak{A}}[s]/x)(a/y)] & \\
& \text{iff} \quad 0 \in B^{\mathfrak{A}}[s(a/y)(t^{\mathfrak{A}}[s]/x)] & \text{(i)} \\
& \text{iff} \quad 0 \in B^{\mathfrak{A}}[s(a/y)(t^{\mathfrak{A}}[s(a/y)]/x)] & \text{(ii)} \\
& \text{iff} \quad 0 \in B(t/x)^{\mathfrak{A}}[s(a/y)] & \text{IH}
\end{array}
$$

As a result, $1 \in B^{\mathfrak{A}}[s'(a/y)]$ iff $1 \in B(t/y)^{\mathfrak{A}}[s(a/y)]$, and $0 \in B^{\mathfrak{A}}[s'(a/y)]$ iff $0 \in B(t/x)^{\mathfrak{A}}[s(a/y)]$, for every $a \in |\mathfrak{A}|$. Therefore, $1 \in A^{\mathfrak{A}}[s']$ iff $1 \in A(t/x)^{\mathfrak{A}}[s]$, and $0 \in A^{\mathfrak{A}}[s']$ iff $0 \in A(t/x)^{\mathfrak{A}}[s]$.

∎

Given these two results, we can now prove that $\mathcal{N}$ is indeed sound with respect to the class of all **LFI1**-structures:

Theorem 10. (Soundness Theorem) Let $\Gamma \cup \{A\} \subseteq Form(\mathcal{S})$. If $\Gamma \vdash A$, then $\Gamma \vDash A$.

Proof: The proof proceeds as usual by induction on the number of nodes of a derivation $\mathcal{D}$ of A from Γ and uses the results proved above at several places. Consider, for example, the case of rule $=E$: if A results from applying $=E$ to $t=u$ and $A(t/u)$, then there are derivations $\mathcal{D}_1$ and $\mathcal{D}_2$ with fewer nodes than $\mathcal{D}$ such that $\mathcal{D}_1$ and $\mathcal{D}_2$ are derivations from Γ of, respectively, $t=u$ and $A(t/u)$. By the induction hypothesis, $\Gamma \vDash t=u$ and $\Gamma \vDash A(t/x)$. Suppose that $\mathfrak{A}, s \vDash \Gamma$. It then follows that $\langle t^{\mathfrak{A}}[s], u^{\mathfrak{A}}[s]\rangle \in =^{\mathfrak{A}}_+$, and therefore $t^{\mathfrak{A}}[s] = u^{\mathfrak{A}}[s]$ (by clause (5) of Definition 2). Now, let x be a variable that occurs nowhere in A, t, or u. Clearly, $A(t/u) = A(x/u)(t/x)$, and therefore $1 \in (A(x/u)(t/x))^{\mathfrak{A}}[s]$ (since $1 \in (A(t/u))^{\mathfrak{A}}[s]$). Notice that since x does not occur in A and t is free for u in A, t is also free for x in $A(x/u)$. By Theorem 9, it follows that $1 \in (A(x/u))^{\mathfrak{A}}[s(t^{\mathfrak{A}}[s]/x)]$. But because $t^{\mathfrak{A}}[s] = u^{\mathfrak{A}}[s]$, we are allowed to conclude that $1 \in (A(x/u))^{\mathfrak{A}}[s(u^{\mathfrak{A}}[s]/x)]$. Finally, by applying Theorem 9 one more time, we infer that $1 \in (A(x/u)(u/x))^{\mathfrak{A}}[s]$, that is, $1 \in A^{\mathfrak{A}}[s]$ (since $A(x/u)(u/x)$ and A are clearly the same formula). Therefore, $\mathfrak{A}, s \vDash A$.

∎

2.2 Replacement Rule and Alphabetic Variants

Before jumping right into the proof of the completeness of **LFI1**, we have some preliminary work to do. Specifically, we need to prove a few of results that will be necessary in the construction of the canonical model to be found in Proposition 21 (and elsewhere before that). These results concern (i) the ability to replace congruent formulas for one another within arbitrary sentential contexts, and (ii) both the deductive and semantic equivalence between formulas that differ only in some of their bound variables, i.e., between formulas that are *alphabetic variants* of one another. Whereas (ii) wouldn't be absolutely necessary if we hadn't formulated the semantics of **LFI1** in terms of the satisfaction relation, the fact that we have done so means that while constructing the canonical model we will also need to consider a "canonical" assignment within that model. And because working with this assignment will require us to apply the Substitution Theorem (Theorem 9) several times, we will also need to be able to move back and forth between certain formulas and their alphabetic variants in order escape the restrictions imposed by the theorem's initial conditions. For example, there will be times when we will want to infer $1 \in A[t/x]^{\mathfrak{A}}[s]$ from $1 \in A^{\mathfrak{A}}[s(t^{\mathfrak{A}}[s]/x)]$, but will be prevented from using the Substitution Theorem to do this due to the lack any guarantees that the term t is free for x in A (which is required by the theorem). Now, because alphabetic variants are, as shall see below, semantically equivalent to one another, we will be able to circumvent this difficulty by applying the theorem to an alphabetic variant of A in which t is indeed free for x, moving back to A once the desired result has been reached. As for (i), besides being used in the proof of (ii), it represents a rather interesting result its own right, providing us with a useful tool for abbreviating derivations in $\mathcal{N}$ that might otherwise turn out to be excessively long.

Unlike the classical bi-implication connective, the bi-implication connective of **LFI1**, $\leftrightarrow$, does not characterize a congruence relation between formulas, given that the set of truth values of $A \leftrightarrow B$ may include the value 1, whilst neither $\neg A \leftrightarrow \neg B$ nor $\circ A \leftrightarrow \circ B$ do – this will be the case whenever $A^{\mathfrak{A}}[s] = \{1\}$ and $B^{\mathfrak{A}}[s] = \{0, 1\}$, or vice versa. As this example suggests, however, we may strengthen $\leftrightarrow$ to enforce that the two formulas have *exactly* the same set of values by taking into account not only the formulas themselves, but also their negations. In other words, we may characterize a congruence relation between formulas in **LFI1** by defining:

$$A \cong B =_{df} (A \leftrightarrow B) \wedge (\neg A \leftrightarrow \neg B)$$

And given this definition, the following result should come as no surprise:

Lemma 11. Let $A, B, C, D \in Form(S)$, and $v \in \mathcal{V}$. Then:

(1) $\vdash A \cong A$
(2) $\vdash A \cong \neg\neg A$
(3) $\vdash A \cong B \rightarrow \neg A \cong \neg B$
(4) $\vdash A \cong B \rightarrow {\circ}A \cong {\circ}B$
(5) $\vdash A \cong B \wedge C \cong D \rightarrow A \wedge C \cong B \wedge D$
(6) $\vdash A \cong B \wedge C \cong D \rightarrow A \vee C \cong B \vee D$
(7) $\vdash A \cong B \wedge C \cong D \rightarrow (A \rightarrow C) \cong (B \rightarrow D)$
(8) $\vdash \forall x (A \cong B) \rightarrow \forall x A \cong \forall x B$
(9) $\vdash \forall x (A \cong B) \rightarrow \exists x A \cong \exists x B$

Once we know that $\cong$ does behave as expected with respect to each of **LFI1**'s connectives and quantifiers, we can easily prove that $\cong$-equivalent formulas are intersubstitutable in every sentential context. And although we shall not make much use of this result in the following pages, except in the proof of Proposition 14 below and in a few other places, it should be pretty clear how valuable being able to replace equivalent formulas for one another in a derivation may sometimes be. The result comes in two forms, as a theorem and as an accompanying rule:

Theorem 12. Let $A, B, C \in Form(S)$ and let $C(B/A)$ be the formula that results from C by replacing all occurrences of A in C by B. Then:

(1) **(Replacement Theorem)** If all the variables that are free in A or B and bound in C are among $x_1, \ldots, x_n$, then $\vdash \forall x_1 \ldots \forall x_n (A \cong B) \rightarrow (C(B/A) \cong C)$.

(2) **(Replacement Rule (RR))** If (i) none of the variables that are free in A or B and bound is C are free in Γ, and if (ii) $\Gamma \vdash A \cong B$, then $\Gamma \vdash C(A/B) \cong C$. In particular, $\vdash A \cong B$ implies $\vdash C(A/B) \cong C$.

Proof: The Replacement Rule is an immediate consequence of the Replacement Theorem, Proposition 6(4), and rule $\rightarrow E$. Hence, we shall only prove (1). The proof proceeds by induction on the complexity of C. Notice first that if B does not occur in C, then $C(A/B)$ and C are the same formula, and the result follows from Lemma 11(1) and Proposition 6(1). If C is an atomic formula, then C is B and $C(A/B)$ is A. Applying rule $\forall E$ n times to the hypothesis yields $\forall x_1 \ldots \forall x_n (A \cong B) \vdash C(A/B) \cong C$. As a result, $\vdash \forall x_1 \ldots x_n (A \cong B) \rightarrow C(A/B) \cong C$ (by Proposition 6(1)). If C is $\neg D$, $\circ D$, $D \wedge E$, $D \vee E$, or $D \rightarrow E$, then the result follows immediately from the induction hypothesis and Lemma 11(2-7). If C is $\forall x D$, notice that since x is bound in C, it is not free in $\forall x_1 \ldots \forall x_n (A \cong B)$ (either it is not free in $A \cong B$ or it is among $x_1, \ldots, x_n$). Consider then the following derivation in **LFI1** (where $\forall \vec{x}$ and F are abbreviations for, respectively, $\forall x_1 \ldots \forall x_n$ and $D(A/B) \equiv D$):

$$\frac{[\forall \vec{x}(A \cong B)]^1 \quad \dfrac{\forall \vec{x}(A \cong B) \rightarrow F}{\dfrac{F}{\forall xF}\forall I} \rightarrow E \qquad \dfrac{\begin{array}{c}\text{IH}\\ \vdots\end{array} \quad \begin{array}{c}\text{Lemma 11(8)}\\ \vdots\end{array}}{\dfrac{\forall xF \rightarrow \forall x(D(A/B)) \cong \forall xD}{\forall x(D(A/B)) \cong \forall xD}\rightarrow E}}{\forall \vec{x}(A \cong B) \rightarrow \forall x(D(A/B)) \cong \forall xD} \rightarrow I_1$$

Finally, if C is $\exists xD$, then the result follows by adapting the derivation above to use Lemma 11(9) instead of Lemma 11(8). ■

Let us move on and prove now a couple of results concerning alphabetic variants that will be useful later on. The intuitive idea is, of course, that two formulas are alphabetic variants of one another if they differ on at most some of their bound variables. This idea can be made a bit more precise by means of the following definition:

Definition 13. (Alphabetic Variants) Let S be a first-order signature and $A, A' \in Form(S)$. A' is an *alphabetic variant* (a.v) of A iff one of the following conditions obtain:

(1) A is an atomic formula and $A' = A$;
(2) A is $\neg B$, A' is $\neg B'$, and B' is an a.v of B;
(3) A is $\circ B$, A' is $\circ B'$, and B' is an a.v of B;
(4) A is $A \wedge B$, A' is $B' \wedge C'$, B' is an a.v of B, and C' is an a.v of C;
(5) A is $A \vee B$, A' is $B' \vee C'$, B' is an a.v of B, and C' is an a.v of C;

(6) A is $A \to B$, A' is $B' \to C'$, B' is an a.v of B, and C' is an a.v of C;

(7) A is $\forall x B$ and either

   (7.1) A' is $\forall x B'$ and B' is an a.v of B; or
   (7.2) A' is $\forall y(B'(y/x))$, y is not free in B', y is free for x in B', and B' is an a.v of B.

(8) A is $\exists x B$ and either

   (8.1) A' is $\exists x B'$ and B' is an a.v of B; or
   (8.2) A' is $\exists y(B'(y/x))$, y is not free in B', y is free for x in B', and B' is an a.v of B.

The restrictions that y not be free in B' and y be free for x in B' are meant to prevent that, say, $\forall x(P(x) \wedge Q(x,y))$ and $\forall y(P(y) \wedge Q(y,y))$, and that $\forall x(P(x) \vee \exists y Q(x,y))$ and $\forall y(P(y) \vee \exists y Q(y,y))$ be regarded as alphabetic variants of one another. After all, we would like the free variables of a formula to remain free in all of its alphabetic variants.

Proposition 14. Let $A, A' \in Form(\mathcal{S})$. If A' is an alphabetic variant of A, then $\vdash A \cong A'$.

Proof: The proof proceeds by induction on the complexity of A. If A is an atomic formula, then A' is A itself and the result follows from Lemma 11(1). If A is $\neg B$, $\circ B$, $B \wedge C$, $B \vee C$, or $B \to C$, then the result follows immediately from the induction hypothesis and RR. If A is $\forall x B$, then there are two cases:

(1) A' is $\forall x B'$ and B' is an alphabetic variant of B: By the induction hypothesis, $\vdash B' \cong B$ and, therefore, $\vdash \forall x B' \cong \forall x B$ (by RR); that is, $\vdash A' \cong A$.

(2) A' is $\forall y(B'(y/x))$, y is not free in B', y is free for x in B', and B' is an alphabetic variant of B: First notice that since y is not free in B' and y is free for x in B', x is free for y in $B'(y/x)$ and $B'(y/x)(x/y) = B'$. Furthermore, since y is not free in B', it is also not free in B (by Lemma 14). Now, by the induction hypothesis, $\vdash B \cong B'$, and so $\vdash \forall x B \cong \forall x B'$ (by rule RR). Since $\cong$ is transitive, to complete the proof it suffices to prove that $\vdash \forall x B' \cong \forall y(B'(y/x))$, which poses no serious difficulties.

Finally, if A is $\exists x B$, then there are two cases as well:

(1) A' is $\exists x B'$ and B' is an alphabetic variant of B: By the induction hypothesis, $\vdash B' \cong B$ and, therefore, $\vdash \exists x B' \cong \exists x B$ (by RR); that is, $\vdash A' \cong A$.

(2) A' is $\exists y B'(y/x)$, y is not free in B', y is free for x is B', and B' is an alphabetic variant of B: By the induction hypothesis, $\vdash B \cong B'$, and thus, by Lemma 11(3), $\vdash \neg B' \cong \neg B'$. By applying the same reasoning as above, it then follows that $\vdash \forall x \neg B \cong \forall y(\neg B'(y/x))$. By Lemma 11(3) once again, $\vdash \neg \forall x \neg B \cong \neg \forall y(\neg B'(y/x))$. And since $\vdash \neg \forall x \neg B \cong \exists x B$ and $\vdash \neg \forall y(\neg B'(y/x)) \cong \exists y(B'(y/x))$ (by Proposition 6(6-7) and Lemma 11(2)), it then follows $\vdash \exists x B \cong \exists y(B'(y/x))$.

∎

The following corollary is an immediate consequence of the proposition above and the soundness of $\mathcal{N}$ (Theorem 10):

Corollary 15. Let $A, A' \in Form(\mathcal{S})$ and let $\mathfrak{A}$ be an $\mathcal{S}$-structure. If A' is an alphabetic variant of A, then $A^{\mathfrak{A}}[s] = A'^{\mathfrak{A}}[s]$, for every $\mathfrak{A}$-assignment s.

As the last piece of preliminary work that needs to be taken care of before we go about proving the completeness of **LFI1**, we have the following technical lemma:

Lemma 16. Let $\mathcal{S} = \langle \mathcal{C}, \mathcal{F}, \mathcal{P} \rangle$ be a first-order signature such that $c \in \mathcal{C}$. Let $\Gamma \cup \{A, B\} \subseteq Form(\mathcal{S})$ be such that c does not occur in Γ and $\Gamma, B \vdash A$. Then there is a variable y that occurs nowhere in $\Gamma \cup \{A, B\}$, such that $\Gamma, B(y/c) \vdash A(y/c)$.

Proof: Suppose that $\Gamma, B \vdash A$ and let $\mathcal{D}$ be a derivation of A from $\Gamma \cup \{B\}$ in $\mathcal{N}$. Let y be the first variable that does not occur in $\mathcal{D}$ and let $\mathcal{D}'$ be the result of substituting every of occurrence of c in $\mathcal{D}$ by y. We shall prove, by induction on $\mathcal{D}$, that $\mathcal{D}'$ is a derivation of $A(y/c)$ from $\Gamma \cup \{B(y/c)\}$ in $\mathcal{N}$: suppose first that $\mathcal{D}$ has only one node. Then, either $A \in \Gamma \cup \{B\}$ or A is the conclusion of an application of PEM or $=I$. If $A \in \Gamma$, then $A(y/c)$ is A itself, given that c does not occur in Γ; and if A is B, then $A(y/c)$ and $B(y/c)$ are also the same formula. In either case, it follows that $\mathcal{D}'$ is a derivation of $A(y/c)$ from $\Gamma \cup \{B(y/c)\}$. Suppose then that $A(y/c)$ results from an application of PEM or $=I$. Clearly, $A(y/c)$ also results from an application of PEM or $=I$ in $\mathcal{D}'$. Finally, let $\mathcal{D}$ be such that it contains $n > 1$ nodes and suppose that the result holds for every derivation with fewer nodes than n. It is immediate to check that if A follows by applying one of the rules of **LFI1** other than PEM and $=I$, then $A(y/c)$ also follows by applying the same rule to the corresponding formulas that occur previously in $\mathcal{D}'$. Therefore, $\Gamma, B(y/c) \vdash A(y/c)$.

∎

2.3 Completeness

Proving the completeness of $\mathcal{N}$ follows, from now on, the standard procedure we are accustomed to see in the proofs of other similar results. The proof involves two main steps: in the first step (Proposition 19) we show that, given a set of formulas Γ that does not derive a certain formula A, Γ can be extended to a set Δ that (i) still does not prove A and (ii) has some desirable syntactic properties. The second step (Proposition 21) consists in showing how to construct a (canonical) model for Δ – which will, of course, be a model of Γ as well. As is customary, Δ's distinctive syntactic properties will allow us to mimic the semantic clauses for the sentential connectives and the quantifiers of **LFI1** in terms of the *membership-in-Δ* relation. In the case of **LFI1**, however, the semantic clauses not only describe the conditions under which a formula is true in a structure (w.r.t. an assignment), but also the conditions under which it is false. Hence, while defining a canonical structure for **LFI1**, we will have to take this aspect into account.

The following two definitions characterize the aforementioned syntactic properties. Later on, in Lemma 20, we will prove that they are all we need in order to construct the canonical model for **LFI1**:

Definition 17. (Henkin Set) Let $\mathcal{S} = \langle \mathcal{C}, \mathcal{F}, \mathcal{P} \rangle$ be a first-order signature and $\Delta \subseteq Form(\mathcal{S})$. Δ is a *Henkin set* iff for every $A \in Form(\mathcal{S})$ and $x \in \mathcal{V}$, there is some $c \in \mathcal{C}$ such that $\Delta \vdash \exists x A$ only if $\Delta \vdash A(c/x)$.

Definition 18. Let $\Delta \cup \{A\} \subseteq Form(\mathcal{S})$. Δ is said to be *maximal* with respect to A iff the following two conditions obtain: (i) $\Delta \nvdash A$; and (ii) for every $B \notin \Delta$, $\Delta, B \vdash A$.

Given these definitions, we can show that every set of formulas Γ that does not prove a certain formula A can be extended to a Henkin set that is maximal with respect to A. As usual, extending Γ to the new set also requires us to expand the language in order to make sure there are enough constants to satisfy the condition expressed in Definition 17.

Proposition 19. Let $\mathcal{S} = \langle \mathcal{C}, \mathcal{F}, \mathcal{P} \rangle$ be a first-order signature and $\Gamma \cup \{A\} \subseteq Form(\mathcal{S})$. If $\Gamma \nvdash A$, then there is a first-order signature $\mathcal{S}^+ = \langle \mathcal{C}^+, \mathcal{F}, \mathcal{P} \rangle$ and a set $\Delta \subseteq Form(\mathcal{S}^+)$ such that $\mathcal{C} \subseteq \mathcal{C}^+$, $\Gamma \subseteq \Delta$, and Δ is a Henkin set that is maximal w.r.t. A.

Proof: Let $\mathcal{C}^+ = \mathcal{C} \cup \{c_i : i \in \mathbb{N}\}$, with $\{c_i : i \in \mathbb{N}\} \cap [\mathcal{S}] = \emptyset$, and adopt a fixed enumeration $B_0, B_1, B_2, \ldots$ of the formulas in $Form(\mathcal{S}^+)$[10]. Define the sequence $\langle j_n \rangle_{n \in \mathbb{N}}$ of natural numbers as follows:

---

[10] For the sake of simplicity, the current proof assumes that $\mathcal{S}$ has no more than $\aleph_0$ non-

(1) j_0 = the least natural number k such that c_k does not occur in B_0.
(2) j_{n+1} = the least natural number k such that c_k does occur in B_k, and for every $i \leq n$, $k \neq j_i$.

Now, let the sequence $\langle \Gamma_n \rangle_{n \in \mathbb{N}}$ of subsets of $Form(\mathcal{S}^+)$ be defined by:

$\Gamma_0 = \Gamma$;

$$\Gamma_{n+1} = \begin{cases} \Gamma_n & \text{if } \Gamma_n, B_n \vdash A \\ \Gamma_n \cup \{B_n\} & \text{if } \Gamma_n, B_n \nvdash A \text{ and } B_n \neq \exists x C \\ \Gamma_n \cup \{B_n, C(c_{j_n}/x)\} & \text{if } \Gamma_n, B_n \nvdash A \text{ and } B_n = \exists x C \end{cases}$$

Finally, let $\Delta = \bigcup_{n \in \mathbb{N}} \Gamma_n$. Clearly, $\Gamma \subseteq \Delta$. We shall now prove that ($*$) for every $n \in \mathbb{N}$, $\Gamma_n \nvdash A$: The proof proceeds by induction on n. By the initial hypothesis, $\Gamma_0 \nvdash A$. Suppose then that $\Gamma_n \nvdash A$ (IH). There are two cases: either $\Gamma_n, B_n \vdash A$ or $\Gamma_n, B_n \nvdash A$. If $\Gamma_n, B_n \vdash A$, then $\Gamma_{n+1} = \Gamma_n$. Thus, $\Gamma_{n+1} \nvdash A$, by (IH). If, on the other hand, $\Gamma_n, B_n \nvdash A$, then either (i) $B_n = \exists x C$, for some $x \in \mathcal{V}$ and $C \in Form(\mathcal{S}^+)$, and $\Gamma_{n+1} = \Gamma_n \cup \{B_n, C(c_{j_n}/x)\}$; or (ii) $\Gamma_{n+1} = \Gamma_n \cup \{B_n\}$. In the latter case, $\Gamma_{n+1} \nvdash A$, since $\Gamma_n, B_n \nvdash A$ (by hypothesis). As for (i), let $B_n = \exists x C$ and $\Gamma_{n+1} = \Gamma_n \cup \{B_n, C(c_{j_n}/x)\}$, and suppose, to reason by contradiction, that $\Gamma_{n+1} \vdash A$. Since c_{j_n} does not occur in $\Gamma_n \cup \{B_n\}$, it follows by Lemma 16 above that there is a new variable y such that $\Gamma_n, B_n, C(c_{j_n}/x)(y/c_{j_n}) \vdash A(y/c_{j_n})$. And since y does not occur in C and c_{j_n} does not occur in A, this yields $\Gamma, B_n, C(y/x) \vdash A$[11]. By Proposition 14 and $\exists E$, we then conclude that $\Gamma_n, B_n \vdash A$:

Proposition 14
$$\frac{\dfrac{\exists x C \cong \exists y C(y/x)}{\dfrac{\exists x C \leftrightarrow \exists y C(y/x)}{\exists x C \to \exists y C(y/x)} \wedge E} \wedge E \quad \exists y C(y/x)}{\exists y C(y/x)} \quad \frac{[B_n] \quad [\exists x C]}{A} \to E \quad \frac{[C(y/x)]^1 \\ \vdots \\ A}{} \exists E_1$$

---

logical symbols – and therefore that the set of formulas generated by $\mathcal{S}$ is denumerable. As the reader will readily notice, however, this assumption will not compromise any of the results to be presented below, nor will she encounter any difficulties while attempting to generalize them to first-order languages of any cardinality.

[11]Notice that due to way the sequence $\langle \Gamma_n \rangle_{n \in \mathbb{N}}$ was defined and because $A \in Form(\mathcal{S})$, c_{j_n} occurs neither in $\Gamma_n \cup \{B_n\}$ nor in A.

But this result contradicts the hypothesis that $\Gamma_n, B_n \nvdash A$. Therefore, $\Gamma_{n+1} \nvdash A$.

Given $(*)$, it is then immediate to prove that Δ is a Henkin set that it is maximal w.r.t. A.

∎

The following result follows almost immediately from definitions 17 and 18 and will be very useful in the proof of Proposition 21 below:

Lemma 20. Let $\Delta \subseteq Form(\mathcal{S})$ be a Henkin set that is maximal w.r.t. A. Then, for every $B, C \in Form(\mathcal{S})$ and $x \in \mathcal{V}$:

(1) $B \in \Delta$ iff $\Delta \vdash B$
(2) $B \wedge C \in \Delta$ iff $B \in \Delta$ and $C \in \Delta$
(3) $B \vee C \in \Delta$ iff $B \in \Delta$ or $C \in \Delta$
(4) $B \to C \in \Delta$ iff $B \notin \Delta$ or $C \in \Delta$
(5) $B \in \Delta$ or $\neg B \in \Delta$
(6) $\neg \neg B \in \Delta$ iff $B \in \Delta$
(7) $\neg(B \wedge C) \in \Delta$ iff $\neg B \in \Delta$ or $\neg C \in \Delta$
(8) $\neg(B \vee C) \in \Delta$ iff $\neg B \in \Delta$ and $\neg C \in \Delta$
(9) $\neg(B \to C) \in \Delta$ iff $B \in \Delta$ and $\neg C \in \Delta$
(10) $\circ B \in \Delta$ iff $B \notin \Delta$ or $\neg B \notin \Delta$
(11) $\neg \circ B \in \Delta$ iff $B \in \Delta$ and $\neg B \in \Delta$
(12) If B' is an alphabetic variant of B, then $B' \in \Delta$ iff $B \in \Delta$
(13) $\exists x B \in \Delta$ iff $B(c/x) \in \Delta$, for some $c \in \mathcal{C}$
(14) $\forall x B \in \Delta$ iff $B(c/x) \in \Delta$, for every $c \in \mathcal{C}$

Proof: The proofs of (1)-(11) are pretty much straightforward. (12) is an immediate consequence of Proposition 14, while (13) follows immediately from rule $\exists I$ and the hypothesis that Δ is a Henkin set. As for (14), its left-to-right direction follows from applying rule $\forall E$ to $\forall x B$, whereas its right-to-left direction can be proven in the following manner: notice first that $(*)$ $\circ \forall x B, \neg \forall x B \vdash \exists x(\circ B \wedge \neg B)$[12]. Suppose then that $\forall x B \notin \Delta$ (i.e., $\Delta \nvdash \forall x B$). By PEM and (3) above, $\Delta \vdash \neg \forall x B$. Since $\vdash \neg \circ \forall x B$ implies $\vdash \forall x B$ (by $\neg \circ E$), it follows that $\Delta \nvdash \neg \circ \forall x B$, and therefore $\Delta \vdash \circ \forall x B$ (by PEM and (3) once again). Thus, $\Delta \vdash \exists x(\circ B \wedge \neg B)$ (by $(*)$ above). Now, since Δ is a Henkin set, there is some $c \in \mathcal{C}$ such that $\Delta \vdash \circ B(c/x) \wedge \neg B(c/x)$, which means that $B(c/x)$ cannot be a member of Δ, for otherwise Δ would

---
[12]This can be proven by proving that $\circ \forall x B, \neg \forall x B, \forall x(\neg \circ B \vee B) \vdash \neg \forall x(\neg \circ B \vee B)$.

be trivial and A would be derivable from Δ. Since we have been reasoning under the hypothesis that $\forall x B \notin \Delta$, we may conclude at this point that if $\forall x B \notin \Delta$, then $B(c/x) \notin \Delta$, for some $c \in \mathcal{C}$. But this claim is equivalent to the claim that if $B(c/x) \in \Delta$, for every $c \in \mathcal{C}$, then $\forall x B \in \Delta$. ∎

Proposition 21 below is the only remaining step to prove the completeness of $\mathcal{N}$. It establishes that every Henkin set $\Delta \subseteq Form(\mathcal{S})$ that is maximal with respect to some formula A has a *model* – i.e., a structure in which all its elements are (at least) true. The proof of this result consists in (i) showing how exactly to build this structure, and in (ii) proving that it is in fact a model of Δ. As usual, this so-called *canonical model* is a "linguistic structure", in the sense that all the parameters of the relevant language are interpreted in terms of the language itself: its domain will be composed of certain equivalence classes of the terms in $Term(\mathcal{S})$, and its function and predicate symbols, interpreted in terms of the *derivability-from-Δ* relation.

Proposition 21. Let $\mathcal{S} = \langle \mathcal{C}, \mathcal{F}, \mathcal{P} \rangle$ be a first-order signature and $\Delta \cup \{A\} \subseteq Form(\mathcal{S})$. If Δ is a Henkin set that is maximal w.r.t. A, then there is an $\mathcal{S}$-structure $\mathfrak{A}$ and an $\mathfrak{A}$-assignment s such that $\mathfrak{A}, s \vDash \Gamma$ and $\mathfrak{A}, s \nvDash A$.

Proof: Let $\sim$ be the relation on $Term(\mathcal{S})$ defined by: $t_1 \sim t_2$ iff $t_1 = t_2 \in \Delta$. By rules $=I$ and $=E$, $\sim$ is an equivalence relation. Let $[t] = \{u : u \in Term(\mathcal{S}) \text{ and } t \sim u\}$ and define the $\mathcal{S}$-structure $\mathfrak{A}$ as follows:

(i) $|\mathfrak{A}| = \{[t] : t \in Term(\mathcal{S})\}$;
(ii) For every $c \in \mathcal{C}$, $c^{\mathfrak{A}} = [c]$;
(iii) For every $f \in \mathcal{F}_n$, $f^{\mathfrak{A}}([t_1], \ldots, [t_n]) = [f(t_1, \ldots, t_n)]$;
(iv) For every $P \in \mathcal{P}_n$, $P^{\mathfrak{A}}$ is the pair $\langle P_+^{\mathfrak{A}}, P_-^{\mathfrak{A}} \rangle$ such that:

$$\langle [t_1], \ldots, [t_n] \rangle \in P_+^{\mathfrak{A}} \text{ iff } P(t_1, \ldots, t_n) \in \Delta;$$

$$\langle [t_1], \ldots, [t_n] \rangle \in P_-^{\mathfrak{A}} \text{ iff } \neg P(t_1, \ldots, t_n) \in \Delta$$

Facts:

(1) For every $f \in \mathcal{F}_n$, the definition of $f^{\mathfrak{A}}$ does not depend on the representatives $t_1, \ldots, t_n$;
(2) For every $P \in \mathcal{P}_n$, the definition of $P^{\mathfrak{A}}$ does not depend on the representatives $t_1, \ldots, t_n$;
(3) $=_+^{\mathfrak{A}} = \{\langle [t], [t] \rangle : t \in Term(\mathcal{S})\}$.

These facts guarantee that $\mathfrak{A}$ is indeed an $\mathcal{S}$-structure. Let s be the $\mathfrak{A}$-assignment defined by $s(x) = [x]$, for every $x \in \mathcal{V}$. By a straightforward induction of the complexity of $t \in Term(\mathcal{S})$, it is immediate to show that $t^{\mathfrak{A}}[s] = [t]$. We will now prove that $\mathfrak{A}, s \vDash \Delta$ and $\mathfrak{A}, s \nvDash A$. In fact, we will prove the following stronger claim:

$$1 \in B^{\mathfrak{A}}[s] \text{ iff } B \in \Delta \quad \text{and} \quad 0 \in B^{\mathfrak{A}}[s] \text{ iff } \neg B \in \Delta$$

for every $B \in Form(\mathcal{S})$. The proof proceeds by induction on the length of B. Because most cases follow almost immediately from the induction hypothesis and the corresponding items in Lemma 20, we shall only consider the cases that involve quantified formulas:

(7) B is $\forall xC$:

(7.1) Suppose that $1 \in B^{\mathfrak{A}}[s]$. Thus, $1 \in C^{\mathfrak{A}}[s([t]/x)])$, for every $t \in Term(\mathcal{S})$. In particular, $1 \in C^{\mathfrak{A}}[s([c]/x)]$, for every $c \in \mathcal{C}$. By Theorem 9 and the induction hypothesis, it follows that $C(c/x) \in \Delta$, for every $c \in \mathcal{C}$. By applying Lemma 20(14), we then conclude that $\forall xC \in \Delta$; that is, $B \in \Delta$. As for the other direction, suppose that $1 \notin B^{\mathfrak{A}}[s]$. Thus, $1 \notin C^{\mathfrak{A}}[s([t]/x)]$, for some $t \in Term(\mathcal{S})$. Let C' be an alphabetic variant of C such that t is free for x in C'. By Corollary 15, it follows that $1 \notin C'^{\mathfrak{A}}[s([t]/x)]$ and, by Theorem 9, that $1 \notin (C'(t/x))^{\mathfrak{A}}[s]$. By the induction hypothesis, $C'(t/x) \notin \Delta$, and because $\forall xC' \in \Delta$ implies $C'(t/x) \in \Delta$, we may infer that $\forall xC' \notin \Delta$. By applying Lemma 20(12), we then get that $\forall xC \notin \Delta$; that is, $B \notin \Delta$.

(7.2) Suppose that $0 \in B^{\mathfrak{A}}[s]$. Thus, $0 \in C^{\mathfrak{A}}[s([t]/x)]$, for some $t \in Term(\mathcal{S})$. Let C' be an alphabetic variant of C such that t is free for x in C'. By Corollary 15, $0 \in C'^{\mathfrak{A}}[s([t]/x)]$ and, by Theorem 9, $0 \in (C'(t/x))^{\mathfrak{A}}[s]$, which implies, by the induction hypothesis, that $\neg C'(t/x) \in \Delta$. Hence, $\exists x \neg C' \in \Delta$, and so $\neg \forall xC' \in \Delta$ (by rules $\exists I$ and $\forall DM$). We then conclude by applying Lemma 20(12) that $\neg \forall xC \in \Delta$; that is, $\neg B \in \Delta$. Suppose now that $\neg \forall xC \in \Delta$. It follows, by rules $\forall DM$ and Lemma 20(13), that $\neg C(c/x) \in \Delta$, for some $c \in \mathcal{C}$. By the induction hypothesis, $0 \in (C(c/x))^{\mathfrak{A}}[s]$, which, by Theorem 9, is equivalent to $0 \in C^{\mathfrak{A}}[s([c]/x)]$. Therefore, $0 \in (\forall xC)^{\mathfrak{A}}[s]$, i.e., $0 \in B^{\mathfrak{A}}[s]$.

(8) B is $\exists xC$:

(8.1) $1 \in B^{\mathfrak{A}}[s]$ iff $1 \in (\neg \forall x \neg C)^{\mathfrak{A}}[s]$ (by Proposition 6(6) and the soundness of $\mathcal{N}$) iff $0 \in (\forall x \neg C)^{\mathfrak{A}}[s]$ iff $\neg \forall x \neg C \in \Delta$ (by (7.2) above) iff $\exists xC \in \Delta$ (by Proposition 6(6)).

(8.2) $0 \in B^{\mathfrak{A}}[s]$ iff $1 \in (\neg \exists xC)^{\mathfrak{A}}[s]$ iff $1 \in (\forall x \neg C)^{\mathfrak{A}}[s]$ (by rule $\exists DM$ and the soundness of $\mathcal{N}$) iff $\forall x \neg C \in \Delta$ (by (7.1)) iff $\neg \exists xC \in \Delta$ (by rule $\exists DM$).

We now have everything we need in order to prove the completeness of $\mathcal{N}$:

Theorem 22. (Completeness Theorem) Let $\mathcal{S} = \langle \mathcal{C}, \mathcal{F}, \mathcal{P} \rangle$ be a first-order signature and $\Gamma \cup \{A\} \subseteq Form(\mathcal{S})$. If $\Gamma \vDash A$, then $\Gamma \vdash A$.

Proof: Suppose that $\Gamma \nvdash A$. By Proposition 19, there is a first-order signature $\mathcal{S}^+ = \langle \mathcal{C}^+, \mathcal{F}, \mathcal{P} \rangle$ and Henkin set $\Delta \subseteq Form(\mathcal{S}^+)$ such that $\mathcal{C} \subseteq \mathcal{C}^+$, $\Gamma \subseteq \Delta$, and Δ is maximal w.r.t. A. By Proposition 21, there is an $\mathcal{S}^+$-structure $\mathfrak{A}$ and an $\mathfrak{A}$-assignment s such that $\mathfrak{A}, s \vDash \Delta'$ and $\mathfrak{A}, s \nvDash A$. Let $\mathfrak{B}$ be the $\mathcal{S}$-reduct of $\mathfrak{A}$ (i.e., the restriction of $\mathfrak{A}$ to $[\mathcal{S}]$). Clearly, $\mathfrak{B}, s \vDash \Gamma$ and $\mathfrak{B}, s \nvDash A$, therefore, $\Gamma \nvDash A$. ∎

3   LFI1 as a Free Logic

In order to demonstrate the value of having at our disposal an objectual interpretation of **LFI1**, together with accompanying soundness and completeness results, we will describe in this section how **LFI1** can be adapted to yield a system of free logic. After presenting the system and proving its completeness, we will then briefly discuss how it can be useful for those willing to apply **LFI1**'s consistency operator, ∘, to formally express certain theses in ontology.

In a nutshell, a *free logic* is a first- or higher-order logical system (i) that allows for empty singular terms and (ii) whose first-order quantifiers are thought of as having existential import. (ii) amounts to interpreting ∀ and ∃ according to Quine's criterion of ontological commitment. As for (i), an *empty singular term* may be characterized as a term that either has no reference whatsoever – e.g., '$\frac{1}{0}$' – or, at least, does not refer to anything existent – e.g., 'Pegasus', 'the round square', etc[13]. Since, as per (ii), the domain of the first-order quantifiers ∀ and ∃ represent the class of all things that do exist (or the class of things that exist according to a certain theory), it follows that if a term t is empty, then its referent, if it has any, cannot lie within the range of ∀ (∃).

Free logics are more easily recognizable by the fact that they invalidate both *universal instantiation* and *existential generalization*:

$$\forall x A \vDash A(t/x) \tag{UI}$$

---
[13] See [Bencivenga, 2002], for a more precise characterization of free logics and for a discussion of their motivations.

$$A(t/x) \vDash xA \qquad \text{(EG)}$$

This is to be expected, for unless we know that t refers to an (existing) object, we are not authorized to conclude that A holds of the referent of t (if any) from the fact that it holds of every (existing) object; nor can we infer that A holds of at least one such object from the fact that $A(t/x)$ is true. However, if t is indeed non-empty, then those two inference schemas should be completely harmless. Free logics countenance this fact by validating the following restricted versions of (UI) and (EI):

$$\forall x A, \mathbf{E}(t) \vDash A(t/x) \qquad \text{(UI}^{\mathbf{E}}\text{)}$$

$$A(t/x), \mathbf{E}(t) \vDash \exists x A \qquad \text{(EG}^{\mathbf{E}}\text{)}$$

Here $\mathbf{E}$ is a special predicate, the *existence predicate*, that holds of every existent object, and such that $\mathbf{E}(t)$ is false whenever t is empty. Although free logics can sometimes be presented without $\mathbf{E}$ as a primitive symbol, we shall henceforth regard it as a logical predicate, in the same way as we have been doing with $=$ all along[14].

Free logics differ among themselves in a myriad of different aspects, most of which have mainly, but not exclusively, to do with the ways that different authors intend to account for the semantics of formulas in which empty terms occur. For example, they may be either *inclusive* or *non-inclusive*, according to whether or not they allow for empty domains of quantification; they may be formulated in terms of a *single-* or a *dual-domain* semantics, according to whether the referents of empty terms are assigned no reference whatsoever or some kind of degenerate reference in a so-called *outer domain*; they may be presented as either an *E-logic* or as an *E^+-logic*, according to whether or not free variables are assigned values in the domain of quantification; and they may be either *positive*, *negative*, or *nonvalent*, according whether some atomic formulas in which empty terms occur are allowed to be true, or whether they must invariably be regarded as false or even truth-valueless[15].

Now, investigating how **LFI1** behaves with respect each of these alternatives would certainly be both important and interesting, but since our aim here is merely to illustrate that one may indeed profit from having

---

[14] See footnote 16 below.

[15] For an overview of the different approaches to free logics that have been presented in the literature, see [Bencivenga, 2002] and [Nolt, 2007]. For the distinction between E- and E^+-logics, see [Troelstra & van Dalen, 1988, Ch. 2, Sec. 2].

an objectual interpretation of **LFI1** at one's disposal, we shall content ourselves with the simplest possible choices; that is, we will present **LFI1** as a positive non-inclusive free E^+-logic with a dual-domain semantics. Before going into more detail about what those choices actually mean, let us first present a natural deduction system for the resulting logic, which we shall from now on call **FLFI1**.

A natural deduction system for **FLFI1** can be obtained by adding the following rule to $\mathcal{N}$:

$$\frac{}{\mathrm{E}(x)}\ I$$

And replacing $\forall E$ and $\exists I$ by:

$$\frac{\forall x A \quad \mathrm{E}(t)}{A(t/x)}\ \forall E^{\mathrm{E}} \qquad \frac{A(t/x) \quad \mathrm{E}(t)}{\exists x A}\ \exists I^{\mathrm{E}}$$

where t must be free for x in A. Let us denote the resulting system and the corresponding consequence relation by, respectively, $\mathcal{N}_{\mathrm{F}}$ and $\vdash_{\mathrm{F}}$. In this system we can easily prove that $\forall x \mathrm{E}(x)$, which expresses the trivial fact that every existent object exists[16]:

$$\frac{\dfrac{}{\mathrm{E}(x)}\ I}{\forall x \mathrm{E}(x)}\ \forall I$$

As for the model theory of **FLFI1**, we shall adopt an inclusive dual-domain semantics: any **FLFI1**-structure will contain two (possibly equal) domains such that one of them, the so-called *inner domain*, will represent the set of existent objects (and so will coincide with both the extension of E and the range $\forall$ and $\exists$), while the other, called the *outer domain*, will represent the class of absolutely every thing, irrespective of whether or not it exists. We shall require the inner domain to be non-empty, which corresponds to the fact that **FLFI1** is a non-inclusive (or exclusive) logic – and, therefore, every language will be assumed to be talking about at least one *existing* thing. Although it is most natural to regard the elements of the outer domain that do not also belong to the inner domain as non-existent entities, i.e., as having some sort of being but lacking the property

---

[16]It is also possible to prove $\exists x(x{=}t) \leftrightarrow \mathrm{E}(t)$ in $\mathcal{N}_{\mathrm{F}}$ (where x does not occur in t). However, unless it is assumed that $\circ \mathrm{E}(x)$ and that $\mathrm{E}(t) \to \circ(t{=}t)$, the same cannot be said about $\neg\exists x(x{=}t) \leftrightarrow \neg\mathrm{E}(t)$. Therefore, we are unable to treat $\mathrm{E}(t)$ as an abbreviation for $\exists x(x{=}t)$ in **FLFI1**, as is often the case in some free logics.

of existence, this is not mandatory. It is quite possible to think of them as representing fictitious entities which we are able to meaningfully talk about even though they do not *really* exist, such as certain error objects of computer programs, or the semantic values of otherwise empty terms that are conventionally assigned to them[17]. Finally, the extensions and the anti-extensions of the predicates (other than **E**) will be allowed include any object belonging to the outer domain – even if it is not an element of the inner domain. As a result, we will end up with a positive free logic, given that atomic formulas in which empty terms occur, and formulas built up from them, will come out as true in some structures. Individual constants and function symbols will be interpreted similarly, and, as in the case of **LFI1**, the extension of the identity predicate will be the set of all pairs $\langle a, a \rangle$, where a is an element of the *outer* domain.

Definition 23. Let $\mathcal{S}_E = \langle \mathcal{C}, \mathcal{F}, \mathcal{P} \rangle$ be a first-order signature such that $\in \mathcal{P}_1$. An $\mathcal{S}_E$-*structure* $\mathfrak{A}$ for **FLFI1** is a function on the set of parameters, $[\mathcal{S}_E]$, of $\mathcal{S}_E$ that satisfies the following conditions:

(1) $\mathfrak{A}(\forall) = \mathfrak{A}(\exists)$ is a pair $\langle |\mathfrak{A}|_i, |\mathfrak{A}|_o \rangle$ such that $|\mathfrak{A}|_i \subseteq |\mathfrak{A}|_o$ and $|\mathfrak{A}|_i \neq \emptyset$;

(2) For every $c \in \mathcal{C}$, $c^{\mathfrak{A}} \in |\mathfrak{A}|_o$;

(3) For every $f \in \mathcal{F}_n$, $f^{\mathfrak{A}}$ is a function $g : |\mathfrak{A}|_o^n \longrightarrow |\mathfrak{A}|_o$;

(4) For every $P \in \mathcal{P}_n$, $P^{\mathfrak{A}}$ is a pair $\langle \mathcal{P}_+^{\mathfrak{A}}, \mathcal{P}_-^{\mathfrak{A}} \rangle$ such that $\mathcal{P}_+^{\mathfrak{A}} \cup \mathcal{P}_-^{\mathfrak{A}} = |\mathfrak{A}|_o^n$;

(5) $=_+^{\mathfrak{A}}$ is the identity relation on $|\mathfrak{A}|_o$ (i.e., $=_+^{\mathfrak{A}}$ is the set $\{\langle a, a \rangle : a \in |\mathfrak{A}|_o\}$);

(6) $\mathbf{E}_+^{\mathfrak{A}} = |\mathfrak{A}|_i$.

In clause (1), $|\mathfrak{A}|_i$ and $|\mathfrak{A}|_o$ are respectively the inner and the outer domain of $\mathfrak{A}$.

In a free logic with a dual-domain semantics, there are two alternative ways of assigning values to variables. The first one, which characterizes an E-logic, is to allow the variables to take values from the entire outer domain, which corresponds to defining an assignment as a function from $\mathcal{V}$ to $|\mathfrak{A}|_o$. Had we opted for this interpretation, we would have had to change the definition above by requiring that $|\mathfrak{A}|_o$, *rather than* $|\mathfrak{A}|_i$, be non-empty; and since $|\mathfrak{A}|_i$ is supposed to be the range of the quantifiers, we would end up with an inclusive free logic version of **LFI1**[18]. The second way is to define an $\mathfrak{A}$-assignment as a function from $\mathcal{V}$ to $|\mathfrak{A}|_i$. This is the interpretation we shall adopt here, which is why **FLFI1** is an $\mathbf{E}^+$-logic – i.e.,

---
[17] For example, we may stipulate that '$\frac{1}{0}$' denotes a certain arbitrary object, say, $*$, even though that term does not actually refer to anything.

[18] We would also have to drop rule E*I* from $\mathcal{N}_\mathbf{F}$ and replace $\forall I$ and $\forall E$ by:

a logic in which free variables are always interpreted as taking existent objects as values (and therefore in which $\mathbf{E}(x)$ is always satisfied)[19]. Once this decision has been made, we can then adopt the same definition of the denotation relation presented in Definition 3, and define the set of truth values of A in an $\mathcal{S}_E$-structure $\mathfrak{A}$ by simply replacing clauses (7) and (8) in Definition 4 by:

(7') $\quad 1 \in (\forall x A)^{\mathfrak{A}}[s] \quad$ iff $\quad 1 \in A^{\mathfrak{A}}[s(a/x)]$, for every $a \in |\mathfrak{A}|_i$;
$\quad\quad 0 \in (\forall x A)^{\mathfrak{A}}[s] \quad$ iff $\quad 0 \in A^{\mathfrak{A}}[s(a/x)]$, for some $a \in |\mathfrak{A}|_i$

(8') $\quad 1 \in (\exists x A)^{\mathfrak{A}}[s] \quad$ iff $\quad 1 \in A^{\mathfrak{A}}[s(a/x)]$, for some $a \in |\mathfrak{A}|_i$;
$\quad\quad 0 \in (\exists x A)^{\mathfrak{A}}[s] \quad$ iff $\quad 0 \in A^{\mathfrak{A}}[s(a/x)]$, for every $a \in |\mathfrak{A}|_i$

After defining the notions of *satisfaction*, *truth* (in a structure), and *logical consequence* for **FLFI1** as we did for **LFI1** (and using $\models_F$ to denote **FLFI1**'s consequence relation), we can prove:

Theorem 24. (Soundness and Completeness) Let $\Gamma \cup \{A\} \subseteq Form(\mathcal{S}_F)$. Then, $\Gamma \vdash_F A$ if, and only if, $\Gamma \models_F A$.

Proof: The reader should encounter no difficulties in adapting the results presented in Section 2 to the case of **FLFI1**. Here are some of the main changes:

Soundness: (i) All derivations presented in Section 2.1 remain as they were, except that wherever $\forall E$ ($\exists I$) was applied to $\forall x A$ (A) to obtain A ($\exists x A$), we shall now use $\forall E^E$ ($\exists I^E$) *in conjunction with* I to reach the same conclusions. (ii) In the statement of the Substitution Theorem (Theorem 9), we must require that $t^{\mathfrak{A}}[s] \in |\mathfrak{A}|_i$, for otherwise $s(t^{\mathfrak{A}}[s]/x)$ would be undefined in some cases. We must prove, in addition, that whenever t and u are such that $t^{\mathfrak{A}}[s] = u^{\mathfrak{A}}[s] \notin |\mathfrak{A}|_i$, then $A^{\mathfrak{A}}[s] = A(t/u)^{\mathfrak{A}}[s]$[20].

$$\frac{\begin{array}{c}[\mathbf{E}(x)]\\ \vdots \\ A\end{array}}{\forall x A} \forall I^E \qquad \frac{\exists x A \quad \begin{array}{c}[\mathbf{E}(x)][A]\\ \vdots \\ B\end{array}}{B} \exists E^E$$

where the restrictions are the same as the ones for $\forall I$ and $\exists E$, except that the free occurrence of x in the hypothesis $\mathbf{E}(x)$ does not prevent us from applying the rules.

[19]As the reader might have noticed, there is something a little bit odd about this way of interpreting the variables; namely, unless an element of $(|\mathfrak{A}|_o - |\mathfrak{A}|_i)$ has a name in the language, we are utterly prevented from expressing anything about it, given that variables, being interpreted in terms of the elements of $|\mathfrak{A}|_i$, cannot be used for this purpose.

[20]This additional result is necessary in the proof of the soundness of rule $=E$.

Completeness: (iii) The proofs of the Replacement Rule (Theorem 12) and of the results concerning alphabetic variants (Proposition 14 and Corollary 15) remain the same except for the modifications described in (i). (iv) The statement of Lemma 16 must be slightly reformulated: if $c \in \mathcal{C}$ does not occur in Γ and $\Gamma, \mathbf{E}(c), B \vdash_\mathbf{F} A$, then there is a variable y that occurs nowhere in $\Gamma \cup \{A, B\}$ such that $\Gamma, B(y/c) \vdash_\mathbf{F} A(y/c)$. This result can then be proven by making some obvious changes in the proof of Lemma 16 presented above. (v) The definition of a Henkin set (Definition 17) must be reformulated as well: Δ is a *Henkin set* if and only if $\Delta \vdash_\mathbf{F} \exists x A$ implies $\Delta \vdash_\mathbf{F} \mathbf{E}(c) \wedge A(c/x)$, for some $c \in \mathcal{C}$. (vi) The construction of the sequence $\langle \Gamma_n \rangle_{n \in \mathbb{N}}$ in the proof of Proposition 19 must be adapted in order to comply with this new definition:

$\Gamma_0 = \Gamma;$

$$\Gamma_{n+1} = \begin{cases} \Gamma_n & \text{if } \Gamma_n, B_n \vdash_\mathbf{F} A \\ \Gamma_n \cup \{B_n\} & \text{if } \Gamma_n, B_n \nvdash_\mathbf{F} A \text{ and } B_n \neq \exists x C \\ \Gamma_n \cup \{B_n, \mathbf{E}(c_{j_n}), C(c_{j_n}/x)\} & \text{if } \Gamma_n, B_n \nvdash_\mathbf{F} A \text{ and } B_n = \exists x C \end{cases}$$

(vii) Items (13) and (14) of Lemma 20 must be modified in the following manner: (13') $\exists x B \in \Delta$ iff $B(c/x) \in \Delta$ for some $c \in \mathcal{C}$ such that $\mathbf{E}(c) \in \Delta$; and (14') $\forall x B \in \Delta$ iff $B(c/x) \in \Delta$ for every $c \in \mathcal{C}$ such that $\mathbf{E}(c) \in \Delta$. Finally, (viii) the canonical structure $\mathfrak{A}$ in the proof of the **FLFI1**-analogue of Proposition 21 will be such that its inner domain, $|\mathfrak{A}|_i$, will be the set:

$$\{[t] : t \in Term(\mathcal{S}_\mathbf{E}) \text{ and } \mathbf{E}(t) \in \Delta\}$$

while its outer domain, $|\mathfrak{A}|_o$, will be the set $\{[t] : t \in Term(\mathcal{S}_\mathbf{F})\}$. Everything else will be as in the original proof of Proposition 21, except for some minor changes in the proofs of items (7) and (8). ∎

Having both ∘ and **E** as logical primitives of **FLFI1** unveils some interesting possibilities when it comes to formally expressing certain ontological theses. First of all, although we have required the extension, $\mathbf{E}_+^\mathfrak{A}$, of **E** to coincide with the inner domain of every **FLFI1**-structure, there are no similar constraints with respect to its anti-extension, $\mathbf{E}_-^\mathfrak{A}$ – and, in particular, no constraints to the effect that it must be the complement of $|\mathfrak{A}|_i$ (relative to $|\mathfrak{A}|_o$). As a result, there can be a structure $\mathfrak{A}$ such that an element a of

$|\mathfrak{A}|_i$ belongs to both $\mathbf{E}_+^{\mathfrak{A}}$ and $\mathbf{E}_-^{\mathfrak{A}}$, and since $\mathbf{E}$ stands for the existence predicate, this translates into the claim that there is an object that does and does not exist. Now, even if one has gone so far as to seriously entertain the possibility of there being predicates with contradictory instances, one may still refrain from maintaining that the existence predicate is among them. One may, that is, argue that it makes no sense to speak of something as simultaneously existing and not existing and so that we must therefore not allow $\mathbf{E}$ to have contradictory instances. Fortunately for those willing to take this route, there is a very simple way of implementing this idea in **FLFI1**, which is to require $\circ \mathbf{E}(t)$ to hold for every term t. Once this further axiom is accepted, it then follows that $\mathbf{E}_+^{\mathfrak{A}} \cap \mathbf{E}_-^{\mathfrak{A}} = \emptyset$, which corresponds proof-theoretically to the claim that any theorem of the form $\mathbf{E}(t) \wedge \neg \mathbf{E}(t)$ will render trivial every theory having $\circ \mathbf{E}(t)$ among its axioms.

That we are able to express the consistency of the existence predicate is indeed a very good thing about **FLFI1**, but the presence of $\circ$ and $\mathbf{E}$ allows us to do a little bit more than that. For we are also capable of expressing such theses as (I) that consistency is a necessary condition for existence – meaning that contradictions cannot hold of certain objects unless those objects do not exist; or (II) that consistency is a sufficient condition for existence – meaning that whenever a certain claim about an object is known to be consistent, then that object must exist. Although (II) is a thesis most of us would immediately reject as false (for we do not normally think of, say, the consistency of the axioms of a certain mathematical theory as sufficient for the existence of its posits), a view along these lines has already been entertained in the philosophy of mathematics by Mark Balaguer [1995] in the form of his *Full-Blooded Platonim*. And in the context of **FLFI1** we could well express it by means of the schema:

$$\circ A \to \mathbf{E}(t_1) \wedge \cdots \wedge \mathbf{E}(t_n)$$

where $t_1, \ldots, t_n$ are all the terms that occur free in A[21].

(I), on the other hand, is a somewhat more palatable thesis than (II), for it corresponds to the idea that even though the world (and the objects

---

[21]Balaguer [1995] argues that the only feasible way of defending *Platonism* (i.e., the claim that mathematical objects are abstract entities that exist independently of us) against the charges of turning mathematical knowledge into a complete mystery, is to embrace the thesis that every mathematical object that could possibly exist does actually exist. Since he takes the notion of possibility that occurs in this formulation in its broadest sense, he means logical possibility (w.r.t. to classical logic). Therefore, full-blooded Platonism entails that all the objects posited by a non-contradictory mathematical theory exist. Notice, however, that Balaguer frames his thesis in terms of the consistency of whole theories, rather than in terms of single formulas. The formal rendering of (II) within **FLFI1** presented above is thus not equivalent to Balaguer's version, but it gets close enough for the purposes of our illustration.

that inhabit it) is itself consistent, admitting thus no contradictory true descriptions, no non-existent entities are subject to this ontological constraint. According to this view, contradictions about non-existents (e.g., the round square) may well be tolerated without one incurring into metaphysical debits one would otherwise face if one were inclined to admit some violations of the *ontological version of the Law of Non-Contradiction*[22]. As in the case of (II), (I) could also be expressed in **FLFI1** by means of a schema – viz., $\mathbf{E}(t_1) \wedge \cdots \wedge \mathbf{E}(t_n) \to {\circ}A$ – but because the consistency-propagation properties of **LFI1** will continue to hold in **FLFI1**, we can express the same idea by requiring $\forall x_1 \ldots \forall x_n {\circ} P(x_1, \ldots, x_n)$ to hold for every predicate P:

Proposition 25. Let $A \in Form(\mathcal{S}_\mathbf{E})$ be such that $t_1, \ldots, t_n$ are all the terms (not belonging to $\mathcal{V}$[23]) that are free in A. Let Δ be the set:

$$\{\forall x_1 \ldots \forall x_n {\circ} P(x_1, \ldots, x_n) : P \text{ occurs in } A\}$$

Then:

$$\Delta \vdash_\mathbf{F} \mathbf{E}(t_1) \wedge \cdots \wedge \mathbf{E}(t_n) \to {\circ}A$$

Proof: The proof proceeds by a straightforward induction on the complexity of A using the consistency-propagation properties of **FLFI1**.
■

As the reader might have expected, by assuming that consistency and existence are related in the way described in (I), one is also capable of fully recovering classical logic whenever all the objects one is reasoning or theorizing about are known to satisfy **E**. This means that anyone subscribing to (I) is in a position to formally express the view that although classical logic is not entirely suited for us to reason about absolutely everything (which is supposed to comprise existent and non-existent objects alike), its inferences are invariably valid whenever the objects concerned all exist. This fact is an immediate consequence of the following **FLFI1**-version of the Derivability Adjustment Theorem (Theorem 7):

Theorem 26. (Derivability Adjustment Theorem (FLFI1 Version)) Let $\Gamma \cup \{A\} \subseteq Form(\mathcal{S}_\mathbf{E})$ be such that ${\circ}$ occurs neither in (any element of) Γ nor in A. If there is a derivation of A from Γ in classical logic, then $\Gamma, \Delta, \Theta \vdash_\mathbf{F} A$, where:

$$\Delta \subseteq \{\forall x_1 \ldots \forall x_n {\circ} P(x_1, \ldots, x_n) : P \text{ is predicate letter of } \mathcal{S}_\mathbf{E}\}$$

and

$$\Theta \subseteq \{\mathbf{E}(t) : t \in Term(\mathcal{S}_\mathbf{E}) \text{ and } t \notin \mathcal{V}\}$$

---

[22] For a fuller discussion of this idea, see Antunes [2018, 2019a].
[23] Since $\vdash_\mathbf{F} \mathbf{E}(x)$, for every variable x that is free in A, there is no need to include $\mathbf{E}(x)$ in the antecedent of $\mathbf{E}(t_1) \wedge \cdots \wedge \mathbf{E}(t_n) \to {\circ}A$.

4 Final Remarks

In this paper we first went over a few motivations for adopting an objectual interpretation of a logic's first-order quantifiers. We then presented the logic **LFI1** within the framework of a Tarskian objectual semantics, proving soundness and completeness results for a corresponding natural deduction system. While proving those theorems, we indicated the steps that would otherwise be unnecessary if one were dealing with a non-objectual semantics instead. Finally, we attempted to demonstrate the value having an objectual interpretation of **LFI1** ready at our disposal by describing how it can be modified to yield a free logic of formal inconsistency, which we dubbed **FLFI1**. After presenting a natural deduction system for **FLFI1** and proving it to be sound and complete with respect to a non-inclusive dual-domain semantics, we briefly illustrated how having both the consistency operator, ∘, and the existence predicate, E, among the logical primitives of **FLFI1** allows us express certain ontological theses about the relationships between those two notions.

Since our incursion into free logics was merely intended here to serve the purpose of an illustration, we have only scratched the surface of what can be done on the behalf of combining them with the logics of formal inconsistency. As in the case of the **LFI**s, free logics also come in all shapes and sizes, and it would certainly be worth the effort to undertake a thorough investigation about how different **LFI**s can be combined with different free logics. Even if we were to focus exclusively on **LFI1**, as we have done so far, several questions would still have to be addressed, such as, for example, whether and how it can be adapted to yield a free logic with a single-domain semantics, whether and how it can adapted to yield a negative free logic, whether a free theory of definite descriptions can be developed within **FLIF1**, and the list goes on.

References

Antunes, H. (2018), 'On existence, Inconsistency, and Indispensability', *Principia: An International Journal of Epistemology* 22, 7–34.

Antunes, H. (2019a), Contradictions for Free: Towards a Nominalistic Interpretation of Contradictory Theories, PhD thesis, University of Campinas. url: http://repositorio.unicamp.br/jspui/bitstream/REPOSIP/334810/1/Almeida_HenriqueAntunes_D.pdf.

Antunes, H. (2019b), 'Enthymematic Classical Recapture', *The Logic Journal of IGPL*. doi: https://doi.org/10.1093/jigpal/jzy061.

Balaguer, M. (1995), 'A Platonist Epistemology', *Synthese* **103**, 303–325.

Bencivenga, E. (2002), Free logics, *in* D. M. Gabbay & F. Guenthner, eds, 'Handbook of Philosophical Logic', 2 ed., Vol. 5, Springer, pp. 147–196.

Carnielli, W. & Coniglio, M. E. (2016), *Paraconsistent Logic: Consistency, Contradiction and Negation*, Springer.

Carnielli, W. & Rodrigues, A. A. (2017), 'An Epistemic Approach to Paraconsistency: A Logic of Evidence and Truth', *Synthese* **196**, 3798–3813. doi: https://doi.org/10.1007/s11229-017-1621-7.

Carnielli, W. & Rodrigues, A. A. (2019), 'On Epistemic and Ontological Interpretations of Intuitionistic and Paraconsistent Paradigms', *Logic Journal of the IGPL*. doi: https://doi.org/10.1093/jigpal/jzz041.

Carnielli, W., Coniglio, M. & Marcos, J. (2007), Logics of Formal Inconsistency, *in* D. M. Gabbay & F. Guenthner, eds, 'Handbook of Philosophical Logic', Vol. 14, Springer, pp. 1–94.

Carnielli, W., Marcos, J. & de Amo, S. (2000), 'Formal Inconsistency and Evolutionary Databases', *Logic and Logical Philosophy* **8**, 115–152.

da Costa, N. C. A. (1974), 'On the Theory of Inconsistent Formal Systems', *Notre Dame Journal of Formal Logic* **11**, 497–510.

de Amo, S., Carnielli, W. & Marcos, J. (2002), A Logical Framework for Integrating Inconsistent Inconsistent Information in Multiple Databases, *in* T. Eiter & K.-D. Schewe, eds, 'IN INTERNATIONAL SYMPOSIUM ON FOUNDATIONS OF INFORMATION AND KNOWLEDGE SYSTEMS', Springer, pp. 67–84.

D'Ottaviano, I. M. L. & da Costa, N. C. A. (1970), 'Sur un Problème de Jaśkowski', *Comptes Rendus de l'Academie de Sciences de Paris* **270A**, 1349–1353.

D'Ottaviano, I. M. L. & da Costa, N. C. A. (1985), 'The Completeness and Compactness of a Three-Valued First-Order Logic', *Revista Colombiana de Matemáticas* **19**, 77–94.

Hand, M. (2007), Objectual and Substitutional Interpretations of the Quantifiers, *in* Jaquette [2007], pp. 649–674.

Jaquette, D., ed. (2007), *Philosophy of Logic*, Handbook of the Philosophy of Logic, 5 ed., North Holland.

Nolt, J. (2007), Free Logics, *in* Jaquette [2007], pp. 1023–1060.

Omori, H. & Waragai, T. (2011), Some Observations on the Systems LFI1 and LFI1*, *in* F. Morvan, A. M. Tjoa & R. Wagner, eds, 'Proceedings of DEXA2011', IEEE computer society, pp. 320–324.

Quine, W. V. O. (1948), 'On What There Is', *The Review of Metaphysics* 2, 21–38.

Quine, W. V. O. (1969), Existence and Quantification, *in* 'Ontological Relativity and Other Essays', pp. 91–113.

Rodrigues, A., Bueno-Soler, J. & Carnielli, W. (2020), 'Measuring Evidence: A Probabilistic Approach to an Extension of Belnap-Dunn Logic', *Synthese*. doi: https://doi.org/10.1007/s11229-020-02571-w.

Sano, K. & Omori, H. (2013), 'An Expansion of First-Order Dunn-Belnap Logic', *Logic Journal of the IGPL* 22, 458–481.

Tarski, A. (1983), The Concept of Truth in Formalized Langues, *in* J. Corcoran, ed., 'Logic, Semantics, Metamathematics: Papers from 1923 to 1938', Hackett Publishing Company, pp. 152–278.

Troelstra, A. S. & van Dalen, D. (1988), *Constructivism in Mathematics*, Vol. I, North Holland.

van Dalen, D. (2013), *Logic and Structure*, 5 ed., Springer.

Remarks on a nice theorem of Monsieur Glivenko[‡]

Itala M. Loffredo D'Ottaviano*   Evandro Luís Gomes[†]

* Centre for Logic, Epistemology and the History of Science
Department of Philosophy
University of Campinas, Campinas, SP, Brazil
itala@unicamp.br

[†] Department of Philosophy
State University of Maringá, Maringá, PR, Brazil

Centre for Logic, Epistemology and the History of Science
University of Campinas, Campinas, SP, Brazil
elgomes@uem.br

Abstract

In this paper, we revisit Glivenko's paper of 1929, in which he introduced his well-known "translation of double negation" from classical into intuitionistic propositional logic. Following, we discuss Glivenko's interpretation from the approach of some general concepts of translations between logics that we have proposed and studied in a series of previous papers. Finally, we question why Kolmogorov's paper of 1925, in which it was proposed the first formalization for intuitionistic logic and the first translation from classical into intuitionistic logic, is not mentioned by Glivenko.

Introduction

Valery Glivenko (1896/97–1940) was a Ukranian Soviet mathematician. He worked on the intersection of algebraic logic and probability theory, foundations of mathematics, real analysis and mathematical statistics.

He taught at the Moscow Industrial Pedagogical Institute until his death at age 43, and most of his work was published in French. There are a few fundamental theorems bearing his name.

[‡]The authors thank to Fábio Maia Bertato, for having suggested them the reading of Troelstra's paper of 1990.

This paper is about Glivenko's "interpretation" of classical propositional logic into intuitionistic propositional logic, introduced in his fundamental paper *Sur quelques points de la logique de M. Brouwer*, presented at the Académie Royale de Belgique by M. De Donder, and published in 1929 by the *Bulletins de la Classe des Sciences* of the Academy (see Glivenko [1929]).

In a previous paper of 1928, *Sur la logique de M. Brouwer*, also presented at the Académie Royale de Belgique (see Glivenko [1928]), Glivenko had presented, according to him, a proof of a "remarkable" Brouwer's theorem, namely that in his logic it is provable the falsity of the falsity of the Principle of the Excluded Third.

In the paper of 1929, Glivenko claims that it is of great interest to study more general relations between classical logic and the logic of M. Brouwer (Luitzen Egbertus Jan Brouwer, 1881–1966). So, the aim of his paper is to prove two results:

- *If a certain expression of the logic of propositions is provable in classical logic, then the falsity of the falsity of this expression is provable in the Brouwerian logic;*

- *If the falsity of a certain expression of the logic of propositions is provable in classical logic, then this falsity is provable in Brouwerian logic.*

The aim of this paper is to analyse Glivenko's interpretation from classical into intuitionistic propositional logic from the approach of some general concepts of translations between logics that we have proposed in a series of previous papers.

During several years we have studied interrelations between logics by the analysis of translations between them. In 1999, da Silva, D'Ottaviano and Sette proposed a very general definition for the concept of translation between logics; in 1997, Feitosa introduced the concept of conservative translation, studied by Feitosa and D'Ottaviano in 2001; in 2007, Carnielli, Coniglio and D'Ottaviano proposed the concept of contextual translation, in order to have a stricter notion of translation and to solve questions related to conservative translations; recently, Moreira, Feitosa and D'Ottaviano have introduced the concept of abstract contextual translation between logics and proved that this new concept is an intermediate concept, wider than the concepts of conservative and contextual translation (see da Silva, D'Ottaviano & Sette [1999]; Feitosa & D'Ottaviano [2001]; Carnielli, Coniglio & D'Ottaviano [2009]; Moreira [2016]).

In this paper we discuss Glivenko's interpretation founded on our general concepts of translations between logics.

In the first section we recall Glivenko's 1928 paper and present, without detailed proofs, Glivenko's 1929 famous interpretation of the double negation from classical into intuitionistic propositional logic.

In the next section we succinctly present some known historical interpretations from classical into intuitionistic logic, contemporary to Glivenko's one, calling special attention to the translation introduced by Kolmogorov in 1925.

Following, after introducing our general concepts of translations between logics, we show that Glivenko's interpretation is not only a conservative translation but also a contextual translation and an abstract contextual translation from classical into intuitionistic propositional logic.

Finally, we discuss the relevance of Glivenko's 1929 paper, we reference some recent extensions of Glivenko's translation, and we question why Glivenko didn't mention, in his two papers, Kolmogorov's original and pioneering translation from classical into intuitionistic logic.

This work was presented at the 16th International Congress on Logic, Methodology and Philosophy of Science and Technology (CLMPST 2019), held in Prague, August 5–10, 2019, in the special Symposium "Proof and Translation – Glivenko's Theorem 90 Years After", organized by Sara Negri and Peter Schuster.

1 Sur quelques points de la logique de M. Brouwer

Glivenko's paper of 1929 is based on his paper of 1928, both published by the *Bulletins de la Classe des Sciences de la Académie Royale de Belgique*.

1.1 Sur la logique de M. Brouwer

In his short paper of 1928, *Sur la logique de M. Brouwer*, Glivenko introduces an axiomatics for what he called the "Brouwerian Logic of Propositions".

By tacitly using a language having as primitive connectives $\sim$, $\wedge$, $\vee$ and $\supset$, respectively corresponding to negation, conjunction, disjunction and implication, Glivenko adopts the following nine axioms as admissible to the Brouwerian logic of propositions, that we denote by BL:

I. $p \supset p$

II. $(p \supset p) \supset ((q \supset r) \supset (p \supset r))$

III. $p \wedge q \supset p$
IV. $p \wedge q \supset q$
V. $(r \supset q) \supset ((r \supset q) \supset (r \supset p \wedge q))$
VI. $p \supset p \vee q$
VII. $q \supset p \vee q$
VIII. $(p \supset r) \supset ((q \supset r) \supset ((p \vee q) \supset r))$
IX. $(p \supset q) \supset ((p \supset \sim q) \supset \sim p)$

Founded on such axioms, the following important results are proved.

Theorem 1.1. *(Glivenko [1928, p. 226]) In the Brouwerian logic of propositions, BL, the falsity of the falsity of the Principle of the Third Excluded is provable. That is:*

$$\vdash_{BL} \;\; \sim\sim (\sim p \vee p)$$

□

Theorem 1.2. *(Glivenko [1928, p. 226–227]) In BL we have the following results:*

$$p \supset \sim\sim p$$
$$(p \supset q) \supset (\sim q \supset \sim p)$$
$$\sim\sim\sim p \supset \sim p$$

□

Theorem 1.3. *If in classical logic, CPL, the falsity of a certain proposition is a consequence of the Principle of the Third Excluded, then the falsity of this proposition also occurs in the Brouwerian logic. That is:*

$$\vdash_{CL} \;\; \sim p \vee p \supset \sim q \;\; \Rightarrow \;\; \vdash_{BL} \;\; \sim q$$

□

1.2 Glivenko's interpretation from classical logic into intuitionistic logic

In his also short paper of 1929, *Sur quelques points de la logique de M. Brouwer*, Glivenko introduces an extended axiomatics for Brouwer's logic.

In addition to the nine axioms presented in his previous paper, he adds four new axioms:

A. $(p \supset (q \supset r)) \supset (q \supset (p \supset r))$
B. $(p \supset (p \supset r)) \supset (p \supset r)$
C. $p \supset (q \supset p)$
D. $\sim q \supset (q \supset p)$

Glivenko says that M. A. Heyting (Arend Heyting, 1898–1980), in a private letter, had suggested him the Axioms C and D, claiming that the Axioms I–IX plus the Axioms A and B would not be sufficient to express Brouwer's logic.

Hence, the axioms assumed by Glivenko for his Brouwerian logic are:

Axioms I–IX plus Axioms A–D.

Following, it is observed that if the Principle of the Excluded Third is added to the 13 above introduced axioms, a complete system of the (Aristotelian) classical propositional logic is obtained. Here, it is mentioned that the concept of a complete axioms system is the one presented in Hilbert & Ackermann [1928].

The two results of the paper are then presented.

Theorem 1.4. *If a certain expression of the logic of propositions is provable in classical logic, CPL, then the falsity of the falsity of this expression is provable in the Brouwerian logic, BL:*

$$\vdash_{CL} \alpha \quad \Rightarrow \quad \vdash_{BL} \sim\sim \alpha$$

Proof. It is proved that the falsity of the falsity of each one of the expressions I–IX and A–D results from

$$p \supset \sim\sim p.$$

We also have, by the Theorem 1.1, that

$$\sim\sim (\sim p \vee p).$$

It is now sufficient to prove that

$$\sim\sim A$$
$$\underline{\sim\sim (A \supset B)}$$
$$\sim\sim B.$$

□

Corollary 1.5. *If the falsity of a certain expression of the logic of propositions is provable in classical logic, then this falsity is provable in Brouwerian logic. That is:*

$$\vdash_{CL} \sim \alpha \;\Rightarrow\; \vdash_{BL} \sim \alpha.$$

□

We observe that, as the converse of Theorem 1.1 also holds, we have that

$$\vdash_{CL} \alpha \;\Leftrightarrow\; \vdash_{BL} \sim\sim \alpha.$$

We may introduce another simple corollary of Glivenko's results.

Corollary 1.6. *Classical propositional logic CPL is inconsistent only if the Brouwerian logic BL is inconsistent.* □

2 Historical translations

In this section, we present some known historical interpretations from classical into intuitionistic logic, one of them anticipating Glivenko's one.

The first known 'translation' involving classical logic and intuitionistic logic was presented by Kolmogorov in 1925.

After Glivenko's papers, Gödel and Gentzen independently introduced two other interpretations from classical into intuitionistic logic. Their interpretations were developed in order to show the relative consistency of classical arithmetic with respect to intuitionistic arithmetic.

2.1 Kolmogorov 1925: two formal systems

Andrei Nikolaevich Kolmogorov (1903–1987) anticipates in his paper of 1925 not only Heyting's formalization of intuitionistic logic, but also results on the translatability of classical mathematics into intuitionistic mathematics (see Kolmogorov [1925]; Kolmogorov [1977]).

The aim of Kolmogorov "is to explain why" the illegitimate use of the Principle of Excluded Middle in the domain of transfinite arguments "has not led to contradictions". He introduces two formal systems: the system B, the intuitionistic Brouwerian calculus; and the system H, the classical propositional Hilbertian calculus.

Kolmogorov's *general logic of judgements* B has the following axioms and rules:

Axiom K_1: $\varphi \to (\psi \to \varphi)$

Axiom K_2: $(\varphi \to (\varphi \to \psi)) \to (\varphi \to \psi)$

Axiom K_3: $(\varphi \to (\psi \to \sigma)) \to (\psi \to (\varphi \to \sigma))$

Axiom K_4: $(\psi \to \sigma) \to ((\varphi \to \psi) \to (\varphi \to \sigma))$

Axiom K_5: $(\varphi \to \psi) \to ((\varphi \to \overline{\psi}) \to \overline{\varphi})$ [Principle of Contradiction]

- Rule of *Modus Ponens*: $\alpha, \alpha \to \beta / \beta$
- Substituition Rule: $\vdash \alpha(p) / \vdash \alpha(p \mid \beta)$

Kolmogorov's system B differs from the system introduced by Heyting in 1930, by:

Axiom K_H: $\varphi \to (\overline{\varphi} \to \psi)$

that substitutes Axiom K_5.

Johansson's *minimal logic*, published in 1936, coincides with Kolmogorov's logic (see Johansson [1936]), where we have that

$$\vdash_B \quad \varphi \to (\overline{\varphi} \to \overline{\psi})$$

The special logic of judgements H is obtained from the system B, by addition of

Axiom K_6: $\overline{\overline{\varphi}} \to \varphi$

Kolmogorov proves that H is equivalent to Hilbert's 1923 formulation of the classical propositional calculus (see Hilbert [1923]).

Given B and H, Kolmogorov defines inductively a function k associating to every formula α of H, a formula α^k of B by adding a double negation in front of every subformula of α.

Kolmogorov proves that, given a set of axioms $A = \{\alpha_1, \ldots, \alpha_n\}$:

$$A \vdash_H \alpha \Rightarrow A^k \vdash_B \alpha^k$$

where

$$A^k = \{\alpha_1^k, \ldots, \alpha_n^k\}.$$

He finally suggests that a similar result can be extended to quantificational systems and, in general, to all known mathematics, anticipating Gödel's and Gentzen's results on the relative consistency of classical arithmetic with respect to intuitionistic arithmetic.

2.2 Gödel 1933

In a paper of 1933, Kurt Gödel (1902–1978) proves that, if α is a theorem of classical propositional logic, then, under a specific interpretation Gd, the interpretation of α, α^{Gd}, is a theorem of the intuitionistic propositional logic IPL (see Gödel [1933]; Gödel [1986]).

Gödel considers the CPC in the primitive connectives $\neg$ and $\wedge$, and the IPL in the primitive connectives $-$ and Δ, for negation and conjunction respectively, in both logics.

The function $Gd : CPC \to IPC$ is defined by:

1. $(p)^{Gd} =_{df} p$
2. $(\neg \varphi)^{Gd} =_{df} -\varphi^{Gd}$
3. $(\varphi \wedge \psi)^{Gd} =_{df} (\varphi^{Gd} \Delta \psi^{Gd})$
4. $(\varphi \vee \psi)^{Gd} =_{df} -(-\varphi^{Gd} \Delta - \psi^{Gd})$
5. $(\varphi \to \psi)^{Gd} =_{df} -(\varphi^{Gd} \Delta - \psi^{Gd})$.

Theorem 2.1. $\vdash_{CPC} \alpha \Rightarrow \vdash_{IPC} \alpha^{Gd}$.

Gödel shows that this result is also valid relatively to intuitionistic arithmetic H' and classical number theory PA, by proving that

$$\vdash_{PA} \alpha \Leftrightarrow \vdash_{H'} (\alpha)^{Gd}. \qquad \square$$

Then Gödel states that the above theorem, extended to the corresponding arithmetics, gives a proof of the relative consistency of classical logic and arithmetic relative to the respective intuitionistic theories.

In the paper, Gödel explicitly mentions Glivenko's result, but he does not mention Kolmogorov's paper.

2.3 Gentzen 1933

Also in 1933, Gerhard Gentzen (1909–1945) presents a rigorous and complete paper, with a simpler translation from CPC into IPC. The aim of Gentzen [1936] is to show that the "applications of the law of double negation in proofs in classical arithmetic can in many instances be eliminated" (see also Gentzen [1969]).

Gentzen introduces the "transformation" Gz from the language of CPC (also extended to the language of classical arithmetic CAR) into IPC (also extended to the language of intuitionistic arithmetic IAR):

1. $(p)^{Gz} =_{df} \neg\neg p$

2. $(\neg\varphi)^{Gz} =_{df} \neg\varphi^{Gz}$
3. $(\varphi \wedge \psi)^{Gz} =_{df} \varphi^{Gz} \wedge \psi^{Gz}$
4. $(\varphi \vee \psi)^{Gz} =_{df} \neg(\neg\varphi^{Gz} \wedge \neg\psi^{Gz})$
5. $(\varphi \to \psi)^{Gz} =_{df} (\varphi^{Gz} \to \psi^{Gz})$
6. $(\forall x\varphi)^{Gz} =_{df} \forall x\varphi^{Gz}$
7. $(\exists x\varphi)^{Gz} =_{df} \neg(\forall x \neg\varphi^{Gz})$
8. other formulas of arithmetic are translated into themselves.

He proves the following theorems.

Theorem 2.2. $\Gamma \vdash_{CPC} \alpha \Leftrightarrow \Gamma^{Gz} \vdash_{IPC} \alpha^{Gz}$. □

Theorem 2.3. $\Gamma \vdash_{CAr} \alpha \Leftrightarrow \Gamma^{Gz} \vdash_{IAr} \alpha^{Gz}$. □

As a consequence, he presents a constructive proof of the consistency of elementary classical arithmetic with respect to intuitionistic arithmetic.

Theorem 2.4. $Cons(IAr) \Rightarrow Cons(CAr)$. □

3 Translations between logics: general definitions

In spite of Kolmogorov, Glivenko, Gödel and Gentzen dealing with interrelations between the studied systems, they are not interested in the meaning of the concept of translation between logics in general. And since then, interpretations between logics have been used to different purposes.

Prawitz & Malmnäs [1968] surveys these historical papers and is the first paper in which a general definition for the concept of translation between logical systems is introduced. Other authors have also proposed definitions for this concept, but we will not discuss their approaches in this paper (see, for instance, Feitosa & D'Ottaviano [2001]).

In this section we introduce our concepts of translation between logics, conservative translation, contextual translation and abstract contextual translation, in order to analyse Givenko's double negation interpretation from our general approaches.

Da Silva, D'Ottaviano and Sette [1999], explicitly interested in the study of interrelations between logic systems in general, propose a general definition for the concept of translation between logics, in order to single out what seems to be in fact the essential feature of a logical translation – logics are characterized as pairs constituted by a set (ignoring the fact that in general a logic deals with formulas of a language) and a consequence operator, and translations between logics are defined as maps preserving consequence relations.

Definition 3.1. *A logic $\mathbb{A}$ is a pair $\langle A, C \rangle$, where the set A is the domain of $\mathbb{A}$ and C is a consequence operator in A, that is, a function $C : \mathcal{P}(A) \to \mathcal{P}(A)$ that satisfies, for $X, Y \subseteq A$:*

(i) $X \subseteq C(X)$

(ii) *If* $X \subseteq Y$ *then* $C(X) \subseteq C(Y)$

(iii) $C(C(X)) \subseteq C(X)$.

The usual concepts and results on closure spaces are here assumed. The general definition of translation between logics is then presented.

Definition 3.2. *A translation from a logic $\mathbb{A}$ into a logic $\mathbb{B}$ is a map $t : A \to B$ such that, for any $X \subseteq A$:*

$$t(C_{\mathbb{A}}(X)) \subseteq C_{\mathbb{B}}(t(X)).$$

Of course, it is possible to consider logics defined over formal languages.

Definition 3.3. *A logic system defined over L is a pair $\mathbf{L} = (L, C)$, where L is a formal language and C is a structural consequence operator in the free algebra $For(L)$ of the formulas of L.*

We will use the symbol $\mathbf{L}$ when the system is over a language with particular logical operators, and A, B for systems over sets only.

If $\mathbf{L}_A$ and $\mathbf{L}_B$ are determined by formal languages with associated syntactic consequence relations $\vdash A$ and $\vdash B$, respectively, then t is a translation if, and only if, for $\Gamma \cup \{\alpha\} \subseteq For(L_A)$:

$$\Gamma \vdash_A \alpha \quad \text{implies} \quad t(\Gamma) \vdash_B t(\alpha). \tag{1}$$

An initial treatment of a theory of translations between logics is presented by da Silva, D'Ottaviano & Sette [1999]. And a very recent survey about our general definitions and results is in D'Ottaviano & Feitosa [2019].

An important subclass of translations, the conservative translations, has been investigated by Feitosa and D'Ottaviano.

Definition 3.4. *Let $\mathbb{A}$ and $\mathbb{B}$ be logics. A conservative translation from $\mathbb{A}$ into $\mathbb{B}$ is a function $t : A \to B$ such that, for every set $X \cup \{x\} \subseteq A$,*

$$x \in C_A(X) \quad \text{if, and only if,} \quad t(x) \in C_B(t(X)).$$

In terms of consequence relations, $t : Form(L_1) \to Form(L_2)$ is a *conservative translation* when, for every $\Gamma \cup \{\alpha\} \subseteq Form(L_1)$,

$$\Gamma \vdash_{C_1} \alpha \quad \text{if, and only if,} \quad t(\Gamma) \vdash_{C_2} t(\alpha).$$

Feitosa and D'Ottaviano have introduced conservative translations involving classical logic, intuitionistic logics, modal logics, Łukasiewicz's and Post's many-valued logics, several paraconsistent logics and predicate logics, dealing with syntactic results, algebraic semantics and matrix semantics (see D'Ottaviano & Feitosa [1999a, 1999b, and 2000], and Feitosa & D'Ottaviano [2001]). And surveys about our general definitions and results may be seen in D'Ottaviano & Feitosa [2007] and D'Ottaviano [2013].

As conservative translations do not preserve the triviality of the source logic, Coniglio & Carnielli [2002] reintroduce the concept of translation and conservative translation from the concept of transfer between two abstract logics over two-sorted languages.

However, in spite of transfers being pretty general, they require that every connective of the source logic must be translated into another connective, and this is too restrictive.

Carnielli, Coniglio & D'Ottaviano [2009] introduce the concept of contextual translation, aimimg to obtain an intermediate concept between translation and conservative translation. Contextual translations are mappings between languages preserving the meta-properties of the source logic, which are defined in a formal first-order meta-language. In order to present the definition of contextual translation, we should introduce several previous formal definitions, for a given language (see D'Ottaviano & Feitosa [2019, p. 11–12]).

Definition 3.5. *A contextual translation (c-translation) $f : L \to L'$ is a mapping $f : L(C, \Sigma) \to L'(C', \Sigma')$ such that L' satisfies the meta-property $\hat{f}(P)$ whenever L satisfies the meta-property (P), where $\hat{f}(P)$ is the adequate image of the components of P.*

We observe that every c-translation is a translation in our sense. But contextual translations and conservative translations showed to be essentially independent concepts – we have studied conservative translations that are not contextual, contextual translations that are not conservative, and translations that are both conservative and contextual.

Rodrigues Moreira [2016] introduces the concept of abstract contextual translation, successfully expecting to obtain the previously mentioned intermediate concept of translation.

Definition 3.6. Let $\mathbb{L}_1 = (L_1, C_1)$ and $\mathbb{L}_2 = (L_2, C_2)$ be two logics. An abstract contextual translation $t : \mathbb{L}_1 \to \mathbb{L}_2$ is a function $t : L_1 \to L_2$ such that, for every set $X_i \cup \{x_i\} \subseteq L_1$, $i \in \{1, 2, \ldots, n\}$, we have that:

If
$$x_1 \in C_1(X_1), x_2 \in C_1(X_2), \ldots, x_{n-1} \in C_1(X_{n-1}) \Rightarrow x_n \in C_1(X_n)$$
then
$$t(x_1) \in C_2(t(X_1)), t(x_2) \in C_2(t(X_2)), \ldots, t(x_{n-1}) \in C_2(t(X_{n-1})) \Rightarrow$$
$$\Rightarrow (x_n) \in C_2(t(X_n)).$$

Abstract contextual translations are particular cases of our general translations between logics and extend our concepts of conservative and contextual translations.

Theorem 3.7. *(1) If t is a contextual translation between Tarskian logics, then t is an abstract contextual translation; (2) if t is a conservative translation, then t is an abstract contextual translation.* $\square$

Gödel [1933] interpretation is not a translation in our sense, even in the propositional level, for, despite preserving theoremhood, it does not preserve derivability (see D'Ottaviano & Feitosa [2011]).

Kolmogorov's and Gentzen's interpretations are conservative and contextual translations from classical into intuitionistic logic.

We have obtained some relevant results to the study of general properties of logic systems from the point of view of translations between them. The proofs can be seen in Feitosa [1997], Feitosa & D'Ottaviano [2001], D'Ottaviano & Feitosa [2007], and Rodrigues Moreira [2016].

4 Glivenko revisited

Let's recall that Glivenko's 1929 function $G : Form(CL) \to Form(BL)$ is such that
$$\vdash_{CL} \alpha \Rightarrow \vdash_{BL} \sim\sim \alpha.$$

According to the concepts that we have introduced in the previous section, we pose some questions concerning Glivenko's function from classical into Brouwerian propositional logic:

- Is it a *translation*?
- Is it a *conservative translation*?

- Is it a *contextual translation*?
- Is it an *abstract contextual translation*?

In this section we will answer such questions positively.

4.1 Glivenko's 'interpretation' is a translation and is conservative

Yes, we can prove, by using algebraic semantics, that G is a conservative translation from CPC into BL. For a detailed proof, see Feitosa [1997], and Feitosa & D'Ottaviano [2004].

The following result is well known.

Proposition 4.1. *Let $H = (H, 1, \Rightarrow, \wedge, \vee, -)$ be a Heyting algebra. If $B = \{--a : a \in H\}$, then $B_{\neg\neg} = (B, \wedge, \vee, -)$ is a Boolean algebra.* □

Now, let $D : H \to B_{\neg\neg}$ such that $D : a \longmapsto --a$.

Lemma 4.2. *Given the function $G : CPC \to IPC$, with $G(\alpha) = \sim\sim \alpha$, we have that, for $\Gamma \cup \{\alpha\} \subseteq Form(CPC)$ and for every valuation v_H from IPC into a Heyting algebra H:*

$$v_H(G(\alpha)) = D \circ v_H \circ G(\alpha).$$

The function $D \circ v_H \circ G$ determines a valuation $v : CPC \to B_{\neg\neg}$, given by

$$D \circ v_H \circ (G(\alpha)) = v_B(\alpha).$$

□

Theorem 4.3. *Glivenko's function G is a conservative translation from CPC into the intuitionistic propositional calculus BL (IPC).*

Proof: ($\Rightarrow$). If $G(\Gamma) \not\vdash G(\alpha)$, as every Heyting algebra is strongly adequate algebraic model for IPC, then there is a valuation $[v_\circ]_H$ into a Heyting algebra H such that, $[v_\circ]_H(G(\alpha)) = 1$, for every $\gamma \in \Gamma$, and $[v_\circ]_H(G(\alpha)) \neq 1$. By the previous Lemma, $v_B(\alpha) = v_H(G(\alpha))$, where v_B takes values in the Boolean algebra $B_{\neg\neg}$, so $[v_\circ]_B(\alpha) = 1$, for every $\gamma \in \Gamma$, and $[v_\circ]_B(\alpha) \neq 1$. Hence, $\Gamma \not\vDash \alpha$, that is, $\Gamma \not\vdash \alpha$. □

Proof: ($\Leftarrow$). If $\Gamma \not\vdash G\alpha$, then there is a valuation $[v_\circ]_B$ into a Boolean algebra B, such that $[v_\circ]_B(\alpha) = 1$, for every $\gamma \in \Gamma$, and $[v_\circ]_B(\alpha) \neq 1_B$. As every Boolean algebra is a Heyting algebra, let us consider $B_{\neg\neg} = \{--a : a \in B\} = B$. By the Lemma, $v_B(G(\alpha)) = v_{B_{\neg\neg}}(\alpha)$, then there is a valuation $[v_\circ]_B$ into a Heyting algebra such that $[v_\circ]_B(G(\gamma)) = 1_{B_{\neg\neg}} = 1_B$, for every $\gamma \in \Gamma$, and $[v_\circ]_B(G(\alpha)) \neq 1_{B_{\neg\neg}} = 1_B$. Hence, $G(\Gamma) \not\vdash G(\alpha)$. □

Hence, we have proved that $\Gamma \vdash_{C_A} \alpha$ implies $t(\Gamma) \vdash_{C_B} t(\alpha)$, and that $\Gamma \vdash_{C_1} \alpha$ if, and only if, $t(\Gamma) \vdash_{C_2} t(\alpha)$. That is, according to Definition 3.1 and Definiton 3.3, respectively, we have proved that G, besides being a translation, is also a conservative translation from classical propositional logic (CPL) into intuitionistic propositional logic (IPL).

4.2 Glivenko's 'interpretation' is a contextual translation

According to Definition 3.5, in order to prove that G is a contextual translation, we have to show that the 'image' (through G) of every deduction rule (meta-property) of CPL is a meta-property that is satisfied by BL.

So, as the only meta-property of CPL is the Rule of Modus Ponens, we have to prove the following result.

Theorem 4.4. *For every formula φ and ψ,*

$$\frac{\neg\neg\varphi \quad \neg\neg(\varphi \to \psi)}{\neg\neg\psi}$$

is satisfied in BL.

Proof.

$$\begin{array}{ll}
\neg\neg\varphi & \text{(premise)} \\
\neg\neg(\varphi \to \psi) & \text{(premise)} \\
\neg\neg\varphi \to \neg\neg\psi & \text{(Theor. of } BL) \\
\neg\neg\psi & \text{(Rule MP of } BL)
\end{array}$$

$\square$

4.3 Glivenko's 'interpretation' is an abstract contextual translation

According to Theorem 3.7, as G is conservative (and contextual), then G is an abstract contextual translation from CPL into BL (IPL).

5 Concluding remarks: Why doesn't Glivenko mention Kolmogorov?

In the previous section, we have proved that Glivenko's double negation interpretation from classical into intuitionistic propositional logic preserves and conserves derivability, and preserves the meta-rules of the source logic.

In spite of Glivenko's function having not been the first translation from classical into intuitionistic logic in the literature, it seems that some of Glivenko's contemporary logicians like for instance Gödel and Gentzen did not know Kolmogorov's paper of 1925. We may consider that it was a breakthrough at the time.

Besides being adequate for the study of interrelations between logics from the approach of our general theory of translations between logics, it can really be applied in many contexts.

It can be recast in order to relate logics other than *CPC* and *IPC*, such as logics intermediate between these and predicate calculi. Several authors, in more recent papers, have revisited Glivenko's theorems and applied his translation to other logics. For instance, we mention, among other papers in the literature, Cignoli & Torrens [2003; 2004], Galatos & Ono [2006], Ono [2009], Ferreira & Oliva [2012], and Farhami & Ono [2012].

So, we do consider that Glivenko's original results remain of interest.

But we have a question that cannot remain silent – why does not Glivenko mention the Kolmogorov [1925] paper in his papers of 1928 and 1929?

Kolmogorov was a little younger than Glivenko, about seven years, for he was born in 1903. In 1920, he began to study at the Moscow State University, where Glivenko had also studied. In 1922, Kolmogorov gained international recognition for constructing a Fourier series that diverges almost everywhere (see Kolmogorov [1923]). In 1929, Kolmogorov earned his Doctor of Philosophy (Ph.D) degree, from the Moscow State University, in 1931 becoming a professor at the Moscow State University.

In spite of being a mathematician and also living in Moscow, did'nt Glivenko know Kolmogorov, didn't he know the Kolmogorov paper?

Looking for some references, for some possible correspondence among Glivenko, Heyting, and Kolmogorov – recall that Glivenko, in his paper of 1929, mentions a suggestion addressed to him by Heyting – very recently we found a beautiful paper by Anne Troelstra (1939–2019), *On the early history of intuitionistic logic*, published in the *Proceedings of the Summer School and Conference on Mathematical Logic honourably dedicated to the Ninetieth Anniversary of Arend Heyting (1898–1980)*, held in Bulgaria in 1988 (see Troelstra [1990]).

In his paper, Troelstra describes the early history of intuitionistic logic, its formalization and the genesis of the so-called Brouwer-Heyting-Kolmogorov interpretation. At the end of the paper some precious source material is presented, in particular letters from Becker, Brouwer, Bernays,

Glivenko and Kolmogorov to Heyting.

Troelstra reproduces six letters from Glivenko to Heyting, dated July 4th 1928, October 13th 1928, October 18th 1928, October 30th 1928, November 13th 1928, and October 24th, 1933.

The letters from 1928 "document the discovery of the Glivenko theorem". According to the letter of July 4th, Glivenko acknowledges the receipt of a letter from Heyting and says that he is pleased to know that their views on the problem of the formalization of the intuitionistic logic and mathematics completely coincide. He reports that, relatively to his *Note* published in Brussels, he had just showed that $\sim\sim (\sim\sim p \to p)$ is not provable in Brouwerian logic.

In October 18th, 1928, Glivenko communicates to Heyting his theorem.

In October 30th, he comments that he intends to publish his result in the *Bulletins de l'Académie Royale de Belgique*, well understood if Heyting's work that is going to appear in the " 'Math. Ann.' contient dejá ce résultat".

However, in November 13th of the same year, Glivenko writes to Heyting that he wants to publish his own result, independently of Heyting's paper.

Almost five years latter, in the letter of October 12th 1933, Glivenko acknowledges the receipt of Heyting's paper and comments other non-related questions.

In October 24th 1933, Glivenko discusses, in French, Kolmogorov's "calculus of problems" interpretation – "His conception is, without doubt, very interesting" – and mentions the possibility of taking falsehood instead of negation as primitive (see Kolmogorov [1932]).

Now, let us call attention to Glivenko's letter of October 13th 1928, sent to Heyting before the publication of his double negation translation of 1929. Glivenko asserts, in French: "In the t. 32 of the "Recueil Mathématique" of Moscow (1925) M. Kolmogoroff, one of the most ingenious Russian mathematicians, published a paper entitled "*Sur le principe de tertium non datur*" (in Russian), where he proposes an axiomatics of the intuitionistic logic of propositions. There, he rejects, not only the principle of the third excluded, but also the principle $(\sim p \to (q \to p))$ (although he admits the principle $(p \to (q \to p))$). It is precisely after this restriction that my conclusion is true, that is, that the falsity of the falsity of the principle has no place."

Troelstra also reproduces three letters from Kolmogorov to Heyting, the last two ones undated. In the first letter, dated October 21th 1931, Kolmogorov points to some critics to Heyting's paper of 1930. In the second one, we have more comments addressed to Heyting's paper.

In his third letter, in German, Kolmogorov mentions his paper of 1925, claiming that he had anticipated Gödel's paper of 1933.

Ergo, founded on the letters from Glivenko to Heyting, we have no doubt that Glivenko, in spite of having not mentioned Kolmogorov in his papers of 1928 and 1929, knew the above mentioned Kolmogorov's papers. In particular, Glivenko knew the paper published by Kolmogorov in 1925, with the translation of "double negations" from the classical propositional Hilbertian calculus H into the intuitionistic Brouwerian calculus B.

Glivenko died prematurely in 1940. It was Kolmogorov who published, in 1941, in Russian, in the *Uspekhi Matem. Nauk*, Glivenko's obituary – *Obituary: Valeri Ivanovich Glivenko (1897–1940)*.

References

Carnielli, W. A., Coniglio, M. E. & D'Ottaviano, I. M. L. (2009), 'New dimensions on translations between logic', *Logica Universalis* **3**, 1–19.

Cignoli, R. & Torrens, A. (2003), 'Hájek basic fuzzy logic and Łukasiewicz infinite-valued logics', *Archive for Mathematical Logics* **4**(2), 361–370.

Cignoli, R. & Torrens, A. (2004), 'Glivenko like theorems in natural expansions of BCK-logic', *Mathematical Logic Quarterly* **50**(2), 111–125.

Coniglio, M. E. & Carnielli, W. A. (2002), 'Transfers between logics and their applications', *Studia Logica* **72**(3), 367–400.

da Silva, J. J., D'Ottaviano, I. M. L. & Sette, A. M. (1999), Translations between logics, *in* X. Caicedo & C. H. Montenegro, eds, 'Models, algebras and proofs', Vol. 203 of *Lectures Notes in Pure and Applied Mathematics*, New York: Marcel Dekker, pp. 435–448.

D'Ottaviano, I. M. L. & Feitosa, H. A. (1999a), 'Conservative translations and model-theoretic translations', *Manuscrito - International Journal of Philosophy* **22**(2), 117–132.

D'Ottaviano, I. M. L. & Feitosa, H. A. (1999b), 'Many-valued logics and translations', *Journal of Applied Non-Classical Logics* **9**(1), 121–140.

D'Ottaviano, I. M. L. & Feitosa, H. A. (2000), 'Paraconsistent logics and translations', *Synthese* **125**, 77–95.

D'Ottaviano, I. M. L. (2013), Translations as representations between logics, *in* E. Agazzi, ed., 'Representation and explanation in the sciences', Milan: FrancoAngeli, pp. 228–243.

D'Ottaviano, I. M. L. & Feitosa, H. A. (2007), Deductive systems and translations, in J.-Y. Béziau & A. Costa-Leite, eds, 'Perspectives on universal logic', Monza: Polimetrica International Scientific Publisher, chapter Deductive systems and translations, pp. 125–157.

D'Ottaviano, I. M. L. & Feitosa, H. A. (2011), On Gödel's modal interpretation of intuitionistic logic, in 'Anthology of Universal Logic: from Paul Hertz to Dov Gabbay', Springer Basel: Studies in Universal Logic.

D'Ottaviano, I. M. L. & Feitosa, H. A. (2019), Translations between logics: a survey, in G. M. Mars, P. Weingartner & B. Ritter, eds, 'The Proceedings of the 14th International Ludwig Wittgenstein Symposium, 2018', Berlin: De Gruyter, pp. 121–140.

Farhami, H. & Ono, H. (2012), 'Glivenko theorems and negative translations in substructural predicate logics', *Archive for Mathematical Logics* **51**(7–8), 695–707.

Feitosa, H. A. (1997), Traduções conservativas [Conservative translations, in Portuguese], phdthesis, Institute of Philosophy and Human Sciences, State University of Campinas.

Feitosa, H. A. & D'Ottaviano, I. M. L. (2001), 'Conservative translations', *Annals of Pure and Applied Logic* **108**, 205–227.

Feitosa, H. A. & D'Ottaviano, I. M. L. (2004), Um olhar algébrico sobre as traduções intuicionistas [an algebraic outlook across the intuitionistic translations, in portuguese], in F. T. Sautter, H. A. Feitosa & W. A. Carnielli, eds, 'Lógica e Filosofia da Lógica [Logic and Philosophy of Logic, in Portuguese]', Vol. 40 of *Coleção CLE*, Campinas: Centro de lógica, Epistemologia e História da Ciência,Unicamp, pp. 49–77.

Ferreira, G. & Oliva, P. (2012), On the relation between various negative translations, in W. Berger & H. Schwichtenberg, eds, 'Logic, Construction, Computation', Vol. 3 of *Mathematical Logic Series*, Ontos-Verlag, pp. 227–258.

Galatos, N. & Ono, H. (2006), 'Glivenko theorems for substructural logics over fl', *Journal of Symbolic Logic* **71**(4), 1353–1384.

Gentzen, G. (1936), 'Die widerspruchsfreiheit der reinem zahlentheorie', *Mathematische Annalen* **112**, 493–565.

Gentzen, G. (1969), On the relation between intuitionist and classical arithmetic, in M. E. Szabo, ed., 'The Collected Works of Gerhard Gentzen', Amsterdam: North-Holland, pp. 53–67.

Glivenko, V. I. (1928), 'Sur la logique de m. brouwer', *Académie Royale de Belgique: Bulletins de la Classe des Sciences* **14**, 225–228. Série 5.

Glivenko, V. I. (1929), 'Sur quelques points de la logique de m. brouwer', *Académie Royale de Belgique: Bulletins de la Classe des Sciences* **15**, 183–188. Série 5.

Gödel, K. (1933), Zur intuitionistischen arithmetik und zahlentheorie, *in* 'Ergebnisse eines mathematischen Kolloquiums', Vol. 4, pp. 34–38.

Gödel, K. (1986), On intuitionistic arithmetic and number theory, *in* S. Feferman et al., eds, 'Collected Works', Vol. 1, Oxford: Oxford University Press, pp. 282–285.

Hilbert, D. (1923), 'Die logischen grundlagen der mathematik', *Mathematische Annalen* **88**, 151–165.

Hilbert, D. & Ackermann, W. (1928), *Grundzüge der theoretischen Logik*, Berlin: Springer.

Johansson, I. (1936), 'Die minimalkalküll, ein reduzierter intuitionistischer formalismus', *Compositio Mathematicæ* **4**, 119–136.

Kolmogorov, A. N. (1923), 'Une série de fourier-lebesgue divergente presque partout [a fourier–lebesgue series that diverges almost everywhere, in french]', *Fundamenta Mathematicae* **4**(1), 324–328.

Kolmogorov, A. N. (1925), 'O printsipe tertium non datur', *Mam. cb* **32**, 646–667.

Kolmogorov, A. N. (1932), 'Zur deutung der intuitionistischen logik', *Math. Z.* **35**, 58–65.

Kolmogorov, A. N. (1977), On the principle of excluded middle, *in* J. Heijenoort, ed., 'From Frege to Gödel: a source book in Mathematical Logic 1879–1931', Cambridge, MA: Harvard University Press, pp. 414–437.

Moreira, A. P. R. (2016), Sobre traduções entre lógicas: relações entre traduções conservativas e traduções contextuais abstratas [On translations between logics: relations between conservative translations and abstract contextual translations, in Portuguese], phdthesis, Institute of Philosophy and Human Sciences, University of Campinas.

Ono, H. (2009), 'Glivenko theorems revisited', *Annals of Pure and Applied Logic* **161**(2), 246–250.

Prawitz, D. & Malmnäs, P. E. (1968), A survey of some connections between classical, intuitionistic and minimal logic, in H. Schimidt et al., eds, 'Contributions to mathematical logic', Amsterdam: North-Holland, pp. 215–229.

Troelstra, A. S. (1990), On the early history of intuitionistic logic, in P. P. Petkov, ed., 'Mathematical Logic (Proceedings of the Summer School and Conference on Mathematical Logic, honourably dedicated to the Ninetieth Anniversary of Arend Heyting (1898–1980), held September 13–23, 1988, in Chaika, Bulgaria)', New York: Plenum Press, pp. 3–17.

The Arithmetization-Free Component of Gödel's Proof

Rodrigo A. Freire

* Department of Philosophy
University of Brasília
rodrigofreire@unb.br

Dedicated to Professor Paulo Augusto Silva Veloso

1 Introduction

There is a widespread understanding according to which (i) a certain amount of elementary arithmetic and the technique of arithmetization are essential components of Gödelian proofs and (ii) the method developed by Gödel is very general and the arithmetic details involved in (i) can vary widely as they do not really matter. For instance, this understanding is stated in the following paragraph of [Raatikainen, 2020].

> The next essential step of Gödel's proof is to take the language of a formal system, which is always precisely defined (this is part of being a formal system), and fix a correspondence of a certain kind between the expressions of that language and the system of natural numbers – a coding, "arithmetization", or "Gödel numbering", of the language. There are many possible ways of accomplishing this, and the details do not really matter (for some more details of one quite standard approach, see the supplementary document Gödel Numbering). The essential point is that the chosen mapping is effective: it is always possible to pass, purely mechanically, from an expression to its code number, and from a number to the corresponding expression. Today, when most of us are familiar with computers and the fact that so many things can be coded by zeros and ones, the possibility of such an arithmetization is hardly surprising.

Although this claim is correct, it is also true that there is a general pattern in the proof that is not concerned with arithmetic or computability. This general pattern has been known for a while, it is presented in

[Tarski et al., 2010], section II.2, but it has been neglected by the specialized discussions on the subject. On the other hand, the role of arithmetization is traditionally praised very high, expounded with great attention to detail in the canonical textbooks, even if those details are not really relevant. Gödel's theorems are a subject of great importance and there are many reasons to detach the arithmetization-free component of Gödel's proof and to make it more widely discussed. Here we give a few.

1. The direct implications of the general pattern in Gödel's proof to the foundations of mathematics are also striking.

2. It streamlines the idea of a metamathematical reconstruction of semantical antinomies, arguably the conceptual core of Gödel's proof.

3. It applies to all consistent first-order theories, decidable and undecidable, complete and incomplete. Hence it belongs to general logic.

4. There is no opposition between the general pattern and arithmetization, they are complementary.

These points justify our interest in the arithmetization-free component of Gödel's proof. We shall see that this arithmetization-free component already shows that a representability failure is present in every first-order theory T: Either T is incapable of representing the diagonalization function, or it is incapable of representing its own validity, or both. We shall analyze to which extent this failure blocks the idea of a formal representation of mathematics as a whole. The remaining points were already clearly emphasized in [Tarski et al., 2010], pages 47 and 48.

> Theorem 1 and its proof represent the metamathematical reconstruction and generalization of arguments involved in various semantical antinomies and, in particular, in the antinomy of the liar. The idea of this reconstruction and the realization of its far-reaching implications is due to Gödel [Gödel, 1931]. The present version of this reconstruction is distinguished by its generality and simplicity. It applies to arbitrary formalized theories, and not only to those in which a comprehensive fragment of arithmetic of natural numbers can be developed; to a large extent it is independent of the way in which the notion of validity has been defined for a given theory, and in particular it

does not involve the notion of a formal proof within this theory; it does not use the apparatus of recursive functions – although this apparatus will play a fundamental role in applications of Theorem 1 to the decision problem.

Theorem 1 in this quote correspond, with some minor variations, to the pair diagonalization lemma and impossibility theorem given below. This splitting is justified for the diagonalization lemma can be applied to obtain other results which are closely related: For instance, the undefinability of truth predicates is a direct consequence.

2 Naming Formulas and Diagonalization

Gödel's technique aims to translate metamathematical assertions about a formal system into statements which can be expressed within the system itself. Traditionally, this is accomplished by an *arithmetization* of the formal system, a clear explanation of which can be found in [Mendelson, 1997], page 192, footnote:

> An *arithmetization* of a theory K is a one-one function g from the set of symbols of K, expressions of K and finite sequences of expressions of K into the set of positive integers. The following conditions are to be satisfied by the function g: (1) g is effectively computable; (2) there is an effective procedure that determines whether any given positive integer m is in the range of g and, if m is in the range of g, the procedure finds the object x such that $g(x) = m$.

Details involved in a particular construction of an arithmetization do not matter, as long as the resulting g satisfies the conditions explained in Mendelson's footnote. The context also requires that K must have closed terms playing the role of numerals. Then, a corresponding closed term $n(x)$ can be assigned to the value $g(x)$, and it is a name of x within K. In particular, a naming of the formulas is obtained, that is, a closed term is assigned to each formula.

Instead of constructing an arithmetization function g, with all the restrictions above explained, then defining a naming of expressions, we shall consider any naming of formulas in a formal theory without further restrictions. That is, many things can be accomplished by considering an *arbitrary* function n from the set of formulas to the set of closed terms of an *arbitrary* first-order theory T. Of course, we are assuming that there are closed terms in T and that T is at least a first-order theory with equality, but that is all.

Let us stick to the individual variable u. Let T and n be as above, an arbitrary first-order theory with equality containing closed terms and an arbitrary naming of formulas, respectively. Consider the *diagonalization* function, the function that assigns $A_u[n(A)]$ to each formula A in the language of T the formula, in which $A_u[n(A)]$ is the formula obtained from A by replacing all free occurrences of u by $n(A)$. This is a very important function assigning a formula of T to each formula of T. The formula A is not required to contain free occurrences of u in the definition of $A_u[n(A)]$, the substitution of the free occurrences of u by $n(A)$. If it does not contain any, the diagonalization of A is A itself.

Let T, n and u be as above. Let D be a formula and v a variable distinct from u. We say that D with u and v represent the diagonalization in the pair (T, n) if, and only if,

$$T \vdash \forall v (D_u[n(A)] \leftrightarrow v = n(A_u[n(A)])),$$

for each formula A. If there are a formula D and a variable v distinct from u such that D with u and v represent the diagonalization in the pair (T, n), then we say that (T, n) is a *diagonalizable pair*.

Theorem 1. [Diagonalization Lemma]

Let T be a first-order theory with equality containing at least one closed term, and let n be an arbitrary function from the set of formulas to the set of closed terms. Assume that (T, n) is a diagonalizable pair. If W is a formula whose only free variable is v, then there is a sentence B such that

$$T \vdash W_v[n(B)] \leftrightarrow B.$$

Proof. From the assumption, there are a formula D and a variable v distinct from u such that D with u and v represent the diagonalization in the pair (T, n).

Let A be $\exists v (D \wedge W)$ and let B be the diagonalization of A. Then, B satisfies our condition. Indeed, B is $\exists v (D_u[n(A)] \wedge W)$, which, from the assumption, is equivalent to $\exists v (v = n(B) \wedge W)$, hence to $W_v[n(B)]$.

More formally, the following sequence provides a deduction of the desired equivalence in T.

1. $B \leftrightarrow \exists v(D_u[n(A)] \wedge W)$ (definition of B)

2. $\forall v(D_u[n(A)] \leftrightarrow v = n(B))$ (representing the diagonalization)

3. $\exists v(D_u[n(A)] \wedge W) \leftrightarrow \exists v(v = n(B) \wedge W)$ (from equivalence 2)

4. $\exists v(v = n(B) \wedge W) \leftrightarrow W_v[n(B)]$ (elimination of equality)

5. $W_v[n(B)] \leftrightarrow B$ (tautological consequence of 1, 3 and 4)

Our choice of B as the diagonalization of A becomes more natural if one follows steps 4, 3 and 1 of the above sequence in this ordering. □

For Peano Arithmetic (PA), there is a function n such that (PA, n) is diagonalizable pair. Indeed, just name the formulas which contain free occurrences of u with the numerals corresponding to odd numbers. If A is one of those formulas, define $n(A_u[A])$ to be the term $n(A) + n(A)$. Name the remaining formulas with the even numerals corresponding to multiples of 4. Let D be the formula of PA expressing that either u is even and $v = u$ or u is odd and $v = u + u$. It is immediate that D with u and v represent the diagonalization in the pair (PA, n).

3 The Impossibility Theorem

We shall see that it is impossible for a diagonalizable pair (T, n) to represent a very important metamathematical property of formulas of T, unless T is inconsistent: The notion of validity in T. The valid formulas can be defined as those for which there is a formal deduction in T, but this definition is not really relevant for what we are going to prove. It is enough to know that validity in T is a metamathematical property of formulas. The validity of A in T is denoted by $T \vdash A$.

Assume that T is framed to formalize and accommodate all mathematics, including its own metamathematics. Then, T must talk about its own formulas and whether they are valid. In order to do that, a naming of formulas n is needed. How could T endowed with a naming n express that a given formula is valid?

Let W be a formula of T and let v be a variable distinct from u. Assume that v is the only variable occurring free in W. We say that W with v represents the notion of validity in the pair (T, n) if for each formula A,

1. if $T \vdash A$, then $T \vdash W_v[n(A)]$ and

2. if $T \not\vdash A$, then $T \vdash \neg W_v[n(A)]$.

If there are W and v representing the validity in (T, n), then we say that the pair (T, n) can represent validity, or, shortly, that (T, n) represents validity. Similarly, representation in (T, n) can be defined in general, for any metamathematical property of formulas.

If T contains two closed terms a and b such that $T \vdash \neg(a = b)$, then there is a naming function n such that (T, n) can represent validity. Indeed, just name the theorems with a, the remaining formulas with b and take W to be $a = v$. In PA we may take n to be a one-one function which sends the theorems to the numerals corresponding to even numbers, the remaining formulas, if there is any, to the numerals corresponding to the odd numbers and take W to be the formula $\exists w(v = 2.w)$.

Nevertheless, theorem 2 shows that for a consistent T there is no naming function n such that (T, n) is diagonalizable and represents validity. So, if T is consistent, no matter what function n from the set of formulas to the set of closed terms we choose, either the representation of the diagonalization fails or the representation of the validity fails, or both.

Theorem 2. [The Impossibility Theorem]

Let T be a consistent first-order theory with equality. There is no function n from the set of formulas to the set of closed terms such that (T, n) is diagonalizable and represents validity.

Proof. Assume that there is such a function. There are W and v representing the validity in (T, n). Since (T, n) is diagonalizable, let B be a sentence, given by theorem 1, such that

$$T \vdash \neg W_v[n(B)] \leftrightarrow B.$$

If $T \vdash B$, then $T \vdash W_v[n(B)]$. From the above equivalence, $T \vdash \neg B$ and T is inconsistent.

If $T \not\vdash B$, then $T \vdash \neg W_v[n(B)]$. From the above equivalence, $T \vdash B$ and T is inconsistent. □

If T is consistent and contains a certain amount of arithmetic, then, through arithmetization, one can define a function n such that the pair (T, n) is diagonalizable and represents all recursive properties. It follows that validity in T is not a recursive property, that is, T is undecidable.

Furthermore, if T is also recursively axiomatizable, then it is incomplete, for completeness would directly imply decidability in this case.

These are well-known consequences of Gödel's first incompleteness theorem. We shall not be concerned with them. Instead, we deal with the arithmetization-free impossibility theorem given above and discuss its foundational relevance.

4 The Foundational Significance of the Impossibility Theorem

Gödel's theorems are considered to be an irremovable obstacle to some formalist conceptions of mathematics, to say the least. For example, it follows from these theorems that is impossible to frame a first-order system of arithmetic in which the quantifiers range only through finitary terms. In particular, the mathematics within a formal system such as PA and the finitary metamathematics of PA are mismatched.

The arithmetization-free impossibility theorem we have proved is a weakened version of Gödel's first incompleteness theorem, but it is still a strong result. For, just like the full version, the impossibility theorem shows that there is a wide gap between mathematics within a formal system and metamathematics. Indeed, assume that T is a consistent first-order theory with equality. Under these conditions, no matter which naming of formulas n is used and how the notion of validity in T and the diagonalization function are defined within T, these definitions do not simultaneously represent the metamathematical notion of validity in T and the metamathematical diagonalization function. At least one representation fails, there is no way to fix this. This implies that the mathematics within T and the metamathematics of T must be mismatched.

Therefore, the impossibility theorem is sufficient to imply that the mathematics within a consistent first-order formal system T and its metamathematics are mismatched. Indeed, there is no way to name the formulas of T making T capable of expressing the metamathematical notions of validity and diagonalization – technically, there is no naming n such that T represents validity and diagonalization under n.

The foundational problem is that, according to a formalist conception of mathematics, the subject matter of a mathematical theory is defined by the axioms of a corresponding formal system. For example, Hilbert is explicit about the definitional role of consistent formal axiomatic systems in a letter to Frege ([Frege, 1980], page 42).

I was very much interested in your sentence: "From the truth of the axioms it follows that they do not contradict one another" because for as long as I have been thinking, writing, lecturing about these things, I have been saying the exact reverse: If the arbitrarily given axioms do not contradict one another, then they are true, and the things defined by the axioms exist. This for me is the criterion of truth and existence.

Let us note the explicit mention to "the things defined by the axioms". If we identify arithmetic with a formal system like PA, assumed to be consistent, then the subject matter of arithmetic is defined by the axioms of PA. However, the subject matter of arithmetic is supposed to be uniquely defined, hence extensionally equivalent to the subject matter of finitary metamathematics. Moreover, if the subject matter defined by PA were extensionally equivalent to the subject matter of finitary metamathematics, then PA and metamathematics could express the same notions of validity in PA and diagonalization under an appropriate numbering (naming) of formulas witnessing this extensional equivalence. The impossibility theorem shows that this cannot be.

The subject matter defined by PA, whatever it may be, is not extensionally equivalent to the subject matter of finitary metamathematics. Therefore, it is not the finitary range as it was generally supposed to be, and this important conclusion is obtained directly from the arithmetization-free impossibility theorem.

References

Frege, G. (1980), *Philosophical and Mathematical Correspondence*, Oxford: Blackwell Publishers.

Gödel, K. (1931), 'über formal unentscheidbare sätze der principia mathematica und verwandter systeme i', *Monatshefte für Mathematik und Physik* **38**, 173–198.

Mendelson, E. (1997), *Introduction to Mathematical Logic*, Chapman and Hall.

Raatikainen, P. (2020), Gödel's incompleteness theorems, *in* E. N. Zalta, ed., 'The Stanford Encyclopedia of Philosophy', Fall 2020 ed., Metaphysics Research Lab, Stanford University.

Tarski, A., Mostowski, A. & Robinson, R. (2010), *Undecidable Theories*, Dover Publications.

A Working Mathematician between Philosophers: On the Logical Analysis of Magnitudes[‡]

Abel Lassalle-Casanave*   Eduardo N. Giovannini[†]

* Federal University of Bahia and
National Council for Scientific and Technological Development (CNPq)
abel.lassalle@gmail.com

[†] University of Vienna and
National Scientific and Technical Research Council (CONICET)
engiovannini@conicet.gov.ar

Much has been written about Hilbert's strive for "purity" in *Foundations of geometry* (1899), especially in relation to the exclusion of arithmetical concepts in the treatment of geometrical problems. But little has been written about another, no less fundamental, aspect of his program, namely the exclusion of the general concept of magnitude, a thesis which bears many resemblances to Dedekind's program for the foundations of arithmetic. In this paper, we will illustrate this aspect of Hilbert's geometrical enterprise by examining the elementary case of straight line segments, their operations, and relations. In particular, we will analyze how Hilbert explicitly avoided the use of principles for general magnitudes, while he proved their validity for the relevant geometrical objects from his geometrical axioms.

Section I presents an informal or "intuitive" description of the distinction between local and global operations and relations for the particular case of straight line segments. This distinction is also illustrated by appealing to some propositions of Euclid's *Elements*. Section II analyzes the treatment of segment congruence carried out by Hilbert in *Foundations* and its relations to the basic principles of equality and addition of segments.

[‡]The first author thanks the support of the CNPq (Brazil) during work on this project, by means of the award of two research grants (numbers 306044/2007-2 and 424126/2018-4). The second author wish to express his gratitude to the *Austrian Science Fund* (FWF) (Project number M 2803-G) and to the *Agencia Nacional de Promoción Científica y Tecnológica* (ANPCyT, Argentina) (Project number PICT 2017-0443), for their support during work on this article.

Section III extends the previous discussion of local and global operations and relations to the case of plane polygonal figures, especially in connection to a proof of the so-called De Zolt's in an abstract setting. Section IV concludes with some reflections on the logical analysis of magnitudes (and something more).

I

In an allegedly "intuitive" description one often encounters that theoretical elements slip in silently. Consider, for example, the following situation: let a segment AC with an interior point B be given (Figure 1). We can talk here of a "join" of the segments AB and BC, and we say that the segment AC is their *local* addition. But suppose we are asked to add any two segments DE and FG. The task will consist now in a generalization of the local operation of addition to a *global* operation, that is, in reducing the case of the two given segments to a local case.

Figure 1: "Local" and "global" addition of segments

Intuitively, the operation of addition consists in the concatenation of segments at a common point on a given straight line. This suggests a kind of "global" operation of addition: *any* two segments can be added by concatenating them at a common point. But in Figure 1 above, a local operation of addition can be defined by means of the following equality: $AC = AB + BC$. Accordingly, the term "local" refers to the fact that the addends AB and BC are already parts of the given segment AC.

The same distinction between "local" and "global" applies to the comparison of segments. Consider now a given segment AB with the interior points C, D, and two any given segments AB and CD (Figure 2). The notion of comparison of straight line segments is based on the notion of inclusion or 'being a part of'. However, Figure 2 describes a *local* notion of segment comparison: the segment CD is already a part of a given segment AB. In turn, segments AB and CD illustrate a global notion of comparison, since one wishes to compare any two straight line segments. The *global* notion of segment comparison is a generalization of the local notion of inclusion.

Figure 2: "Local" and "global" comparison of straight line segments

To sum up, first one can distinguish between a local and a global conceptions of the operation of addition of straight line segments. Second, one can also identify local and global notions of the comparison of straight line segments. While this will prove instructive in the analysis of Hilbert's practice, let us first motivate the distinction with a preliminary comparison with Euclid's practice.

If our main goal is to compare Hilbert's and Euclid's geometrical practices, we must note that while the former considers infinite straight lines, the latter only accepts "finite straight lines" (straight line segments), which can always be drawn between two points (by Postulate 1) or produced continuously "in straight line" (by Postulate 2). Moreover, and equally important, Euclid does not deal with *any* two segments, but only with given segments. Leaving this conceptual issue aside, for Euclid to add two segments is to place at a given point a segment equal to a given segment in the prolongation in straight line of the other given segment.

As is well known, Euclid proceeds by steps. In Proposition I.1, he constructs an equilateral triangle with sides equal to a given segment (Figure 3). In Proposition I.2, he places a segment at a given point to another given segment (Figure 4). Then, in Proposition I.3, given two unequal segments, Euclid removes the smaller segment from the greater one (Figure 5).[1] The global operation of addition is just a simple adjustment of this last proposition.[2] Some remarks about Euclid's resolution of these problems contribute to identifying other differences with the Hilbertean practice, which is the main focus of this paper.

Less than the construction of an equilateral triangle, Proposition I.1 allows to put two segments equal to a given one in such a way that each of their extremities coincides with the extremity of the given segment, and the other extremities coincide with each other. The formulation of the first proposition of the *Elements* abbreviates the formulation of this problem. This is suggested by the use of the resolution of I.1 in I.2. More precisely, the construction of an equilateral triangle is used to remove

---

[1] For Euclid's proof see Heath [1956].
[2] For details see [Lassalle Casanave & Panza, 2012].

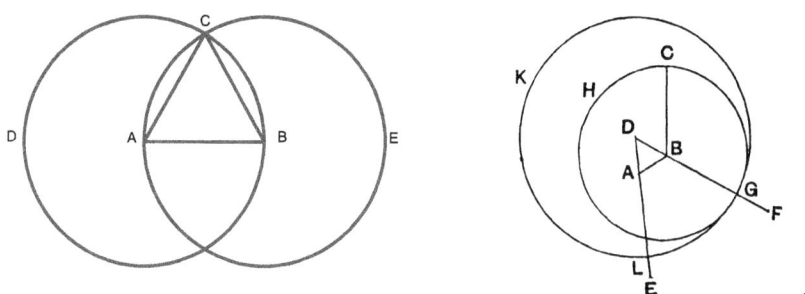

Figure 3: *Elements*, Proposition I.1   Figure 4: *Elements*, Proposition I.2

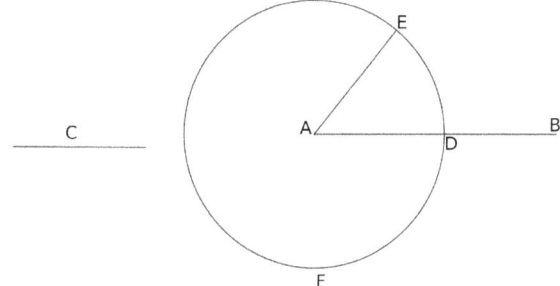

Figure 5: *Elements*, Proposition I.3

equals from equals by Common Notion 3 ("If equals be subtracted from equals, the remainders are equal"). One should notice that it is not the first time that a common notion is used, for Proposition I.1 appeals essentially to Common Notion 1 ("Things which are equal to the same thing are also equal to one another"). Moreover, although Proposition I.3 can be understood as the subtraction of a segment lesser than a greater one, it can also be considered differently from another perspective, namely to put a given segment (lesser) on another given segment (greater).

Let us return to the intuitive description of the situation exhibited in Figure 1. As we already mentioned, this situation can be used to illustrate a local relation of 'lesser'. In fact, since the straight line segment AB is a part of AC, we have the lesser relation as the local relation of inclusion. Now, the global relation of lesser is a generalization of the local relation: Proposition I.3 reduces the global relation to the local one, that is, if a given segment is lesser than another given segment, then the first is a part of second.

In the cases we have examined, the notion of equality of straight lines segments occurs systematically, not only in Common Notion 2 ("If equals be added to equals, the wholes are equals") and Common Notion 3, but also in Common Notion 4 ("Things which coincide with one another are equal to one another"). But as it can be noticed from the resolution of Propositions I.1-I.3, in order to prove that two segments coincide, Euclid uses the definition of a circle as a figure whose radii are equal as well as the common notions.

Finally, another aspect that is worth mentioning here is the problem of the uniqueness of the constructed segments. From a modern perspective, it is usually claimed that Postulate 1 states the existence as well as the *uniqueness* of a segment constructed from any two given points. However, it should be noted that this interpretation of Postulate 1 is not the only possible one. For example, some early modern commentators of the *Elements* found necessary to add a new principle in the proof of the Proposition I.4 –the congruence criterion called 'side–angle–side'–according to which two segments do not enclose a space. This proposition was incorporated in modern versions of the *Elements* as Postulate 6.

II

The geometrical relation of *congruence* naturally provides the criterion of equality for straight line segments. Let us analyze how Hilbert's treatment of segment congruence in *Foundations* is related to the basic principles of equality in an abstract theory of magnitudes.

In Hilbert's axiomatic system, straight line segments are introduced *by definition* as sets of points on a straight line. Segment congruence is a *primitive* quaternary relation, $C(A, B, A', B')$, read as "two segments are congruent, in symbols, $AB \equiv A'B'$". Hilbert lays down three axioms of congruence for line segments or *linear axioms*, which are sufficient to introduce the notion of segment congruence in an adequate way. The first axiom is the following:

Axiom III. 1 *If A, B are two points on a line a, and A' is a point on the same or on another line a', then it is possible to find a point B' on a given side of the line a' through A' such that the segment AB is congruent or equal to the segment $A'B'$. In symbols: $AB \equiv A'B'$.* [Hilbert, 1971, p. 10]

Axiom III.1 asserts the possibility of the construction of any straight line segment or, more precisely, the existence of a segment congruent to a given one. Given any straight line segment and a point on the same or on a different line, this axiom guarantees that a segment congruent to the given segment can be constructed from the given point.³ It should be noted that the uniqueness of this segment, which is essential for the comparison, is not assumed by Hilbert, but later proved.⁴ The second axiom of congruence for line segments is the following:

Axiom III. 2 *If a segment $A'B'$ and a segment $A''B'''$, are congruent to the same segment AB, then the segment $A'B'$ is also congruent to the segment $A''B'''$, or briefly, if two segments are congruent to the third one they are congruent to each other. [Hilbert, 1971, p. 10]*

In a sense, this axiom resembles Euclid's Common Notion 1: "Things which are equal to the same thing are also equal to one another". It should be mentioned that during the nineteenth century, it was usual to identify this axiom with the transitive property; however, it expresses a weaker proposition. These two linear congruence axioms suffice *to prove* that this relation satisfies the reflexive, symmetric, and transitive properties, namely that segment congruence is an equivalence relation. For example, the reflexive property can be proved as follow: let a segment AB be constructed on a ray that it is congruent to another segment $A'B'$ (axiom III.1), that is, $AB \equiv A'B'$. Then, from the congruences $AB \equiv A'B'$, $AB \equiv A'B'$, it follows by axiom III.2 that $AB \equiv AB$. q.e.d.⁵

The fact that segment congruence can be considered a criterion of equality for line segments is a consequence of the first two axioms of congruence. On the other hand, the third axiom introduces a novel element:

Axiom III. 3 *On a line a let AB and BC be two segments which except for B have no point in common. Furthermore, on the same or on another line a' let $A'B'$ and $B'C'$ be two segments which except for B' also have no point in common. In that case, if $AB \equiv A'B'$ and $BC \equiv B'C'$, then $AC \equiv A'C'$. [Hilbert, 1971, p. 11]*

This axiom establishes the possibility of the addition of segments, in particular, the additive property of equality. It also recalls to Euclid's Common Notion 2: "If equals be added to equals, the wholes are equals". Now, the conception of the addition of segments that is involved in this axiom

---

³Although Hilbert occasionally uses the traditional terminology to talk about geometrical constructions, he does not employ this 'constructive language' in a classical way. This is particularly evident in Hilbert's further analysis of the constructive power of different geometrical instruments, such as the ruler, the protractor of segments [*Streckenübertrager*], and the scale [*Eichmaß*].

⁴Cf. Hilbert [1971, p. 13].

⁵Hilbert does not provide the proofs for the symmetric and transitive properties in *Foundations*. For a modern treatment, see Hartshorne [2000].

is not what we have called above a "global" operation, that is, the concatenation of any two segments at a common point. On the contrary, Hilbert deals here with a "local" operation of addition, as defined by the equality "$AC = AB + BC$" (see Figure 6).[6]

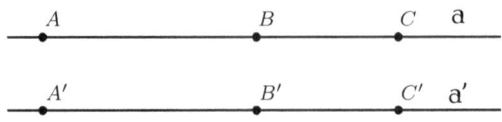

Figure 6: Hilbert's third axiom of segment congruence

Based on axiom III.1, one can easily generalize this local operation of addition to a global operation. More precisely, let AB and CD be two segments on two given lines l and l'. By axiom III.1, there exist a point E on line l such that $CD \equiv BE$. We then define the segment AE as the addition of the segments AB and CD, that is $AE = AB + CD$ (Figure 7).[7]

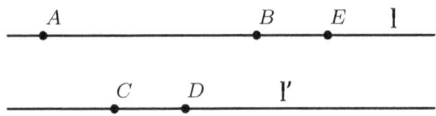

Figure 7: Generalization of "local" addition

Now, we have mentioned that axiom III.1 establishes the possibility of constructing a line segment congruent to any given segment from a given point on a line. The uniqueness of this construction is a central requirement, but does not need to be postulated axiomatically, for it can be derived

---

[6]This definition of addition is formulated by Hilbert in his construction of a segment arithmetic:

> If A, B, C are three points on a line and B lies between A and C the *sum* of the two segments $a = AB$ and $b = BC$ is denoted by $c = AC$ and is expressed as $c = a + b$. [Hilbert, 1971, p. 51]

[7]For details see Hartshorne [2000, p. 84].

from other axioms of congruence (in Hilbert's case, the axioms of congruence for angles).[8] In particular, this uniqueness condition bears important relations to other "principles" for general magnitudes. Consider the following proposition:

Proposition 1 *Let A, B, C be three points on a line, such that B lies between A and C, and let A', B', C' be three points on the same or on another line, such that B', C' line on one side from A'. If $AB \equiv A'B'$ and $AC \equiv A'C'$, then B' lies always between A' and C' and $BC \equiv B'C'$. (Figure 8)*

Figure 8: Removal of segments

This proposition follows immediately from the uniqueness condition for the construction of congruent segments. More interestingly, in a sense it is the "geometrical" analog of the Euclid's Common Notion 3: "If equals be subtracted from equals, the remainders are equal". Then, one could argue that in Hilbert's axiomatic system, the uniqueness condition of the construction of congruent segments plays the role of the fundamental Common Notion 5: "And the whole is greater than the part". More precisely, the former proposition could be interpreted as meaning that if B lies between A and C, then AB cannot be congruent to AC, and this follows from the uniqueness condition, because since B and C are on the same line from A, if $AB \equiv AC$, then B and C should be equal, which is impossible.

Let us consider now the introduction of a relation of (total) order for line segments. As mentioned above, the global notion of comparison is a generalization of the local notion of inclusion. Let A, B, C be three points on a line, we say that $AB < AC$ if and only if B lies between A and C. The linear axioms of betweenness –that is, the axioms of the order of points on a line– along with the axioms of congruence for line segments ground a relation of total ordering for line segments. Again, notice that the following definition of a relation of order refers to a local criterion of

---

[8]Other axiomatizations include this condition as an axiom. See Hartshorne [2000].

inequality: the segment AB is already a part of the segment AC. However, it is easy to generalize this definition to a global criterion of comparison. Consider, for example, the following definition:

Definition 1 *Let AB and CD be any two given line segments. We define $AB < CD$ (read that AB is less than CD), if there exists a point E which lies between C and D such that $AB \equiv CE$. Similarly, in this case we also say that $CD > AB$ (read as CD is greater than AB).*[9] *(Figure 9)*

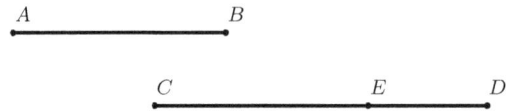

Figure 9: Global comparison of straight line segments

On the basis of axiom III.1, we can perform the construction to establish this inequality. Take any of the two segments, say AB. By the latter axiom, we know that there exists a unique point E such that AE is congruent to CD. If E lies between A and B, then $AB > CD$. On the contrary, if B lies between A and E, then $AB < CD$.

One can also easily obtain the following two properties of order, which shows that the above definition introduces a total order for line segments:

a) Given two line segments $AB \equiv A'B'$ and $CD \equiv C'D'$, then $AB < CD$ if and only if $A'B' < C'D'$.

b) Transitivity: If $AB < CD$, and $CD < EF$, then $AB < EF$.

c) Total or Trichotomy: Given two line segments AB, CD, one and only one of the three following conditions holds:

$$AB < CD, \qquad AB \equiv CD, \qquad AB > CD.$$

To sum up, Hilbert's treatment of comparison, equality, and addition of straight line segments is grounded on purely geometrical axioms of incidence, betweenness, and congruence. These axioms put forward a local

---

[9] See also Hartshorne [2000].

conception of addition and comparison, but which can be easily generalized to a global conception. Moreover, these relations and operations among segments not only do not assume any kind of numerical concepts, but also do not appeal to any kind of principle for general magnitudes. However, one can obtain from these axioms the usual properties of the concept of magnitude, namely, the structure of an *ordered commutative semi-group* [Stein, 1990, see, e.g.,]. Schematically, in this "standard" approach one starts by postulating a relation of equivalence $\sim$ for magnitudes which satisfies the usual properties (reflexivity, symmetry, and transitivity). Then, one introduces an operation of addition + of magnitudes which satisfies the associative and commutative properties, but also the following comparability property: for any magnitudes a and b, there exists a magnitude c such that either $a = b + c$ or $b = a + c$ or $a = b$. The first two properties provide the structure of a commutative or Abelian semigroup. If the third property is also satisfied, we obtain an ordered commutative semigroup.[10]

III

The distinction between local and global operations and relations might seem unmotivated or even pointless in the case of the addition and comparison of straight line segments, for the operations we designated as global can effortlessly be reduced to local operations. But this situation changes entirely when we attempt to extend these notions to plane polygonal figures. Intuitively, one can understand the global operation of addition of any two polygons as juxtaposing any of their common sides; similarly, the global relation of comparison can be conceived as the inclusion of one polygon into the other. However, it is not the case that in general any two polygons can always be added, for there are polygons that cannot be juxtaposed. For example, consider a regular star pentagon and a regular decagon with sides equal or greater than the distance of two consecutive vertices of the pentagon (see Figure 10). It is immediately clear that these two polygons cannot have two points in common at their edges, without also have common points in their *interiors*. Neither is the case that any two polygons can be *directly* compared, since simply it might not be possible to embed one into the other (see Figure 11). Moreover, although we did not need to resort to the distinction between a straight line segment and its *length*, we must now introduce the notion of the *area* of a polygon, but without appealing to numerical considerations, just as in the case of segments. Nevertheless, the task of adding areas leads us again

---

[10]In fact, one defines that $a < b$ if and only if there exists a c such that $b = a + c$.

to the problem of avoiding a general theory of magnitudes, and to a new instance of the reduction of the global operation to a local case.

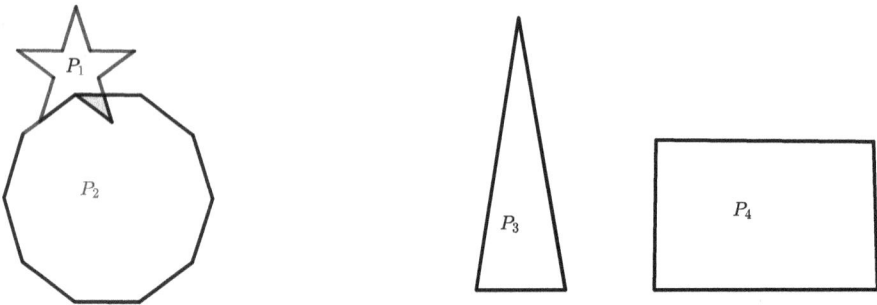

Figure 10: Non-juxtaposable polygons

Figure 11: Non-embeddable polygons

There are, however, two central theorems in the geometrical theory of equivalence which could allow us *in principle* to bypass these initial difficulties with the global operation of addition and the local relation of comparison of polygons. First, every polygon can be decomposed into a finite number of triangles. A proof of this theorem requires the use of induction on the number of sides of the polygon. Second, every polygon can be transformed into an equivalent (viz. equidecomposable[11]) rectangle with a given altitude. This theorem has a more ancient pedigree, since it was proved by Euclid in the crucial proposition I.45 of the *Elements*. Then, given that any two polygons can always be transformed into two equivalent rectangles with the same altitude, they can be effortlessly added, just by concatenating them at a common adjacent side; and similarly, they can be easily ordered, just by comparing their respective bases. These theorems would allow us to circumvent the problem of adding and comparing any two polygons, by reducing the global operation and relation to a local case, in which one has to add and compare only "compatible" plane figures.

But consider now *two different transformations* of a polygon into two equivalent rectangles with a given base. Is it an immediate consequence of the notions of equidecomposition and addition of plane polygons that these rectangles must be also congruent, that is, that their bases must also coincide? This is where the so-called De Zolt's postulate in the theory of plane area makes its entrance. In the formulation provided by Hilbert in *Foundations*, the postulate goes as follows: "If a rectangle is decom-

---

[11]Two simple polygons are said to be *equidecomposable* if they can be decomposed into a finite number of triangles that are congruent in pairs.

posed by lines into several triangles and one of these triangles is omitted then it is impossible to fill out the rectangle with the remaining triangles" [Hilbert, 1971, p. 69]. De Zolt's postulate – Theorem 52 in Hilbert's axiomatic presentation – guarantees that a polygon cannot be equivalent (viz. equidecomposable) with a proper part. This shows that this central geometrical proposition is related to a *local* operation of addition of polygons, which consists in removing the common or adjacent side of a pair of triangles obtained from an adequate triangulation of a polygon, that is, a triangulation which produces no overlaps and any pair of triangles is disjoint (they have only one vertex in common, or one side, and other points in common). In a nutshell, all the addends are already a part of a given plane polygon.

From an algebraic point of view, this situation can be translated into a theory of magnitudes restricted to, what might be called, "compatible magnitudes". This theory does not include the commutative of addition as a basic property, but it is capable of deriving De Zolt's postulate. Moreover, in this theory the latter postulate has to be distinguished from Common Notion 5, i.e., the whole is greater than the part. In what follows we illustrate the fundamental concepts of this proof, by appealing to our previous remarks on addition and comparison of segments.[12] We conclude with a reflection on the logical analysis of the concept of magnitude.

With an eye to the aim of proving De Zolt's postulate, in a treatment of magnitudes one can restrict the operation of addition to compatible magnitudes. This restriction aims at characterizing in an abstract way the fact that addition is conceived as a local operation. As we saw above, the local aspect of addition means that the operation is not performed on any two magnitudes, but on magnitudes which are already part of a given magnitude. Compatible magnitudes are those which result from considering in an adequate way certain decomposition of a previously given magnitude. In a geometrical setting, the idea of compatibility can be illustrated by considering two parts of a given magnitude which have a common element already contained in the latter. For example, the local addition and comparison of straight line segments (Figure 1 and Figure 2) or two polygonal parts of a polygon P with a common side.

The addition of compatible (α_p) magnitudes satisfies the usual properties of associativity, monotonicity, cancellations, and particularly domination: for every pair of compatible magnitudes, their addition is equal or greater to any of the addends.[13] One may also consider special kinds of

---

[12]For a detailed exposition of the proof see Giovannini et al. [2019].

[13]Let $\sim$ and $\preceq$ denote a relation of equivalence and comparison of magnitudes, respectively. Moreover, let ; denote the operation of addition. Then, the domination property of

magnitudes, namely, trivial and proper magnitudes, and use this distinction to characterize *strict parts*.[14] This provides a precise formulation of Common Notion 5, in terms of strict parts.

One might further introduce magnitude lists and truncations (a truncation of a list being a sublist consisting of all its elements but one), as well *partitions* as proper compatible magnitude lists.[15] These concepts are instrumental in formulating and deriving abstract versions of De Zolt's postulate involving global and local decompositions of a magnitude m: a partition a with $\sum[a] \sim m$, for the former, and $\sum[a] = m$, for the latter. These versions involve a partition p and a truncation q of p. They are as follows.

($\sim$) If p is a global decomposition of m, then q is not a global decomposition of m.

($=$) If p is a local decomposition of m, then q is not a local decomposition of m.

Finally, one can derive these versions from the following result on partition truncation:

> Given a partition p, consider a truncation q of p. Then, either q is not compatible or $\sum[q] \prec \sum[p]$.

One should notice that proper cases of Common Notion 5 can be derived from this proposition, thereby establishing tight connections between the above abstract version of De Zolt's postulate and "the whole is greater than the part".

---

addition says, in symbols, that "for all a, b, with a cmp b: $a \preceq a; b \succeq b$".

[14]Consider a magnitude m. Call m *left trivial* iff $m; b \preceq b$, for some magnitude b with m cmp b. Call m *right trivial* iff $a; m \preceq a$, for some magnitude a with a cmp m. Magnitude m is *trivial* (m ∈ Trv) iff m is left or right trivial. Magnitude m is *proper* (m ∈ Prp) iff m is not trivial (i. e. m ∉ Trv).

[15]Consider a magnitude list $\underline{a} = \langle a_1, \ldots, a_n \rangle$.

- List $\underline{a} = \langle a_1, \ldots, a_n \rangle$ is *proper* iff $a_1, \ldots, a_n$ are all proper.
- List $\underline{a} = \langle a_1, \ldots, a_n \rangle$ is *compatible* iff a_1 cmp a_2 cmp $\ldots$ cmp a_{n-1} cmp a_n.

By a *partition* we mean a proper compatible magnitude list.

Given a compatible list $\underline{a} = \langle a_1, \ldots, a_n \rangle$, its *sum* is the sum of all its elements, namely the magnitude $\sum[\underline{a}] = a_1; \ldots; a_n$. For $n = 1$, we take $\sum[\langle a_1 \rangle] = a_1$.

IV

The authors of this note in honor of Professor Veloso were working on a paper about the geometrical practices of Euclid and Hilbert in relation to the notion of plane area, when they faced some difficulties to distinguish between Common Notion 5 and De Zolt's postulate. In our view, the latter postulate was more similar to a principle introduced in early modern times as a genuine geometrical principle, namely that 'the whole is equal to the sum of the parts', than to the Euclidean Common Notion 5. Incidentally, we thought that the Euclidean notion should be understood as what one might call, anachronistically, a principle of the theory of magnitudes. We had then the excellent idea of presenting our hesitations to Professor Veloso, on the occasion of the XXI *Colóquio Conesul de Filosofia de las Ciencias Formales*, in October 2016. Some days later, we received a couple of pages with a proof sketch of a version of De Zolt's postulate in an abstract setting. We decided to ask permission to Professor Veloso to include his name as a co-author, and to include those pages as an Appendix of the paper entitled "De la Práctica Euclidiana a la Práctica Hilbertiana: las Teorías del Área Plana" [Giovannini et al., 2017].

Shortly afterward, we received from Professor Veloso a more detailed proof, consisting of about 30 pages. Unlike the previous version, in this new proof the distinction 'local/global' played a key role. In our opinion, this distinction constitutes a central aspect of what we may call a *logical analysis* of the concept of magnitude. Usually, magnitudes are defined as commutative or Abelian ordered semi-groups. This definition already assumes that the operations are global, and allows to distinguish later between Archimedean and non-Archimedean theories of magnitudes.[16] But in his analysis of De Zolt's postulate, Professor Veloso takes a step back, which leads to the notion of a theory of *compatible* magnitudes. From an abstract viewpoint, the salient difference of this theory is that it does not assume the commutative law of addition.

Between the first sketch and the new version of the proof, we came across a (rough) English translation of a short communication by the mathematician Antoni Łomnicki (1881-1941), written originally in Polish. This communication analyzed De Zolt's postulate in the context of a general theory of magnitudes, in the spirit of the Tarskian school, and served as an inspiration to Professor Veloso. The second proof was subject to several modifications, partially due to our doubts and inquiries, but mostly

---

[16]Hale [2000] refers to an Archimedean theory of magnitudes as a *normal* quantity domain. For a historical study of the emergence of systems of non-Archimedean magnitudes see Ehrlich [2006].

because of the successive refinements introduced by him. Notwithstanding, the remained material still contained very valuable insights, and it was adapted and later published under the title "On Comparison, equivalence and addition of magnitudes" [Veloso et al., 2019]. An interesting aspect of this work are some peculiarities of the theory of magnitudes presented there, when compared to the standard approaches already mentioned. In particular, this theory departs from a relation of comparison as a primitive relation, instead of using as customary the relations of equality and *strict* comparison as primitives. However, a possible "drawback" of this approach is that this relation of comparison does not have a "natural" geometrical interpretation.

Lastly, in July 2019, we published the final version of the work with the title "De Zolt's Postulate: An Abstract Approach", combining mathematical, historical, and conceptual elements.[17] By briefly recounting this history of collaboration, we wish to record our fantastic professional experience of accompanying the work of Professor Veloso in assisting errant philosophers. Truly, a working mathematician between philosophers.

References

Ehrlich, P. (2006), 'The rise of non-archimedean mathematics and the roots of a misconception i: The emergence of non–archimedean systems of magnitudes', *Archive for History of Exact Sciences* **60**(1), 1–121.

Giovannini, E., Haeusler, E. H., Lassalle-Casanave, A. & Veloso, P. (2019), 'De zolt's Postulate: An Abstract Approach', *The Review of Symbolic Logic*, First View.

Giovannini, E., Lassalle Casanave, A. & Veloso, P. (2017), 'De la práctica euclideana a la práctica hilbertiana: las teorías del área plana', *Revista Portuguesa de Filosofia* **73**(2), 1263–1294.

Hale, B. (2000), 'Reals by abstraction', *Philosophia Mathematica* **8**(3), 100–123.

Hartshorne, R. (2000), *Geometry: Euclid and Beyond*, Springer, New York.

Heath, T. (1956), *The Thirteen Books of Euclid's Elements (second edition, vols. I–III*, Dover Publications, New York.

---

[17]The final version of the paper was prepared also in collaboration with Edward Hermann Haeusler. The article a contains a polished English translation of the short communication by Łomnicki, mentioned above.

Hilbert, D. (1971), *Foundations of Geometry*, Open Court, La Salle.

Lassalle Casanave, A. & Panza, M. (2012), 'Sobre el significado del Postulado 2 de los *Elementos*', *Notae Philosophicae Scientiae Formalis* 1(2), 103–115.

Stein, H. (1990), 'Eudoxos and Dedekind: on the ancient Greek theory of ratios and its relation to modern mathematics', *Synthèse* 84, 163–211.

Veloso, P., Lassalle-Casanave, A. & Giovannini, E. (2019), 'On comparison, equivalence and addition of magnitudes', *Principia. An International Journal of Epistemology* 23(2), 153–173.

Boolean Real Semigroups

F. Miraglia* and Hugo R. O. Ribeiro[†]

* Institute of Mathematics and Statistics
University of São Paulo, Brazil
miraglia@ime.usp.br

[†] Institute of Mathematics and Statistics
University of São Paulo, Brazil
hugorafaelor2@gmail.com

Dedicated to Paulo Veloso, dear friend and colleague of the first author

Our purpose here is fourfold: firstly to give, employing the languages of special groups (SG) and real semigroups (RS), new, oftentimes conceptually different and clearer, proofs of the characterization of RSs whose space of characters is Boolean in the natural Harrison (or spectral) topology, originally appearing in section 7.6 and 8.9 of [Marshall, 1996], therein named zero-dimensional abstract real spectra and here called Boolean Real Semigroups. Secondly, to give a natural *Horn-geometric* axiomatization of Boolean RSs (in the language of RSs) and establish the closure of this class by certain important constructions: Boolean powers, arbitrary filtered colimits, products, reduced products and RS-sums and by surjective RS-morphisms (in particular, quotients). Thirdly, to characterize morphisms between Boolean RSs. Fourthly, to give a characterization of quotients of Boolean RSs. Hence, the present paper considerably extends the original work by which it was motivated, namely the references in [Marshall, 1996] mentioned above.

Recall that if G is an RS, $G^\times = \{u \in G : u^2 = u\}$ is the pre-reduced special group of units in G, and $\mathrm{Id}(G) = \{e \in G : e^2 = e\}$ is the distributive lattice of idempotents in G.

In section 1 we recall, for the benefit of the reader, the basic properties of reduced special groups (RSG) and real semigroups. Section 2 discusses the factorization of a reduced special group into a direct product. Section 3 presents our account of the characterization of Boolean RSs, showing in particular that $G^\times$ is a RSG. The third paragraph at the beginning of section 3 explains the differences between our approach and that in [Marshall, 1996]. In section 4 we show that if G_1, G_2 are Boolean RSs, there is a natural bijective correspondence, preserving composition and isomorphism, between $Hom_{RS}(G_1, G_2)$ and compatible pairs $\langle f, h \rangle$, where $f : G_1^\times \longrightarrow G_2^\times$ is a RSG-morphism and $h : \mathrm{Id}(G_1) \longrightarrow \mathrm{Id}(G_2)$ is a bounded lattice morphism (cf. Definition 4.6).

For the notions of geometric and Horn-geometric formulas and theories in a first-order language with equality we refer to reader to section 2 of Chapter 1 in [Dickmann & Miraglia, 2015].

1 Fundamentals

In this section we collect the basic facts on Reduced Special Groups (RSG) and Real Semigroups (RS) to be used in the sequel. The fundamental reference for Special Groups is [Dickmann & Miraglia, 2000], while for Real Semigroups are [Dickmann & Petrovich, 2004], [Dickmann & Petrovich, 2008] and [Dickmann & Petrovich, 2016].

A. Reduced Special Groups

1.1 Notation. a) Write $\mathbb{Z}_2 = \{\pm 1\}$ for the multiplicative subgroup units in the integers, $\mathbb{Z}$. Write $\mathbb{F}_2$ for the two-element field.

b) If S, T are subsets of a group K, set $ST = \{ab \in K : a \in S \text{ and } b \in T\}$. When $S = \{a\}$, write aT for ST.

c) If K is a group of exponent 2 ($x^2 = 1$, $\forall\ x \in K$) with a distinguished element -1, write $-a$ for $-1 \cdot a$, $a \in K$. Clearly, the subgroup $\{1, -1\}$ of K is isomorphic to $\mathbb{Z}_2$ iff $1 \neq -1$ in K.

d) Let A be a set and $\equiv$ a binary relation on $A \times A$. We extend $\equiv$ to a binary relation $\equiv_n$ on A^n, by induction on $n \geq 2$ as follows :

(i) $\equiv_2\ =\ \equiv$

(ii) $\langle a_1, \ldots, a_n \rangle \equiv_n \langle b_1, \ldots, b_n \rangle$ iff there are $x, y, z_3, \ldots, z_n$ in A such that $\langle a_1, x \rangle \equiv \langle b_1, y \rangle$, $\langle a_2, \ldots, a_n \rangle \equiv_{n-1} \langle x, z_3, \ldots, z_n \rangle$ and $\langle b_2, \ldots, b_n \rangle \equiv_{n-1} \langle y, z_3, \ldots, z_n \rangle$.

Whenever clear from the context, we abuse notation and indicate the aforedescribed extension of $\equiv$ by the same symbol. Thus, if $a_1, \ldots, a_n$, $b_1, \ldots, b_n \in A$, we write $\langle a_1, \ldots, a_n \rangle \equiv \langle b_1, \ldots, b_n \rangle$ to mean that these two sequences are related by $\equiv_n$.

e) If $f: A \longrightarrow B$ is a set-mapping and $W \subseteq A$, $f \upharpoonright W : W \longrightarrow B$ denotes the restriction of f to the subset W of A. ∎

Definition 1.2 (Def. 1.2, [Dickmann & Miraglia, 2000]) *A **pre-Special Group (pSG)** is a group K of exponent 2, together with a distinguished element -1 and a binary relation $\equiv_K$ on K^2, called **binary isometry**, verifying for all $a, b, c, d \in K$,*

[SG 0] : *$\equiv_K$ is an equivalence relation.*

[SG 1] : *$\langle a, b \rangle \equiv_K \langle b, a \rangle$.*   [SG 2] : *$\langle a, -a \rangle \equiv_K \langle 1, -1 \rangle$.*

[SG 3] : *$\langle a, b \rangle \equiv_K \langle c, d \rangle$ implies $ab = cd$.*

[SG 4] : *$\langle a, b \rangle \equiv_K \langle c, d \rangle$ implies $\langle a, -c \rangle \equiv_K \langle -b, d \rangle$.*

[SG 5] : *$\langle a, b \rangle \equiv_K \langle c, d \rangle$ implies $\langle xa, xb \rangle \equiv_K \langle xc, xd \rangle, \forall\, x \in K$.*

*$\langle K, \equiv_K, -1 \rangle$ is a **reduced pre-special group (pRSG)** if $-1 \neq 1$ and $\equiv_K$ satisfies*

[red] (reduction) :   *$\forall\, a \in K, \langle a, a \rangle \equiv_K \langle 1, 1 \rangle$ iff $a = 1$.*

*A pre-special group is a **special group (SG)** if it satisfies*

[SG 6] (3-transitivity) : *The extension of $\equiv_K$ to K^3, (cf. 1.1.(d)) is a transitive relation.*

Axioms [SG 0] – [SG 6] imply the extension of $\equiv_K$ to K^n is transitive for all $n \geq 3$ (cf. Theorem 1.2, p. 16, [Dickmann & Miraglia, 2000]). ∎

All of the classic terminology of quadratic forms adapts to the present context:

Definition 1.3 (Def. 1.3, [Dickmann & Miraglia, 2000]) *Let K be a pSG. A **form** φ on K is an n-tuple of $\langle a_1, \ldots, a_n \rangle$ of elements of K; the number n is called the **dimension** of φ, $\dim(\varphi)$; we also call φ an n-**form**. By convention, two forms of dimension 1 are isometric iff they have the same coefficients.*

If $\varphi = \langle a_1, \ldots, a_n \rangle$ is a form on K, define

(1) **The set of elements represented by** φ *as*

$$D_G(\varphi) = \{b \in G : \exists\, z_2, \ldots, z_n \in B \text{ such that } \varphi \equiv_G \langle b, z_2, \ldots, z_n \rangle\}.$$

If $a \in G$, then $D_G(\langle a \rangle) = \{a\}$.

(2) **The discriminant** *of φ as $d(\varphi) = \prod_{i \leq n} a_i$.*

For forms $\varphi = \langle a_1, \ldots, a_n \rangle$ and $\psi = \langle b_1, \ldots, b_m \rangle$ we define the

(3) **direct sum** *as $\varphi \oplus \psi = \langle a_1, \ldots, a_n, b_1, \ldots, b_m \rangle$*

(4) **tensor product** *as $\varphi \otimes \psi = \langle a_1 b_1, \ldots, a_i b_j, \ldots, a_n b_m \rangle$*

If $a \in K$, the product $\langle a \rangle \otimes \varphi$ is written $a\varphi$.

(5) *A **Pfister form** is one of the type $\mathcal{P} = \bigotimes_{i=1}^{n} \langle 1, a_i \rangle$, with $a_i \in K$, $1 \leq i \leq n$. The integer n is the **degree** of $\mathcal{P}$; clearly, $\dim(\mathcal{P}) = 2^n$.* ∎

Lemma 1.4 (Lemma 1.5, [Dickmann & Miraglia, 2000]) *Let $\langle K, \equiv_K, -1 \rangle$ be a pSG. Let a, b, c, d be elements of K and φ, ψ be n-forms on K. Then*
a) $\langle a, b \rangle \equiv_K \langle c, d \rangle$ *iff* $ab = cd$ *and* $ac \in D_K(1, cd)$. *Further,* $c \in D_K(1, a)$ *iff* $\langle c, ac \rangle \equiv \langle 1, a \rangle$.
b) $\varphi \equiv_K \psi$ *implies* $d(\varphi) = d(\psi)$.
c) *If K is a SG, then* $\varphi \equiv_K \psi$ *implies* $D_K(\varphi) = D_K(\psi)$. ∎

Lemma 1.5 *Let K be a SG and let $x, a, b, c, d \in K$.*
a) (Transitivity) $x \in D_K(a, b)$ *and* $b \in D_K(c, d) \Rightarrow$ *there is* $y \in D_K(a, c)$ *so that* $x \in D_K(y, d)$.
b) (Exchange) $D_K(a, b) \cap D_K(c, d) \neq \emptyset \Rightarrow D_K(a, -c) \cap D_K(-b, d) \neq \emptyset$.
c) In fact, (a) and (b) are equivalent.

Proof. a) The antecedent in (a) implies, respectively, $\langle x, xab \rangle \equiv_K \langle a, b \rangle$ and $\langle b, bcd \rangle \equiv_K \langle c, d \rangle$. Then, Prop. 1.6.(a) in [Dickmann & Miraglia, 2000] yields

$$\langle a, c, d \rangle \equiv_K \langle a, b, bcd \rangle \quad \text{and} \quad \langle a, b, bcd \rangle \equiv_K \langle x, xab, bcd \rangle,$$

and the 3-transitivity of isometry obtains $\langle x, xab, d \rangle \equiv_K \langle a, b, d \rangle$, i.e., $x \in D_K(a, b, d)$. Now, Prop. 1.6.(c) in [Dickmann & Miraglia, 2000] yields $y \in D_K(a, b)$ such that $x \in D_K(y, d)$, as needed.
b) If $x \in D_K(a, b) \cap D_K(c, d)$, [SG4] in 1.2 gives $-b \in D_K(a, -x)$ and $-x \in D_K(-c, -d)$. By (a), there is $z \in D_K(a, -c)$ so that $-b \in D_K(z, -d)$; hence, $z \in D_K(a, -c)$ and $-z \in D_K(b, -d)$, and $z \in D_K(a, -c) \cap D_K(-b, d)$.
c) The proof of (b) shows transitivity implies exchange. For the converse, if $x \in D_K(a, b)$ and $b \in D_K(c, d)$. Then, $-b \in D_K(a, -x) \cap D_K(-c, -d)$ and the rule of exchange yields $t \in D_K(a, c) \cap D_K(x, -d)$. Hence, $t \in D_K(a, c)$ and $-x \in D_K(-t, -d)$, i.e. $x \in D_K(t, d)$, as needed. ∎

The multiplicative subgroup $\mathbb{Z}_2 = \{\pm 1\}$ of $\mathbb{Z}$ has a **unique** structure of RSG, with binary isometry defined by $\langle a, b \rangle \equiv_2 \langle c, d \rangle$ iff $a + b = c + d$ (sum in $\mathbb{Z}$).

1.6 Morphisms and Space of Orders. a) If K_1, K_2 are pSGs, a map $f : K_1 \longrightarrow K_2$, is a **SG-morphism** if it is a group morphism, preserving -1 and binary isometry, that is, by 1.4, for all $a, b \in K_1$, $b \in D_{K_1}(1, a)$ implies $f(b) \in D_{K_2}(1, f(a))$.
b) If K is a SG, a SG-morphism, $\sigma : K \longrightarrow \mathbb{Z}_2$ is called a **SG-character or order** of K. Note that for all $a \in K$, σ is a SG-character of K iff it satisfies
[ker] $\qquad a \in \ker \sigma$ implies $D_K(1, a) \subseteq \ker \sigma$.

If K is **reduced**, then X_K is not empty, being called the **space of orders** of K. It carries a natural Boolean space topology, having as a sub-basis
$\{[\![x = 1]\!] : x \in K\}$, where $[\![x = 1]\!] = \{\sigma \in X_K : \sigma(x) = 1\}$.
Note that $[\![-y = 1]\!] = [\![y = -1]\!]$, for all $y \in K$.

c) If $K_1 \xrightarrow{f} K_2$ is SG-morphism and φ, ψ are forms of the same dimension on K_1, then $\varphi \equiv_{K_1} \psi$ implies $f \star \varphi \equiv_{K_2} f \star \psi$, where, for $\theta = \langle a_1, \ldots, a_n \rangle \in K_1^n$, $f \star \theta = \langle f(a_1), \ldots, f(a_n) \rangle$. ∎

1.7 Pfister's local-global principle. If K is a RSG, $\varphi = \langle a_1, \ldots, a_n \rangle$ is a form over K and $\sigma \in X_K$, **the signature of φ at σ**, is defined by $sgn_\sigma(\varphi) = \sum_{k=1}^n \sigma(a_k)$ (sum in $\mathbb{Z}$). A very important result is the following (cf. Prop. 3.7, [Dickmann & Miraglia, 2000])

Theorem 1.8 (Pfister's local-global principle) *Let K be a RSG and let φ, ψ be forms of the same dimension over K. Then,*
$$\varphi \equiv_K \psi \quad \text{iff} \quad \forall \, \sigma \in X_K, \; sgn_\sigma(\varphi) = sgn_\sigma(\psi).$$ ∎

1.9 Duality. We shall not describe the theory of Abstract Order Spaces (AOS), referring the reader to [Marshall, 1996]. We recall Thm. 3.19 in [Dickmann & Miraglia, 2000], establishing a **natural duality** between the categories of Reduced Special Groups and Abstract Order Spaces, allowing the transfer of certain results between these categories. ∎

If K is a RSG, then $D_K(1, b)$ is a subgroup of K and the relation $a \in D_K(1, b)$ is a **partial order** ($a \preceq b$). It is proven in Prop. 1.2 of [Dickmann, Marshall & Miraglia, 2005] that this idea leads to an first-order axiomatization of the theory **reduced** SGs, that will be very useful in what follows:

Proposition 1.10 *A structure $\langle K, \cdot, \preceq, 1, -1 \rangle$ for the first-order language $\{\cdot, 1, -1, \preceq\}$, where $\cdot$ is a binary operation, $\preceq$ is a binary relation, and 1, -1 are constants, is a reduced special group iff it satisfies the following set of axioms:*
[R 0] $\langle K, \cdot, 1 \rangle$ *is a group of exponent two, with $1 \neq -1$.*
[R 1] (Partial order): $\preceq$ *is partial order on K, with first element 1 and last element -1.*
[R 2] (Involution): *For all $a, b \in K$, $a \preceq b \Rightarrow -b \preceq -a$.*
[R 3] (Subgroup): *For all $b \in K$, $\{x : x \preceq b\}$ is a subgroup of K, that is, $x, y \preceq b$ implies $xy \preceq b$.*

[R 4] (**Weak Compatibility**): *For all $a, b, c, d \in K$, $a \preceq b$ and $bd \preceq cd$ imply $\exists\, e \in K$ so that $e \preceq d$ and $ae \preceq ce$.* ∎

Remarks 1.11 a) By 1.5.(a), every RSG satisfies axiom [R 4].

b) In the case of a RSG, K, the transitivity of $a \in D_K(1, b)$ is tantamount to this subgroup being saturated (which may fail if K is not reduced).

c) Given a structure, $\langle K, \cdot, \preceq, 1, -1 \rangle$, satisfying the conditions in 1.10, binary isometry may be obtained, as suggested by Lemma 1.4.(a), by $\langle a, b \rangle \equiv_K \langle c, d \rangle$ iff $ab = cd$ and $ac \preceq cd$. ∎

1.12 Saturated Subgroups. This theme is discussed in some detail in Chapter 2 of [Dickmann & Miraglia, 2000]. Here we register, for the convenience of the reader, only the most basic facts. Let K be a RSG.

A subgroup, Δ, of K is **saturated** if $a \in \Delta$ implies $D_K(1, a) \subseteq \Delta$. By 1.6.(b), the kernel of any SG-character is a saturated subgroup of K. If $-1 \in \Delta$, we say Δ is **improper**, and proper otherwise (i.e., $\Delta \neq K$). Clearly, the intersection and union of a upwards directed family of saturated subgroups is saturated.

a) (Lemma 2.4.(b), [Dickmann & Miraglia, 2000]) A subgroup Δ of K is saturated iff for all Pfister forms $\mathcal{P}$ with coefficients in Δ, $D_K(\mathcal{P}) \subseteq \Delta$.

b) If $\mathcal{P}$ is a Pfister form over K, the $D_K(\mathcal{P})$ is a saturated subgroup of K.

Cor. 2.29 and Thm. 2.11 in [Dickmann & Miraglia, 2000] yield the following important

Theorem 1.13 (Separation) *Let K be a RSG, let Δ be a saturated subgroup of K and let a be an element of G such that $a \notin \Delta$. Then, there is $\sigma \in X_K$ so that $\Delta \subseteq \ker \sigma$ and $\sigma(a) = -1$. In particular, for $a, b \in K$, $b \in D(1, a)$ iff for all $\sigma \in X_G$, $\sigma(a) = 1$ entails $\sigma(b) = 1$.* ∎

c) Let $S \subseteq K$ and write $\mathfrak{s}(S)$ for the saturated subgroup of K generated by S, called the **saturation** of S (in [Dickmann & Miraglia, 2000] this is written $\overline{S}$). The Separation Theorem 1.13 and item (a) yield

$$\mathfrak{s}(S) = \bigcap \{\ker \sigma : \sigma \in X_K \text{ and } S \subseteq \ker \sigma\}$$
$$= \bigcup \{D_K(\otimes_{a \in F} \langle 1, a \rangle) : F \text{ is a finite subset of } S\}.$$

d) Saturated subgroups of a RSG K classify all RSG-quotients of K (cf. Prop. 2.28, [Dickmann & Miraglia, 2000]). If Δ is a saturated subgroup of K, write $K/\Delta = \{a/\Delta : a \in K\}$ for the quotient RSG and $p_\Delta : K \longrightarrow K/\Delta$, for the canonical projection. Note that $a/\Delta = \{x \in K : ax \in \Delta\}$. Def. 2.27 and Prop. 2.28 in [Dickmann & Miraglia, 2000] show that for $a, b, c, d \in K$, the following are equivalent:

(i) $\langle a/\Delta, b/\Delta \rangle \equiv_{K/\Delta} \langle c/\Delta, d/\Delta \rangle$;
(ii) There are $a', b', c', d' \in K$ so that $aa', bb', cc', dd' \in \Delta$ and $\langle a', b' \rangle \equiv_K \langle c', d' \rangle$;
(iii) There is a Pfister form $\mathcal{P}$ over Δ so that $\langle a, b \rangle \otimes \mathcal{P} \equiv_K \langle c, d \rangle \otimes \mathcal{P}$.

e) Recall that a **subspace** of X_K (in the sense of Abstract Order Spaces (AOS)) is a subset of the form $\bigcap_{i \in I} [\![a_i = 1]\!]$, $a_i \in K$, $i \in I$ (cf. Definition in section 2.4, p. 32 of [Marshall, 1996]). Write $\mathrm{Sub}(X_K)$ for the family of subspaces of X_K and $\mathrm{Sat}(K)$ for the set of saturated subgroups of K. Consider the maps:

$$\Sigma : \mathrm{Sub}(X_K) \longrightarrow \mathrm{Sat}(K) \quad \text{and} \quad S : \mathrm{Sat}(K) \longrightarrow \mathrm{Sub}(X_K),$$

where $\Sigma(X) = \bigcap_{\sigma \in X} \ker \sigma$ and $S(\Delta) = \bigcap_{a \in \Delta} [\![a = 1]\!]$.

The Separation Theorem 1.13 guarantees Σ and S to be inverse, order reversing, bijective correspondences between $\mathrm{Sub}(X_K)$ and $\mathrm{Sat}(K)$. ∎

B. Real Semigroups

Definition 1.14 (Def. 1.1, [Dickmann & Petrovich, 2004]) *A* **ternary semigroup (TS)** *is a structure* $S = \langle S, 1, 0, -1, \cdot \rangle$, *such that* $\forall\, x \in S$:

[TS1] S is a commutative semigroup with unit 1;   [TS2] $x^3 = x$;
[TS3] $-1 \neq 1$ and $(-1)(-1) = 1$;   [TS4] $x \cdot 0 = 0$;
[TS5] $x = (-1) \cdot x \Rightarrow x = 0$.

Write $-x$ for $(-1) \cdot x$. *A* **TS-morphism** *is a map preserving product and the constants* 0, 1, −1. *Note that for all* $x \in S$, $x^4 = x^2$.

Definition 1.15 a) (Def. 2.1, [Dickmann & Petrovich, 2004]) *A* **real semigroup (RS)** *is a ternary semigroup* G, *together with a ternary relation* D *on* G, *to be written* $a \in D(b, c)$, *called* **binary representation**, *and with* **transversal binary representation**, D^t, *defined by*

[t-rep] $a \in D^t(b, c) \Leftrightarrow a \in D(b, c)$ and $-b \in D(-a, c)$ and $-c \in D(b, -a)$,

satisfying, for all $a, b, d, e \in G$:

[RS 0] $c \in D(a, b) \Leftrightarrow c \in D(b, a)$;   [RS 1] $a \in D(a, b)$;
[RS 2] $a \in D(b, c) \Rightarrow ad \in D(db, dc)$;
[RS 3] *(Strong associativity)* $a \in D^t(b, c)$ and $c \in D^t(d, e) \Rightarrow \exists\, x \in D^t(b, d)$ s.t. $a \in D^t(x, e)$;
[RS 4] $e \in D(c^2 a, d^2 b) \Rightarrow e \in D(a, b)$;
[RS 5] $ad = bd$, $ae = be$ and $c \in D(d, e) \Rightarrow ac = bc$;
[RS 6] $c \in D(a, b) \Rightarrow c \in D^t(c^2 a, c^2 b)$;

[RS 7] (Reduction) $D^t(a, -b) \cap D^t(b, -a) \neq \emptyset \Rightarrow a = b$.

[RS 8] $a \in D(b, c) \Rightarrow a^2 \in D(b^2, c^2)$.

b) *Representation and transversal representation (t-representation) are extended, by induction, to forms of dimension $n \geq 3$: if $\varphi = \langle a_1, \ldots, a_n \rangle$ is a form over a real semigroup, G, and $x \in G$,*

$$x \in D_G(\varphi) \Leftrightarrow \exists u \in D_G(a_2, \ldots, a_n) \text{ s. t. } x \in D_G(a_1, u).$$

Similarly, for $D_G^t(\varphi)$: $x \in D_G^t(\varphi) \Leftrightarrow \exists u \in D_G^t(a_2, \ldots, a_n)$ s. t. $x \in D_G^t(a_1, u)$.

c) *Write $\mathcal{L}_{RS} = \langle \cdot, 1, 0, -1, D \rangle$ for the first-order language of real semigroups; a structure for this language satisfying all the axioms in (a), with the possible exception of [RS3], is called a **pre-Real Semigroup** (**pRS**).*

d) *If G is a pRS, write $G^\times = \{u \in G : u^2 = 1\}$ for the **group of units** in G.*

Remark 1.16 If G is a pRS, Prop. I.2.10 in [Dickmann & Petrovich, 2016] establishes the following equivalence:

(1) G is a RS (i.e., satisfies axiom [RS3]);

(2) [RS 3'] : For all $a, b, c, d \in G$,

$$D_G^t(a, b) \cap D_G^t(c, d) \neq \emptyset \Rightarrow D_G^t(a, -c) \cap D_G^t(-b, d) \neq \emptyset.$$

The proof for RSs is entirely analogous to that of Lemma 1.5. ∎

Remark 1.17 The ternary semigroup $\mathbf{3} = \{1, 0, -1\}$ (with product induced by the integers) has a unique structure of RS, with representation and t-representation given by (cf. Cor. 2.4, p. 109, [Dickmann & Petrovich, 2004]):

(D) $\begin{cases} D_3(0, 0) = \{0\}; \quad D_3(0, 1) = D_3(1, 0) = D_3(1, 1) = \{0, 1\}; \\ D_3(0, -1) = D_3(-1, 0) = D_3(-1, -1) = \{0, -1\}; \\ D_3(1, -1) = D_3(-1, 1) = \mathbf{3}. \end{cases}$

(D^t) $\begin{cases} D_3^t(0, 0) = \{0\}; \quad D_3^t(0, 1) = D_3^t(1, 0) = D_3^t(1, 1) = \{1\}; \\ D_3^t(0, -1) = D_3^t(-1, 0) = D_3^t(-1, -1) = \{-1\}; \\ D_3^t(1, -1) = D_3^t(-1, 1) = \mathbf{3}. \end{cases}$ ∎

Definition 1.18 a) *A map $f : G \longrightarrow H$, where G, H are pRSs, is a **RS-morphism** if it is a TS-morphism preserving representation (or equivalently, t-representation), i.e., for all $a, b, c \in G$, $a \in D(b, c) \Rightarrow f(a) \in D(f(b), f(c))$ (or the corresponding statement with D replaced by D^t).*

b) *If G is a pRS, write X_G for the set of RS-morphisms from G to $\mathbf{3}$, called the **space of characters** of G. For $g \in G$ and $j \in \mathbf{3} = \{0, 1, -1\}$, set $[\![g = j]\!] = \{\sigma \in X_G : \sigma(g) = j\}$.*

Some of the basic consequences of the axioms concerning forms over a RS appear in Propositions 2.3 (p. 107) and 2.7 (p. 110) of [Dickmann & Petrovich, 2004]; to ease presentation and reference, we register the Lemma and the Proposition that follow, giving proofs only for those items not established in the above mentioned references. These basic properties of representation and t-representation in RSs, will be used frequently, sometimes without explicit mention.

Lemma 1.19 *Let G be a pRS. For all $a, b, c, d \in G$, we have*
(0) $D^t(a,b) = D^t(b,a) \subseteq D(a,b)$;
(1) $a \in D^t(b,c) \Rightarrow -b \in D^t(-a,c)$.
(2) $0 \in D(a,b)$;
(3) $a \in D^t(b,c) \Rightarrow ad \in D^t(bd, cd)$;
(4) $a \in D(0,1) \cup D(1,1) \Rightarrow a = a^2$; (5) $d \in D(ca, cb) \Rightarrow d = c^2 d$;
(6) $a^2 \in D(1,b)$; in particular, $D(1,1) = \mathrm{Id}(G) = \{a \in G : a = a^2\}$;
(7) $a \in D^t(b,b) \Rightarrow a = b$; (8) $a \in D(0,0) \Rightarrow a = 0$; (9) $1 \in D^t(1,a)$;
(10) $D^t(1,-1) = G$; (11) $ab \in D(1,-a^2)$;
(12) $0 \in D^t(a,b) \Rightarrow a = -b$; (12.a) $a \in D^t(b,0) \Rightarrow a = b$.
(13) $c \in D(a,b) \Leftrightarrow c \in D^t(c^2 a, c^2 b)$. In particular,
(13.a) D and D^t are interdefinable; (13.b) $D(a,b) \cap G^\times = D^t(a,b) \cap G^\times$.
(14) $D(a,-a) = D^t(a,-a) = a^2 \cdot G$.
  (14.a) $0 \in D^t(a,b) \Leftrightarrow b = -a$; (14.b) $a \in D^t(b,0) \Leftrightarrow a = b$.
(15) $\{a, -a\} \subseteq D^t(b,c) \Rightarrow c = -b$.

Proof. Item (0) follows immediately from the [RS 0] and the definition of D^t (cf. [t-rep] in Definition 1.15). Items (1) – (12) are proven in Proposition 2.3, pp. 107-108, in [Dickmann & Petrovich, 2004], upon noting that the arguments given therein do not employ axiom [RS 3] (strong associativity). For (12.a), note that its hypothesis entails $0 \in D^t(-a, b)$ and so (12) yields $a = b$.

(13) The implication ($\Rightarrow$) follows immediately from [RS 6] in 1.15; for the converse, item (0) and [RS 4] in 1.15 yield $c \in D^t(c^2 a, c^2 b) \subseteq D(c^2 a, c^2 b)$, and so $c \in D(a,b)$, as needed. The statement in (13.a) is clear, while if $c \in G^\times$, then (13) guarantees that $c \in D(a,b)$ iff $c \in D^t(a,b)$.

(14) Given $x \in G$, by item (9) we have $1 \in D^t(1, -ax)$, and so (3) entails $a \in D^t(a, -a^2 x)$, whence, from (1) we obtain $a^2 x \in D^t(a, -a)$, showing that $a^2 \cdot G \subseteq D^t(a, -a)$. Now, from $x \in D(a, -a) = D(a \cdot 1, a \cdot (-1))$, (5) yields

$x = a^2x$, whence $D(a, -a) \subseteq a^2 \cdot G$. But then, item (0) and the preceding argument imply $a^2 \cdot G \subseteq D^t(a, -a) \subseteq D(a, -a) \subseteq a^2 \cdot G$, establishing (14). Notice that (14) guarantees that $0 \in D^t(x, -x)$ for all $x \in G$. But then, (14.a), (14.b) and are direct consequences of (1) together with of (14), (12), and (14), (12.a), respectively.

(15) By (3), $-a \in D^t(b, c)$ entails $a \in D^t(-b, -c)$ and so $a \in D^t(b, -(-c)) \cap D^t(-c, -b)$; from axiom [RS 7] (reduction) we obtain $c = -b$. ∎

Proposition 1.20 *Let H be a RS, $\varphi = \langle x_1, \ldots, x_n \rangle$, ψ be forms over H and let $a, b, c \in H$.*

a) $D_H(\varphi)$ and $D_H^t(\varphi)$ do not depend on the order of their entries, i.e, for any permutation σ of those entries, $D_H(\varphi) = D_H(\varphi^\sigma)$ and $D_H^t(\varphi) = D_H^t(\varphi^\sigma)$.

b) (1) $a \in D_H(\varphi) \Rightarrow ac \in D_H(c\varphi)$ and $a \in D_H^t(\varphi) \Rightarrow ac \in D_H^t(c\varphi)$;
(2) $a \in D_H(c\varphi) \Rightarrow a = ac^2$;
(3) $a \in D_H(\varphi) \Rightarrow a \in D_H^t(a^2\varphi)$. In particular, $D_H(\varphi) \cap H^\times = D_H^t(\varphi) \cap H^\times$.

c) $a \in D_H(\varphi \oplus \psi) \Leftrightarrow$ there are $b \in D_H(\varphi)$, $c \in D_H(\psi)$ so that $a \in D_H(b, c)$. A similar statement holds replacing D_H by D_H^t.

d) If a is an entry in φ, then $a \in D_H(\varphi)$.

e) $a \in D_H(\varphi)$ and $b \in D_H(\psi) \Rightarrow ab \in D_H(\varphi \otimes \psi)$. A similar statement holds for D_H^t.

f) For $b \in G$ and $n \geq 1$, $D_G^t(\underbrace{b, \ldots, b}_{n\times}) = \{b\}$.

g) $D_H^t(\langle 1, a^2 \rangle \otimes \varphi) = D_H^t(\varphi)$.

j) If $H \xrightarrow{f} H'$ is a RS-morphism (cf. 1.18.(a)), then $a \in D_H(\varphi) \Rightarrow f(a) \in D_{H'}(f \star \varphi)$, where $f \star \varphi = \langle f(x_1), \ldots, f(x_n) \rangle$. Similarly, replacing D_H by D_H^t.

h) (1) $a \in D_H(\varphi) \Rightarrow a \in D_H(\varphi \oplus \psi)$; (2) $a \in D_H^t(\varphi) \Rightarrow a \in D_H^t(a^2\varphi \oplus a^2\psi)$.

Proof. See Prop. 2.7, pp. 110 − 112 in [Dickmann & Petrovich, 2004]. ∎

Remark 1.21 If G is a **real semigroup**, Theorems 4.3 and 4.4. (p. 116) of [Dickmann & Petrovich, 2004] guarantee that its space of **RS-characters**, X_G, separates points in G. Moreover, the representation relations induced by X_G in G coincide with the original ones carried by G, i.e., for all $a, b, c \in G$, we have the following equivalences:

[D_G] $\quad a \in D_G(b, c) \quad \Leftrightarrow \quad \forall \, \sigma \in X_G, \, \sigma(a) \in D_3(\sigma(b), \sigma(c))$,

$[D_G^t]$ $a \in D_G^t(b, c)$ ⇔ $\forall\, \sigma \in X_G,\ \sigma(a) \in D_3^t(\sigma(b), \sigma(c))$.   ■

1.22 Topologies on X_G. If G is a RS, the collection $\{[\![g = 1]\!] : g \in G\}$ (cf. 1.18.(b)) constitutes sub-basis for a (completely normal or root system, cf. [Dickmann, Schwartz & Tressl, 2019], p. 290) spectral topology on X_G. For the constructible topology on X_G, with which X_G is a Boolean space, we have:

Fact 1.23 Let G be a RS and let $a, b, c \in G$.

a) (Cor. III.7.3, [Dickmann & Petrovich, 2016] and Prop. 6.1.5, [Marshall, 1996]) There is a unique $z \in G$, such that $z = z^2$ and $z \in D_G^t(a^2, b^2)$. Moreover, $[\![z = 0]\!] = [\![a = 0]\!] \cap [\![b = 0]\!]$.

b) The sets of the form $[\![b = 0]\!] \cap \bigcap_{k=1}^{n} [\![a_k = 1]\!]$, for $a_1, \ldots, a_n, b \in G$ constitute a basis for the constructible (*or patch*) topology on X_G.   □

Hence, the spectral topology on X_G is Boolean iff for all $z \in G$, the closed set $[\![z = 0]\!]$ is **clopen** in the spectral topology of X_G.   ■

1.24 Duality. By Thm. 4.1 in [Dickmann & Petrovich, 2004] (cf. also Thm. I.5.1, [Dickmann & Petrovich, 2008]), the category **RS** is **isomorphic** to **ARS**[op], where **ARS** is the category of abstract real spectra in the sense of M. Marshall. Note that the category **RS** is Horn-geometrically axiomatizable, while **ARS** is not. Clearly, this duality allows the transfer of statements between these categories.   ■

1.25 The group of units of a RS. Let G be a RS and let $G^\times$ be its group of units (1.15.(d)). For any $a, b \in G$, 1.19.(13.b) entails $D_G^t(a, b) \cap G^\times = D_G(a, b) \cap G^\times$.

Define a binary relation on $G^\times$ by $a \preceq b$ iff $a \in D_G(1, b) = D_G^t(1, b)$

Lemma 1.26 (cf. 1.2.11, [Dickmann & Petrovich, 2016]) *Notation as above, the structure $\langle G^\times, \preceq, 1, -1 \rangle$ satisfies the axioms* [R 0], [R 1], [R 2] *and* [R3] *in the statement of Proposition 1.10. In particular, $G^\times$ is a pSG.*

Proof. Clearly $G^\times$ is a group of exponent 2 ([R 0]). To show $\preceq$ is a partial order ([R 1]), if $a \in G^\times$, then $a \preceq a$ follows from axiom [RS 1] and, as observed above, $D_G \cap G^\times = D_G^t \cap G^\times$. If $a \in D_G(1, b)$ and $b \in D_G(1, a)$, then, 1.19.(1) entails $-1 \in D_G^t(-a, b) \cap D_G^t(-b, a)$ and the reduction axiom [RS 7] of RSs implies $a = b$. For the transitivity of $\preceq$, let $a \in D_G^t(1, b)$ and

$b \in D_G^t(1, c)$. By axiom [RS 3], there is $z \in G$ such that $z \in D_G^t(1, 1)$ and $a \in D_G^t(z, c)$. Now, 1.19.(7) yields $z = 1$, whence, $a \in D_G^t(1, c)$.

Axiom [RS 1] yields $1 \in D_G(1, a)$, i.e, $1 \preceq a$, while 1.19.(10) gives $D_G^t(1, -1) = G$ and so $a \preceq -1$, for all $a \in G^\times$. Axiom [R 3] in 1.10 is immediate from 1.19.(1): $a \in D_G^t(b, c) \Rightarrow -b \in D_G^t(-a, c)$.

It remains to establish [R 3]: for $a \in G^\times$, $\{x \in G^\times : x \preceq a\}$ is a subgroup of $G^\times$. If $y, z \in D_G^t(1, a)$, then 1.20.(e) entails, $xy \in D_G^t(\langle 1, a \rangle \otimes \langle 1, a \rangle) = D_G^t(1, a, a, a^2) = D_G^t(1, 1, a, a)$. By 1.20.(c), there are $t \in D_G^t(1, 1)$ and $z \in D_G^t(a, a)$ so that $xy \in D_G^t(t, z)$. As above, we obtain $t = 1$ and and $z = a$. ∎

When treating Boolean RSs we shall use Proposition 1.10 to show that their group of units is a RSG. By 1.26, it suffices to verify the weak associativity axiom ([R 4]) associated to the partial order $\preceq$ defined above. ∎

1.27 Idempotents in Real Semigroups. Let G be a RS and let $\mathrm{Id}(G) = \{a \in G : a^2 = a\} = G^2$ be the set of **idempotents** in G; for $f, g \in \mathrm{Id}(G)$ define $f \leq g$ iff $fg = g$. Then, $\leq$ is a partial order, endowing $\mathrm{Id}(G)$ with a distributive lattice structure, wherein

$f \wedge g =$ the unique element in $D_G^t(f, g)$ (cf. 1.23.(a))   and $f \vee g = fg$,

whose bottom element is 1 and top element is 0 (Prop. I.6.8, [Dickmann & Petrovich, 2016]). Note: this is the *opposite* of the usual lattice structure associated to idempotents in a *commutative unitary ring*. ∎

2 Direct Sum Decompositions of Reduced Special Groups

In this section, K is a reduced special group; notation is as in 1.12.

Definition 2.1 *Let K be a RSG. A saturated subgroup of K, Δ, is a* **direct sum subgroup (DSS)** *of K if there is a saturated subgroup, $\Delta^\perp$, such that the natural map $\gamma_\Delta : K \longrightarrow K/\Delta \times K/\Delta^\perp$, given by $a \longmapsto (a/\Delta, a/\Delta^\perp)$ is a RSG-isomorphism, where $K/\Delta \times K/\Delta^\perp$ has its canonical product structure. K and $D_K(1,1) = \{1\}$ are both also considered as DSS, the* **trivial DSS** *(although $K/K = \{1\}$ is not a RSG).*

Remark 2.2 Notation as 2.1, Δ is a DSS iff $\Delta^\perp$ is a DSS; since γ_Δ is injective, we get $\Delta \cap \Delta^\perp = \{1\}$ ($\gamma_\Delta(x) = (1, 1)$ iff $x \in \Delta \cap \Delta^\perp$ iff $x = 1$). ∎

Here is an improved special group version of Lemma 8.9.1 in [Marshall, 1996]:

Proposition 2.3 *For a saturated subgroup Δ of K, the following are equivalent:*
(1) Δ *is a direct sum subgroup of* K;
(2) *There is* $a \in K$ *so that:*
  (i) $\Delta = D_K(1,a)$ (and $\Delta^\perp = D(1,-a)$);
  (ii) *For all $b \in K$, there is $c \in K$ so that* $\langle 1,b \rangle \otimes \langle 1,a \rangle \equiv_K \langle 1,1 \rangle \otimes \langle 1,c \rangle$.

Proof. Notation is as in 2.1 and 2.2. To ease presentation, write $D(\cdot,\cdot)$ for $D_K(\cdot,\cdot)$ and T for $K/\Delta \times K/\Delta^\perp$.

(1) $\Rightarrow$ (2) : Since γ_Δ is a RSG-isomorphism, there is $a \in K$ so that $\gamma_\Delta(a) = (1,-1)$ in T. Hence, $a \in \Delta$ and so $D(1,a) \subseteq \Delta$. If $x \in \Delta$, then $\gamma_\Delta(x) = (1, x/\Delta^\perp)$; since isometry and representation in T are coordinatewise defined, we get $\gamma_\Delta(x) \in D_T(\langle 1,1 \rangle, (1,-1)) = D_T(\gamma_\Delta(1), \gamma_\Delta(a))$, whence $x \in D(1,a)$, establishing 2.(i). Note that $\gamma_\Delta(-a) = (-1,1) \in T$ and reasoning as above obtains $\Delta^\perp = D(1,-a)$. For 2.(ii), if $b \in K$, there $c \in K$ such that $\gamma_\Delta(c) = (b/\Delta, -1) \in T$. Thus,

(*) $\qquad cb \in \Delta = D(1,a)$ and $-c \in D(1,-a)$.

To show $\langle 1,a \rangle \otimes \langle 1,b \rangle = \langle 1,a,b,ab \rangle \equiv_K 2\langle 1,c \rangle = \langle 1,1,c,c \rangle$, we compare signatures (Pfister's local-global principle for RSGs, 1.8): if $\sigma(a) = 1$, then the first relation in (*) yields $\sigma(cb) = 1$ and so $\sigma(b) = \sigma(c)$, which in turn entails the equality of the signatures of $\langle 1,a \rangle \otimes \langle 1,b \rangle$ and $2\langle 1,c \rangle$ at σ; if $\sigma(-a) = 1$, the second relation in (*) yields $\sigma(c) = -1$ and the signatures of $\langle 1,a \rangle \otimes \langle 1,b \rangle$ and $2\langle 1,c \rangle$ at σ are both 0, establishing (1) $\Rightarrow$ (2).

(2) $\Rightarrow$ (1) : Since $D(1,a) \cap D(1,-a) = \{1\}$, it is clear that $\gamma_\Delta : K \longrightarrow T$ is injective (cf. 2.2.(a)) and a RSG-morphism. To show it is an embedding, assume, for $x, y \in K$, $\gamma_\Delta(x) \in D_T(\langle 1,1 \rangle, \gamma_\Delta(y))$. Hence,

$$x/\Delta \in D_{K/\Delta}(1, y/\Delta) \quad \text{and} \quad x/\Delta^\perp \in D_{K/\Delta^\perp}(1, y/\Delta^\perp).$$

By the equivalence in item 1.12.(d), there are Pfister forms, $\mathcal{P}$ and $\mathcal{P}^\perp$ over Δ and $\Delta^\perp$, respectively such that

(**) $\qquad \langle x, xy \rangle \otimes \mathcal{P} \equiv_K \langle 1,y \rangle \otimes \mathcal{P}$ and $\langle x, xy \rangle \otimes \mathcal{P}^\perp \equiv_K \langle 1,y \rangle \otimes \mathcal{P}^\perp$.

Now if $\sigma(a) = 1$, since $\mathcal{P}$ has entries in $D(1,a)$, the first relation in (**) yields $2^n(\sigma(x) + \sigma(xy)) = 2^n(1 + \sigma(y))$ and the signatures of $\langle x, xy \rangle$ and $\langle 1,y \rangle$ are the same for all $\sigma \in [\![a = 1]\!]$; similarly, if $\tau \in [\![a = -1]\!] = [\![-a = 1]\!]$, the second relation in (**) entails the signature of $\langle x, xy \rangle$ and $\langle 1,y \rangle$ to be the same for all $\tau \in [\![a = -1]\!]$. Hence, $\langle x, xy \rangle$ and $\langle 1,y \rangle$ have the same total signature and 1.8 entails $\langle x, xy \rangle \equiv_K \langle 1,y \rangle$ and $x \in D_T(\langle 1,y \rangle)$, establishing that γ_Δ is a RS-embedding.

It remains to show γ_Δ is surjective. Let $u, v \in K$; then, there is $c \in K$ so that $\langle 1, -uv \rangle \otimes \langle 1, a \rangle \equiv_K \langle 1, 1 \rangle \otimes \langle 1, c \rangle = \langle 1, c \rangle \oplus \langle 1, c \rangle$. We claim that

(***) $\quad -(cv)v = -c \in D(1, -a)$ and $(-cv)u = -cuv \in D(1, a)$.

By Prop. 2.2 in [Dickmann & Miraglia, 2000], $a \in D(\langle 1, -uv \rangle \otimes \langle 1, a \rangle)$ = $D(\langle 1, c \rangle \oplus \langle 1, c \rangle)$, and so Prop. 1.6.(e) in [Dickmann & Miraglia, 2000] and 1.19.(d) yield $a \in D(1, c)$. Thus, $-c \in D(1, -a)$; if $\tau \in X_K$ is such that $\tau(a) = 1$, the isometry $\langle 1, -uv \rangle \otimes \langle 1, a \rangle \equiv_K 2 \langle 1, c \rangle$ and Theorem 1.8 entail $2\tau(-uv) = 2\tau(c)$ and so $\tau(-cuv) = 1$, for each $\tau \in [\![a = 1]\!]$, whence $-cuv \in D(1, a)$ (by 1.13), establishing (***). Thus, $-cv/\Delta = u/\Delta$, $-cv/\Delta^\perp = v/\Delta^\perp$ and therefore $\gamma_\Delta(-cv) = (u/\Delta, v/\Delta^\perp) \in T$, completing the proof. ∎

Remarks 2.4 (1) In the terminology of section 8.9 in [Marshall, 1996], a subset A of X is **complemented or a direct summand** if both A and $X \setminus A$ are subspaces of X and the natural injection

$$K \longrightarrow K{\upharpoonright}A \times K{\upharpoonright}X \setminus A, \text{ given by } A \longmapsto \langle a{\upharpoonright}A, a{\upharpoonright}X \setminus A \rangle,$$

is an isomorphism. Given the duality between the categories of RSGs and AOSss (cf. 1.9), in the language of RSG and with notation as in items (c) and (d) of 1.12, this can be equivalently rephrased as follows: write Δ for $\Sigma(A)$ and $\Delta^\perp$ for $\Sigma(X \setminus A)$; then, the natural map

$$K \longrightarrow K/\Delta \times K/\Delta^\perp, \text{ given by } a \longmapsto \langle a/\Delta, a/\Delta^\perp \rangle$$

is an isomorphism. In fact, the AOS $\langle K{\upharpoonright}A, A \rangle$ corresponds, by duality, to the RSG-quotient $K \xrightarrow{p_\Delta} K/\Delta$. Hence, all complemented subspaces, A, of X are of the form $[\![a = 1]\!]$, for some $a \in K$ (and so $\Sigma(A) = D_K(1, a)$) and satisfy condition 2.(ii) of 2.3. In fact, Proposition 2.3 and Lemma 8.9.1 in [Marshall, 1996] entail the existence of a bijective correspondence between DSSs of K and direct summands of X_K.

(2) Sometimes, it is easier to deal with certain statements in the language of AOSss; an example follows. For $A \subseteq X$ and $\mu \in \{0, 1\}$, set

$$A^\mu = \begin{cases} A & \mu = 0; \\ \neg A = X \setminus A & \text{if } \mu = 1;. \end{cases}$$

Fact 2.5 [Note (3), p. 179, [Marshall, 1996], without proof] If $A_1, \ldots, A_n$ are direct summands of $\langle K, X \rangle$, set $E_\mu = \bigcap_{i=1}^n A_i^{\mu(i)}$, $\mu \in \{0, 1\}^n$; let

$$\mathfrak{q} = \{\mu \in \{0, 1\}^n : E_\mu \neq \emptyset\},$$

be the set of indices corresponding to the non-empty atoms of the BA generated by $\{A_1, \ldots, A_n\}$. Let $\Delta_\mu = \Sigma(E_\mu)$, $\mu \in \mathfrak{q}$. Then, all E_μ are direct summands of X and the natural map from K to $\prod_{\mu \in \mathfrak{q}} K/\Delta_\mu$, $a \longmapsto \langle a/\Delta_\mu \rangle_{\mu \in \mathfrak{q}}$, is an isomorphism.

Proof. It is enough to deal with the case $n = 2$ and proceed by induction. Let A, B be direct summands of X; we may assume all four atoms, $A \cap B$, $A \cap \neg B$, $B \cap \neg A$, $\neg A \cap \neg B$ to be non-empty. First, we prove that the natural map from $K \upharpoonright B$ to $[K \upharpoonright (A \cap B) \times K \upharpoonright (B \cap \neg A)]$ is an isomorphism. Given $u, v \in K$, since A is a direct summand of $\langle K, X \rangle$, there is $x \in K$ such that $x \upharpoonright A = u \upharpoonright A$ and $x \upharpoonright \neg A = v \upharpoonright \neg A$. Hence,
$$\begin{cases} x \upharpoonright (A \cap B) = (x \upharpoonright A) \upharpoonright B = (u \upharpoonright A) \upharpoonright B = u \upharpoonright (A \cap B); \\ x \upharpoonright (B \cap \neg A) = = (x \upharpoonright \neg A) \upharpoonright B = (v \upharpoonright \neg A) \upharpoonright B = v \upharpoonright (B \cap \neg A), \end{cases}$$
as desired. Similarly, one shows $K \upharpoonright \neg B$ and $[K \upharpoonright (\neg B \cap A) \times K \upharpoonright (\neg B \cap \neg A)]$ to be isomorphic; thus, all atoms above are direct summands and K is naturally isomorphic to the product of the ensuing four factors. ■

3 The Structure of Boolean Real Semigroups

Definition 3.1 *A real semigroup, G, is* **Boolean** *if its space of RS-characters, X_G, endowed with the spectral topology, is Boolean, i.e, if the spectral and constructive topologies on X_G coincide.*

Example 3.2 There are two important examples of Boolean RSs:
(1) If K is a reduced special group, let G be K together with a new absorbing element, 0. By Remark 2.2.4 in [Dickmann & Petrovich, 2004] (see also Corollary 2.5, [Dickmann & Petrovich, 2004]), $G = K \cup \{0\}$ can be made into a RS, whose space is RS-characters is the Boolean space X_K, the space of orders of K;
(2) Post algebras of order 3, i.e., the structure of all continuous maps from a Boolean space X to $3 = \{1, 0, -1\}$, the latter endowed with the discrete topology (cf. section 3 in [Dickmann & Petrovich, 2008]). If $P = \mathbb{C}(X, 3)$, then $X_P = X$.
The forthcoming discussion exhibits a significant number of other examples of Boolean RSs. ■

In this section, among other things, we give new proofs of the following results in [Marshall, 1996]:
• Thm. 7.6.4 corresponds to the first statement in item (3) of Theorem 3.6, below; our proof employs Proposition 1.10, an alternative axiomatic for RSGs;
• Thm. 8.9.2 corresponds to Proposition 3.10, below, which is phrased using clopen direct summands of the space of orders of the RSG K (spaces of orders do not have zero sets);

- Theorem 8.9.3 corresponds to Theorem 3.13, below, but our proof uses the exchange principle in RSs (cf. Remark 1.16).

We also describe a Horn-geometric axiomatization of Boolean RSs, prove the preservation of this class of RSs by important constructions (Theorem 3.18) and characterize the rings whose associated RS is Boolean (Proposition 3.16).

3.3 Notation and Remarks. As usual, if G is a RS, $a \in G$ and $\varepsilon \in 3$:
a) By Proposition 1.10 and Lemma 1.26, $G^\times$ is a reduced pre-special group.
b) $[\![a = \varepsilon]\!] = \{\tau \in X_G : \tau(a) = \varepsilon\}$. If $\varepsilon = 0$, we may write $Z(a)$ for $[\![a = 0]\!]$.
c) $\mathbb{Z}_2 = \{1, -1\} = 3^\times$ is the RSG of units in $3 = \{0, 1, -1\}$. ∎

For a proof of the following beautiful result in the language of ARSs, see Thm. 6.8.1, p.129 of [Marshall, 1996] (the same proof applies to RSs):

Theorem 3.4 (Hörmander-Łojasiewcz Inequality) *Let G be a RS and let $a, b \in G$. Let Y be a closed subspace of X_G such that $Y \cap Z(a) \subseteq Z(b)$. Then, there is $c \in D_G^t(a, b)$ so that $c \!\restriction\! Y = a \!\restriction\! Y$ (i.e., for $\sigma \in Y$, $\sigma(c) = \sigma(a)$).* ∎

Lemma 3.5 *Let G be a Boolean RS.*
a) For each $a \in G$, there is $\nabla a \in G$ such that
(1) $[\![\nabla a = 1]\!] = [\![a = 1]\!]$ and $[\![\nabla a = -1]\!] = Z(a) \cup [\![a = -1]\!]$. In particular, $\nabla a \in G^\times$.
(2) $a = \nabla a \cdot a^2$, and so $G = G^\times \cdot \mathrm{Id}(G)$. (3) $a, a^2 \in D_G^t(1, \nabla a)$.
b) For all $u \in G^\times$, $\nabla u = u$.
c) For all $a, b, c \in G$, $a \in D_G^t(b, c) \Rightarrow \nabla a \in D_G^t(\nabla b, \nabla c) \cap G^\times = D_{G^\times}(\nabla b, \nabla c)$.
d) For all $a \in G$ and $\tau \in X_G$,
(1) $Z(a) = [\![\nabla(a^2) = -1]\!]$; (2) $\tau \in [\![\nabla(a^2) = 1]\!]$ iff $\tau(a) = \tau(\nabla a)$.

Proof. a) For $a \in G$, since X_G is Boolean, $[\![a = 1]\!]$ is clopen in X_G. Now, observe that $-1 \in G$ satisfies
$$[\![a = 1]\!] \cap Z(a) = \emptyset = Z(-1).$$
By Theorem 3.4, there is $\nabla a \in G$ so that $\nabla a \in D_G^t(a, -1)$ and $a \!\restriction\! [\![a = 1]\!] = \nabla a \!\restriction\! [\![a = 1]\!]$. Now notice that:
- For $\tau \in Z(a)$, $\tau(\nabla a) \in D_3^t(\tau(a), -1) = D_3^t(0, -1) = \{-1\}$, and $\tau(\nabla a) = -1$;

- For $\tau \in [\![a = -1]\!]$, the same argument as above shows that $\tau(\nabla a) = -1$.
Consequently, the equalities in (1) are verified and $\nabla a \in G^\times$. Items (2) and
(b) follow easily from the equalities in (1) (note: $\forall \tau \in X_G$, $\tau(\nabla a \cdot a^2) = \tau(a)$; clearly, $a^2 \in \mathrm{Id}(G)$). For (3), since $\nabla a \in D_G^t(-1, a)$, it follows that $-a \in D_G^t(-1, -\nabla a)$, and so, scaling by -1 obtains $a \in D_G^t(1, \nabla a)$; since $D_G^t(1, \nabla a)$ is a subsemigroup of G, the preceding relation yields $a^2 \in D_G^t(1, \nabla a)$.

c) We employ the equalities in item (a.1); for $\tau \in X_G$, we discuss two cases:
(i) $\tau(\nabla a) = 1$: Then, $\tau(a) = 1$ and so, since $\tau(a) \in D_3^t(\tau(b), \tau(c))$, there are only the following possibilities:

(*) $\tau(b) = \tau(c) = 1$; $\tau(b) = 1, \tau(c) = 0$; $\tau(b) = 0, \tau(c) = 1$; or $\tau(ab) = -1$.

In the first case in (*), $\tau(\nabla b) = \tau(\nabla c) = 1$, while in the others $\tau(\nabla b)\tau(\nabla c) = -1$ and so $\tau(\nabla a) \in D_3^t(\tau(\nabla b), \tau(\nabla c))$.

(ii) $\tau(\nabla a) = -1$: Then, either $\tau \in Z(a)$ or $\tau \in [\![a = -1]\!]$; in the former case, either $\tau(b) = \tau(c) = 0$ and so $\tau(\nabla b) = \tau(\nabla c) = -1$, or else, $\tau(b) = -\tau(c) \neq 0$ (i.e., the last alternative in (*) above), yielding $\tau(\nabla b)\tau(\nabla c) = -1$, hence $\tau(\nabla a) = -1 \in D_3^t(\tau(\nabla b), \tau(\nabla c))$. In the latter case, $\tau(a) \in D_3^t(\tau(b), \tau(c))$, leads to a list of possibilities as in (*), with -1 replacing 1 in the first three alternatives, whilst the fourth remains the same; hence we obtain either $\tau(\nabla b) = \tau(\nabla c) = -1$ or $\tau(\nabla b)\tau(\nabla c) = -1$, and thus $\tau(\nabla a) \in D_3^t(\tau(\nabla b), \tau(\nabla c))$, as needed.

d) Item (1) is immediate from (a.1); for (2), (a.1) entails $[\![\nabla a^2 = 1]\!] = [\![a^2 = 1]\!]$. If $\tau \in [\![a^2 = 1]\!]$, then $\tau \notin Z(a)$ and either $\tau \in [\![a = 1]\!] = [\![\nabla a = 1]\!]$ or $\tau \in [\![a = -1]\!] = [\![\nabla a = -1]\!] \setminus Z(a)$, whence $\tau(a) = \tau(\nabla a)$, as needed. ∎

Theorem 3.6 *For a real semigroup G, the following are equivalent:*
(1) For all $x \in G$, there is $u \in G^\times$ so that $x = ux^2$ and $u \in D_G^t(-1, x)$.
(2) The spectral topology on X_G is Boolean;
(3) $G^\times$, with the representation relation induced by G, is a RSG, and the restriction map $\tau \in X_G \longmapsto \tau \upharpoonright G^\times \in X_{G^\times}$ is a homeomorphism [1].

Proof. Item (a.2) in Lemma 3.5 yields (2) $\Rightarrow$ (1). Moreover, since the space of orders of any RSG is Boolean, it is clear that (3) $\Rightarrow$ (2). To complete the proof it suffices to establish (1) $\Rightarrow$ (2) and (2) $\Rightarrow$ (3).

(1) $\Rightarrow$ (2). Fix $a \in G$ and let $u \in G^\times$ satisfy the conditions in (1) with $x = a^2$. For $\tau \in X_G$:
- If $\tau(a) = 0$, then $u \in D_G^t(-1, a^2)$ entails $\tau(u) \in D_3^t(-1, 0)$ and $\tau(u) = -1$;
- If $\tau(u) = -1$, then $\tau(a^2) = -\tau(a^2)$, hence $\tau(a^2) = 0 = \tau(a)$,

---
[1] $X_{G^\times}$ is the space of orders of the reduced special group $G^\times$.

and so $[\![a = 0]\!] = [\![a^2 = 0]\!] = [\![u = -1]\!]$; thus, the closed set $[\![a = 0]\!]$ in the spectral topology of X_G is, in fact, <u>clopen</u> in this topology. Hence, (cf. 1.23.(b)), the spectral and constructible topologies in X_G coincide, establishing (2).

(2) $\Rightarrow$ (3). We start by proving:

$G^\times$ is a RSG. By Lemma 1.26 it suffices to verify $G^\times$ satisfies [R4] in Proposition 1.10. Since $a, b, c, d \in G^\times$, the hypothesis in [R4] is equivalent to

(I) $\qquad\qquad a \in D_{G^\times}(1, b)$ and $b \in D_{G^\times}(c, d)$.

Since $D_G^t \cap G^\times = D_{G^\times}$, (I) yields

(II) $\qquad\qquad a \in D_G^t(1, b)$ and $b \in D_G^t(c, d)$.

But then, (II) together with items (a) and (c) of Proposition 1.20 yield

(III) $\qquad\qquad a \in D_G^t(1, c, d) = D_G^t(1, d, c)$.

Hence, (III) and 1.20.(c) yield $z \in D_G^t(1, d)$, such that $a \in D_G^t(z, c)$. Now, items (b) and (c) of Lemma 3.5 imply $\nabla z \in D_{G^\times}(1, d)$ and $a \in D_{G^\times}(\nabla z, c)$. Just take $y = \nabla z \in G^\times$ to establish the consequent of [R 4], completing the proof that $G^\times$ is a reduced special group.

X_G is (naturally) homeomorphic to $X_{G^\times}$. Set $\eta : X_G \longrightarrow X_{G^\times}$ given by $\eta(\tau) = \tau\restriction G^\times$; clearly, $\eta(\tau)$ is a group morphism from $G^\times$ to $\{1, -1\}$, taking -1 to -1 and respecting binary representation (because the same is true of τ). Moreover:

- η is continuous: For $u \in G^\times$, just note that $\eta^{-1}\left[[\![u = 1]\!]_{X_{G^\times}}\right] = [\![u = 1]\!]_{X_G}$;
- η is injective: If $\tau \neq \tau'$ in G, there is $a \in G$ such that $\tau(a) \neq \tau'(a)$.

Case 1. If $\tau(a) = 1$, then $\tau'(a) \in \{0, -1\}$, then 3.5.(a.1) implies $\tau(\nabla a) = 1$ and $\tau'(\nabla a) = -1$;

Case 2. If $\tau(a) = -1$ and $\tau'(a) = 0 \in \{0, 1\}$, then $\tau(-a) = 1$, while $\tau'(-a) \in \{0, -1\}$ and so 3.5.(a.1) yields $\tau(\nabla -a) = 1$ and $\tau'(\nabla -a) = -1$,

therefore, $\tau \neq \tau'$ implies $\eta(\tau) = \tau\restriction G^\times \neq \tau'\restriction G^\times = \eta(\tau')$, as claimed.

It remains to show:

- η is surjective. Since compact sets in a Hausdorff space are closed, we shall establish the surjectivity of η by showing that its image is dense in $X_{G^\times}$. For this, it is enough to check that Im η meets every non-empty basic clopen in $X_{G^\times}$, that is, for $u_1, \ldots, u_n \in G^\times$,

(IV) $\bigcap_{j=1}^n [\![u_j = 1]\!]_{X_{G^\times}} \neq \emptyset \Rightarrow \exists \tau \in X_G$, so that $\tau(u_j) = 1$, $1 \leq j \leq n$.

To that end we introduce the following notation: if $\varphi = \langle a_1, \ldots, a_n \rangle$ is a form over G, write $\nabla\varphi$ for the n-form over $G^\times$ given by $\langle \nabla a_1, \ldots, \nabla a_n \rangle$. We also need:

Fact 3.7 Let φ be a form of dimension ≥ 2 over G and let $a \in G$. Then,
$$a \in D_G^t(\varphi) \quad \Rightarrow \quad \nabla a \in D_{G^\times}(\nabla \varphi).$$

Proof. By Prop. 1.6.(c) in [Dickmann & Miraglia, 2000], if K is special group, then for all forms θ_1, θ_2 over K and $x \in K$,

(I) $x \in D_K(\theta_1 \oplus \theta_2) \Leftrightarrow \exists\, u \in D_K(\theta_1)$ and $v \in D_K(\theta_2)$ so that $x \in D_K(u, v)$.

For $\dim(\varphi) = 2$, the statement is 3.5.(c); we proceed by induction on the dimension of φ. Assume the result true for $\dim(\psi) = m$ and let $c \in G$. If $a \in D_G^t(\langle c \rangle \oplus \psi)$, then 1.20.(c) yields $x \in D_G^t(\psi)$ such that $a \in D_G^t(c, x)$. Hence, 3.5.(c) and the induction hypothesis entail $\nabla a \in D_{G^\times}(\nabla c, \nabla x)$ and $\nabla x \in D_{G^\times}(\nabla \psi)$. Since $G^\times$ is a RSG, (I) above yields $\nabla a \in D_{G^\times}(\langle \nabla c \rangle \oplus \nabla \psi)$, completing the induction step. □

Consider the Pfister form $\mathcal{P} = \bigotimes_{i=1}^n \langle 1, u_j \rangle$; in $G^\times$, we have $-1 \notin D_{G^\times}(\mathcal{P})$, because, by assumption, there is $\sigma \in X_{G^\times}$ sending all u_j to 1. [2]

Now notice that $-1 \notin D_G(\mathcal{P})$; otherwise, by 1.20.(b.3), we would get $-1 \in D_G^t((-1)^2 \mathcal{P}) = D_G^t(\mathcal{P})$, and Fact 3.7 would then entail, since $\mathcal{P} = \nabla \mathcal{P}$ (by 3.5.(b)), $-1 \in D_{G^\times}(\mathcal{P})$, an impossibility.

By Prop. 2.7.(6) in [Dickmann & Petrovich, 2004] and Cor. 4.7 in [Dickmann & Petrovich, 2008], $D_G(\mathcal{P})$ is a saturated subsemigroup of G, containing $\{u_1, \ldots, u_n\}$, but not -1. Now, Thm. 4.2 in [Dickmann & Petrovich, 2004] furnishes a RS-character, τ, whose values in $D_G(\mathcal{P})$ are in $\{0, 1\}$. Since the $u_i \in G^\times$, we must have $\tau(u_i) = 1$, $1 \leq i \leq n$, establishing the density of $\mathrm{Im}\, \eta$ in $X_{G^\times}$, as needed. ∎

Remark 3.8 Theorem 3.6.(1) yields, together with the axioms for RSs in 1.15, a **Horn-geometric** axiomatization for the class of Boolean RSs in the language real semigroups, $\langle \cdot, 1, -1, D \rangle$ (D^t is defined from D by a conjunction of atomic formulas, cf. [t-rep] in 1.15.(a)). ∎

Before complementing the results in Theorem 3.6, we set down

3.9 Remarks and Notation. a) Let G be a Boolean RS and let $\eta: X_G \longrightarrow X_{G^\times}$ be the homeomorphism given, as above, by $\tau \in X_G \longmapsto \tau \!\upharpoonright\! G^\times$. For each $a \in G$, set
$$L(a) = [\![\nabla a^2 = -1]\!]_{X_{G^\times}} \subseteq X_{G^\times} \text{ and let } L(G^\times) = \{L(a) : a \in G\}.$$
In particular, for e in $\mathrm{Id}(G)$, $L(e) = [\![\nabla e = -1]\!]_{X_{G^\times}}$. Fix $e \in \mathrm{Id}(G)$; note that for each $\tau \in X_G$,

---

[2] In any RSG, -1 is not represented by $2^n = \bigotimes_{i=1}^n \langle 1, 1 \rangle$.

$$\tau(e) = 0 \text{ iff } \tau(\nabla e) = -1 \text{ iff } \tau{\restriction}G^\times(\nabla e) = -1,$$

whence $\eta[Z(e)] = [\![\nabla e = -1]\!]_{X_{G^\times}}$. Since η is a homeomorphism, image by η preserves unions and intersections, and so $L(G^\times)$ is a distributive sublattice of the BA of clopens of $X_{G^\times}$ (isomorphic to the BA of clopens of X_G). Hence, the lattices of idempotents in G, that of zero sets in X_G and $L(G^\times)$ are all isomorphic. Note that $L(G^\times)$ is a lattice of **clopens in $X_{G^\times}$**.

b) For $a, b \in G$,

$(\sigma) \qquad \sigma \in \neg L(a) \cap \neg L(b) = \neg L(ab) \Rightarrow \sigma(\nabla(ab)) = \sigma(\nabla a)\sigma(\nabla b).$

To see this, let $\mu \in X_G$ be such that $\mu{\restriction}G^\times = \sigma$. Then, $\mu \in [\![\nabla(a^2) = 1]\!]_{X_G} \cap [\![\nabla(b^2) = 1]\!]_{X_G} = [\![\nabla(a^2b^2) = 1]\!]_{X_G}$ and 3.5.(d) yields $\mu(x) = \mu(\nabla x)$, for $x \in \{a, b, ab\}$. Hence, $\mu(\nabla(ab)) = \mu(ab) = \mu(a)\mu(b) = \mu(\nabla a)\mu(\nabla b)$; since $\mu{\restriction}G^\times = \sigma$ and $\nabla x \in G^\times$, the preceding equality obtains (σ), as needed. ∎

Proposition 3.10 *If the equivalent conditions in Theorem 3.6 hold, then for all $e \in \mathrm{Id}(G)$ and $u \in G^\times$, $\langle 1, u \rangle \otimes \langle 1, \nabla e \rangle \equiv_{G^\times} \langle 1, 1 \rangle \otimes \langle 1, \nabla(ue) \rangle$. In particular* (cf. Proposition 2.3):

(1) $D_{G^\times}(1, \nabla e)$ *and* $D_{G^\times}(1, -\nabla e)$ *are DSSs in the RSG $G^\times$;*

(2) $L(G^\times) = \{L(a) \subseteq X_{G^\times} : a \in G\}$ *is a lattice of direct summands of $X_{G^\times}$* (isomorphic to Id_G).

Proof. Fix $e \in \mathrm{Id}(G)$ and $u \in G^\times$. To show $\varphi = \langle 1, u \rangle \otimes \langle 1, \nabla e \rangle \equiv_{G^\times} \langle 1, 1 \rangle \otimes \langle \nabla(ue) \rangle = \psi$, it suffices to check their total signatures to be the same; taking into account the homeomorphism η in Theorem 3.6, it suffices to verify the signatures of φ and ψ to be the same at each $\tau \in X_G$ (by Theorem 1.8). For $\tau \in X_G$:

• If $\tau(\nabla e) = -1$, then $\tau \in [\![e = 0]\!] = [\![ue = 0]\!]$ ($u \in G^\times$) and so $\tau(\nabla(ue)) = -1$, and the signatures of φ and ψ are both 0 at τ;

• If $\tau(\nabla e) = 1$, then $\tau \in [\![e = 1]\!]$, whence $0 \neq \tau(u) = \tau(ue) = \tau(\nabla(ue))$ (recall: by 3.5.(d.2), $\tau \notin Z(ue)$ entails $\tau(ue) = \tau(\nabla(ue))$); thus,

$$\tau(\varphi) = \tau(1) + \tau(u) + \tau(\nabla e) + \tau(u\nabla e) = 2 + 2\tau(u) = 2 + 2\tau(\nabla(ue))$$
$$= \tau(\psi),$$

as needed. Item (1) is an immediate consequence of Proposition 2.3, while (2) follows from Remark 2.4.(1), recalling that for $e \in \mathrm{Id}(G)$, $L(e) = [\![-\nabla e = 1]\!]_{X_{G^\times}}$ is the subspace associated to the DSS $D_{G^\times}(1, -\nabla e)$, ending the proof. ∎

We shall now show how given a RSG, K, and a non-empty bounded sublattice, L, of direct summands of X_K (equivalently, DSSs of K, cf. 2.4.(1)), one can obtain a real semigroup, $\mathcal{G} := \mathcal{G}(K, L)$, such that $X_\mathcal{G}$ is naturally homeomorphic to X_K, $\mathcal{G}^\times$ is isomorphic to K and $\mathrm{Id}(\mathcal{G})$ is isomorphic to L (with the partial order defined in 1.27).

3.11 Construction To simplify exposition, write X for X_K.

a) Let $P = \mathbb{C}(X, 3)$, the Post algebra of continuous maps from X to 3, where 3 is endowed with the discrete topology. Consider the map

$(\gamma)\ \gamma : K \times L \longrightarrow P$, given by $\langle v, A \rangle \longmapsto \gamma(v, A)(\sigma) = \begin{cases} \sigma(v) & \text{if } \sigma \notin A; \\ 0 & \text{if } \sigma \in A. \end{cases}$

For $A \in L$ and $a \in K$, set

- $e_A : X \longrightarrow 3$ given by $e_A(\sigma) = \begin{cases} 0 & \text{if } \sigma \in A; \\ 1 & \text{if } \sigma \notin A. \end{cases}$
- $\widehat{a} : X \longrightarrow 3$, given by $\widehat{a}(\sigma) = \sigma(a)$.

b) With notation as in (a), note that:

(1) for $A, B \in L$, $e_A e_B = e_{(A \cup B)}$. Moreover, $e_\emptyset = 1$ and $e_X = 0$.

(2) For each $\langle a, A \rangle \in K \times L$, $\gamma(a, A) = \widehat{a} e_A$.

(3) To ease notation, write the elements of $\mathrm{Im}\ \gamma \subseteq P$ as $a e_A$, instead of $\widehat{a} e_A$ ($a \in K$, $A \in L$).

c) It is straightforward to check that $a e_A \cdot b e_B = a b e_A e_B = a b e_{A \cup B}$. Hence,

$$\mathrm{Im}\ \gamma := \mathcal{G}(K, L) = \{ a e_A \in P : \langle a, A \rangle \in K \times L \},$$

is a ternary subsemigroup of P, with $1 = e_\emptyset$, $0 = e_X$ and $-1 = -1 e_\emptyset$. As usual, write $-v$ for $-1 \cdot v$.

Endow $\mathcal{G} := \mathcal{G}(K, L)$ with the representation and transversal representation induced by the Post algebra P; therefore, for $a, b, c \in K$ and $A, B, C \in L$, $D_\mathcal{G}^t$ is given by:

$a e_A \in D_\mathcal{G}^t(b e_B, c e_C)$ iff for all $\sigma \in X$, $a e_A(\sigma) \in D_3^t(b e_B(\sigma), c e_C(\sigma))$.

Note: The values in 3 of $\sigma \in X$ at each element of $\mathcal{G}$ appearing in the preceding expression is described by formula (γ) above. ∎

Since all axioms of RSs, with the exception of [RS 3], are universal, and $\mathcal{G} \subseteq P$ (a real semigroup, cf. 3.2.(b)), $\mathcal{G}$ **is pre-real semigroup.** To show it is, in fact, a real semigroup, we will employ an equivalent to [RS 3], namely, the exchange principle (cf. 1.16).

Let $A, B, C, D \in L$ and $a, b, c, d \in K$ and set

$$\mathfrak{a} = a e_A, \quad \mathfrak{b} = b e_B, \quad \mathfrak{c} = c e_C \quad \text{and} \quad \mathfrak{d} = d e_D.$$

For $W \in L$, write $\neg W$ for $X \setminus W$.

We say that the *exchange principle holds for* $\langle \mathfrak{a}, \mathfrak{b}; \mathfrak{c}, \mathfrak{d} \rangle$ if

(exch) $D_\mathcal{G}^t(\mathfrak{a}, \mathfrak{b}) \cap D_\mathcal{G}^t(\mathfrak{c}, \mathfrak{d}) \neq \emptyset \Rightarrow D_\mathcal{G}^t(\mathfrak{a}, -\mathfrak{c}) \cap D_\mathcal{G}^t(-\mathfrak{b}, \mathfrak{d}) \neq \emptyset.$

We start with the following observations:

Lemma 3.12 *With notation as above,*

a) $ae_A = be_B$ *iff* $A = B$ *and for all* $\sigma \in \neg A = \neg B$, $\sigma(a) = \sigma(b)$. *In other words,* $\mathfrak{a} = \mathfrak{b}$ *iff* $A = B$ *and* $a/\Delta = b/\Delta$, *where* $\Delta = \Sigma(\neg A)$, *the saturated subgroup associated to* $\neg A$ *(cf. 1.12.(e)).*

b) $\mathrm{Id}(\mathcal{G}) = \{e_A : A \in L\}$. *Moreover, the map* $A \in L \longmapsto e_A \in \mathrm{Id}(\mathcal{G})$ *is a lattice isomorphism between* L *(cf. 1.27) and* $\mathrm{Id}(\mathcal{G})$

c) We may identify $\mathcal{G}^\times$ *with* K, *that is,* $ae_A \in \mathcal{G}^\times$ *iff* $A = \emptyset$ *(and so* $ae_A = a$).

d) For $\mathfrak{a} = ae_A \in \mathcal{G}$, *there is* $u \in K = \mathcal{G}^\times$ *so that* $\mathfrak{a} = u\mathfrak{a}^2$ *and* $u \in D_{\mathcal{G}}^t(-1, \mathfrak{a})$. *Thus, the pre-real semigroup* $\mathcal{G}$ *satisfies condition (1) in Theorem 3.6.*

e) $D_{\mathcal{G}}^t(\mathfrak{a}, \mathfrak{b}) \neq \emptyset$.

f) If any one among $\mathfrak{a}, \mathfrak{b}, \mathfrak{c}, \mathfrak{d}$ *is equal to 0, then the exchange principle holds for* $\langle \mathfrak{a}, \mathfrak{b}; \mathfrak{c}, \mathfrak{d} \rangle$.

g) If $0 \in D_{\mathcal{G}}^t(\mathfrak{a}, \mathfrak{b}) \cap D_{\mathcal{G}}^t(\mathfrak{c}, \mathfrak{d})$, *then the exchange principle holds for* $\langle \mathfrak{a}, \mathfrak{b}; \mathfrak{c}, \mathfrak{d} \rangle$.

Proof. a) If $\sigma \in A$, then $\sigma(ae_A) = 0$ iff $\sigma \in A$ (recall: $a, b \in K$). It is then clear that $\mathfrak{a} = \mathfrak{b}$ implies $A = B$ and $\sigma(a) = \sigma(b)$ for all $\sigma \notin A = B$. The converse is clear. The second statement in (a) is just a rephrasing of the proven equivalence. Item (b) is straightforward.

c) Clearly, $ae_A \in \mathcal{G}^\times$ iff its value at each $\sigma \in X$ is non-zero, i.e., $A = \emptyset$.

d) Let $\mathfrak{a} = ae_A \in \mathcal{G}$; note that $\mathfrak{a}^2 = e_A$ ($a \in K$). Let $\Delta = \Sigma(A)$ and $\Delta^\perp = \Sigma(\neg A)$ be the saturated DSSs associated to A and $\neg A$ (cf. 1.12.(e)). Since $x \in K \longmapsto (x/\Delta, x/\Delta^\perp)$ is an isomorphism, there is $u \in K$ so that $u/\Delta = -1/\Delta$ and $u/\Delta^\perp = a/\Delta^\perp$. Hence, for all $\sigma \in X$:

(1) $\sigma \in \neg A \Rightarrow \sigma(u) = \sigma(a)$;   (2) $\sigma \in A \Rightarrow \sigma(u) = -1$.

Note that (1) and item (a) entail $u\mathfrak{a}^2 = ue_A = ae_A = \mathfrak{a}$. Moreover, (1) and (2) obtain $u \in D_{\mathcal{G}}^t(-1, \mathfrak{a})$. Indeed, for $\sigma \in X_G$, we have:

- If $\sigma \in A$, then $\sigma(u) = -1 \in D_3^t(-1, 0)$;
- If $\sigma \in \neg A$, then $\sigma(u) = \sigma(a)$; since $u, a \in K$, then $-\sigma(u)^2 = \sigma(a)^2 = 1$. Whence, by 1.19.(14), $1 \in D_3^t(-\sigma(u), \sigma(a)) = D_3^t(-\sigma(a), \sigma(a)) = 3$ and 1.19.(1)) entails $\sigma(u) \in D_3^t(-1, \sigma(a))$.

e) Let $E_1 = A \cap B$, $E_2 = A \cap \neg B$, $E_3 = \neg A \cap B$ and $E_4 = \neg A \cap \neg B$ be the four atoms of the BA generated by A and B in $B(X)$. Let $\Delta_k = \Sigma(E_k)$ be the associated saturated subgroup of K, $1 \leq k \leq 4$ (as in 1.12.(e)). For $u, v \in K$ and $1 \leq k \leq 4$, the expression "$u = v$ in E_k" stands for $u/\Delta_k = v/\Delta_k$. We now consider the conditions required to construct a witness for the claim in (e) in each of the atoms E_k:

(1) $E_1 = A \cap B$. In this case, we take $y_1 = 1$ in E_1;

(2) $E_2 = A \cap \neg B$. If $\sigma \in E_2$, then a witness t for our claim must satisfy $\sigma(t) \in D_3^t(0, \sigma(\mathfrak{b}))$, i.e., $\sigma(t) = \sigma(\mathfrak{b})$. In this case, we take $y_2 = \mathfrak{b}$ in E_2;

(3) $E_3 = \neg A \cap B$. With the same reasoning as in (3), we take $y_3 = \mathfrak{a}$ in E_3;

(4) $E_4 = \neg A \cap \neg B$. If $\sigma \in E_4$, then a witness t for (e) must verify $\sigma(t) \in D_3^t(\sigma(\mathfrak{a}), \sigma(\mathfrak{b}))$; in this case, we take $y_4 = \mathfrak{a}$ in E_4.

Since K is isomorphic to $\prod_{k=1}^4 K/\Delta_k$, there is $v \in K$ so that $v/\Delta_k = y_k/\Delta_k$, $1 \le k \le 4$. Then,

(*) $$t = v e_{A \cap B} \in D_{\mathcal{G}}^t(\mathfrak{a}, \mathfrak{b}).$$

To prove (*), it suffices to show that for each $\sigma \in X$, we have $\sigma(t) \in D_3^t(\sigma(\mathfrak{a}), \sigma(\mathfrak{b}))$. Since $X = \bigcup_{k=1}^4 E_k$, we prove (*) holds for σ in each of the atoms E_k, taking into account the selections made in (1) – (4) above.

(1*) If $\sigma \in E_1$, then $\sigma(t) = 0 \in D_3^t(\sigma(\mathfrak{a}), \sigma(\mathfrak{b})) = D_3^t(0, 0)$;

(2*) If $\sigma \in E_2$, then $\sigma(t) = \sigma(v) = \sigma(\mathfrak{b}) \in D_3^t(\sigma(\mathfrak{a}), \sigma(\mathfrak{b})) = D_3^t(0, \sigma(\mathfrak{b}))$, as needed; a similar reasoning applies if $\sigma \in E_3$;

(4*) If $\sigma \in E_4$, then $\sigma(t) = \sigma(v) = \sigma(\mathfrak{a}) \in D_3^t(\sigma(\mathfrak{a}), \sigma(\mathfrak{b})) = D_3^t(\sigma(\mathfrak{a}), \sigma(\mathfrak{b}))$, which holds because $\mathfrak{a}, \mathfrak{b} \in K$ and so $\sigma(\mathfrak{a}), \sigma(\mathfrak{b}) \in \{1, -1\}$,

establishing (*) and completing the proof of (e).

f) Without loss of generality, we may assume $\mathfrak{a} = 0$; if $t \in \mathcal{G}$ satisfies $t \in D_{\mathcal{G}}^t(0, \mathfrak{b}) \cap D_{\mathcal{G}}^t(\mathfrak{c}, \mathfrak{d})$, then $t = \mathfrak{b}$ [3]. But $\mathfrak{b} \in D_{\mathcal{G}}^t(\mathfrak{c}, \mathfrak{d})$ entails $-\mathfrak{c} \in D_{\mathcal{G}}^t(-\mathfrak{b}, \mathfrak{d})$, whence $-\mathfrak{c} \in D_{\mathcal{G}}^t(0, -\mathfrak{c}) \cap D_{\mathcal{G}}^t(-\mathfrak{b}, \mathfrak{d})$, as needed.

g) If $0 \in D_{\mathcal{G}}^t(\mathfrak{a}, \mathfrak{b}) \cap D_{\mathcal{G}}^t(\mathfrak{c}, \mathfrak{d})$, then (cf. 1.19.(12)), $\mathfrak{b} = -\mathfrak{a}$ and $\mathfrak{d} = -\mathfrak{c}$, and the exchange principle leads to $D_{\mathcal{G}}^t(\mathfrak{a}, -\mathfrak{c}) \ne \emptyset$, that is guaranteed by (e). ∎

We now have

Theorem 3.13 $\mathcal{G} = \mathcal{G}(K, L)$ *is a Boolean real semigroup, with* $\mathcal{G}^\times = K$, $\mathrm{Id}(\mathcal{G})$ *lattice isomorphic to* L *and* $X_{\mathcal{G}}$ *naturally isomorphic to* X_K, *via the restriction map* $\tau \mapsto \tau \upharpoonright K$.

Proof. In view of Lemma 3.12 and the equivalences in Theorem 3.6, it remains only to establish that $\mathcal{G}$ is a RS. Moreover, again due to 3.12, in this proof we may assume, for $\mathfrak{a}, \mathfrak{b}, \mathfrak{c}$ and $\mathfrak{d} \in \mathcal{G}$:

(!) There is $t \in D_{\mathcal{G}}^t(\mathfrak{a}, \mathfrak{b}) \cap D_{\mathcal{G}}^t(\mathfrak{c}, \mathfrak{d})$, and $\mathfrak{a}, \mathfrak{b}, \mathfrak{c}, \mathfrak{d}, t$ are all distinct from 0.

To ease presentation, we introduce the following

---

[3] Recall (1), (14a) and (14b) in 1.19.

3.14 Notation and Remarks. a) If $t \in \mathcal{G} \setminus \{0\}$, to ease the discussion that follows, write t^* for a unit in K (cf. 3.12.(a)) that determines t, i.e., $t = t^* e_W$, for some $W \in L$. Let $\mathfrak{a} = a e_A$, $\mathfrak{b} = b e_B$, $\mathfrak{c} = c e_C$ and $\mathfrak{d} = d e_D$. Note that

(&) $$(A \cap B) \cup (C \cap D) \subseteq W,$$

since if σ belongs to this union, then $\sigma(t) = 0$.

b) If G is a RS, $x \in G^\times$ and $y \in G$, it follows easily from [RS 1] and [RS 6] in 1.15.(a) that $x \in D_G^t(x, y)$.

c) We shall construct two tables, the first corresponding to case in which $\sigma \in \neg W$ and the second for $\sigma \in W$ (and so $\sigma(t) = 0$).

For $Z \in L$ and $x, y, z \in K$, the tables below uses the following conventions, with $\Delta = \Sigma(Z)$ (cf. 1.12.(e)):

- "$x = y$ in Z" stands for $x/\Delta = y/\Delta$ or equivalently, for all $\sigma \in Z$, $\sigma(x) = \sigma(y)$.
- "$x \in D(y, z)$ in Z" stands for $x/\Delta \in D_{K/\Delta}(y/\Delta, z/\Delta)$;
- A "1" in a column means that we are <u>outside</u> that set, while a "0" means we are in that set. For instance, the sequence "1 0 1 0" in the columns marked A, B, C, D corresponds to $\neg A \cap B \cap \neg C \cap D$, which in first table is the line E_6 and line E_{22} in Table 2.
- The column marked "$\bigcap$" has either a $\sqrt{}$, meaning that the atom E_k is not necessarily empty, or $\emptyset$ (whose meaning is obvious).
- We assume there is a witness, $t \in \mathcal{G}$, for the antecedent in [RS 3']; its possible values and the constraints it imposes on the coefficients in each of the atoms E_k, $1 \leq k \leq 32$, of the BA generated by A, B, C, D, W is registered in the corresponding column in each table.
- The column corresponding to $y \in K$ yields the values and conditions for it to be a witness of the consequent of [RS 3']. For instance, the first line of the first table indicates that <u>in E_1</u>, $D(a, b) \cap D(c, d) \neq \emptyset$, with $a, b, c, d \in K$. But then 1.5.(b) yields $D(a, -c) \cap D(-b, d) \neq \emptyset$ in E_1, and so it is possible select y_1 in this intersection <u>in E_1</u>.
- In the table for W, the expression "impossible" in the last column indicates that the condition on the column "constraints (in E_k)" cannot hold; thus, the antecedent of our implication is false in E_k.
- **Note:** The intersection of the sets in any line containing a unique 1 (all other entries are 0) <u>must be empty</u>: e.g., consider $E_8 = \neg A \cap B \cap C \cap D \cap \neg W$; if $\sigma \in E_8$, then we would have

$$\sigma(t) \in D_3^t(\sigma(\mathfrak{a}), \sigma(\mathfrak{b})) \cap D_3^t(\sigma(\mathfrak{c}), \sigma(\mathfrak{d})) = D_3^t(\sigma(a), 0) \cap D_3^t(0, 0),$$

that is impossible: $D_3^t(0, 0) = \{0\}$, while $D_3^t(\sigma(a), 0) = \{\sigma(a)\} \subseteq \{1, -1\}$. ∎

Table 1. $\sigma \in \neg W$ and for $1 \leq k \leq 16$, we assume $E_k \subseteq \neg W$; recall (&) in 3.14.(a).

	A	B	C	D	∩	t^* and constraints (in E_k)	A	C	B	D	y_k and constraints (in E_k)
E_1	1	1	1	1	✓	$t^* \in D(a,b) \cap D(c,d)$	1	1	1	1	$y_1 \in D(a,-c) \cap D(-b,d)$
E_2	1	1	1	0	✓	$t^* = c \in D(a,b)$	1	1	1	0	$y_2 = -b \in D(a,-c)$
E_3	1	1	0	1	✓	$t^* = d \in D(a,b)$	1	0	1	1	$y_3 = a \in D(-b,d)$
E_4	1	1	0	0	∅	——	1	0	1	0	——
E_5	1	0	1	1	✓	$t^* = a \in D(c,d)$	1	1	0	1	$y_5 = d \in D(a,-c)$
E_6	1	0	1	0	✓	$t^* = a = c$	1	1	0	0	$y_6 = 1; a = -(-c) = c$
E_7	1	0	0	1	✓	$t^* = a = d$	1	0	0	1	$y_7 = a = d$
E_8	1	0	0	0	∅	——	1	0	0	0	——
E_9	0	1	1	1	✓	$t^* = b \in D(c,d)$	0	1	1	1	$y_9 = -c \in D(-b,d)$
E_{10}	0	1	1	0	✓	$t^* = b = c$	0	1	1	0	$y_{10} = -c = -b$
E_{11}	0	1	0	1	✓	$t^* = b = d$	0	0	1	1	$y_{11} = 1; b = -(-b) = d$
E_{12}	0	1	0	0	∅	——	0	0	1	0	——
E_{13}	0	0	1	1	∅	——	0	1	0	1	——
E_{14}	0	0	1	0	∅	——	0	1	0	0	——
E_{15}	0	0	0	1	∅	——	0	0	0	1	——
E_{16}	0	0	0	0	∅	——	0	0	0	0	——

Table 2. $\sigma \in W$ and so $\sigma(t) = 0$; for $16 \leq k \leq 32$, we assume $E_k \subseteq W$.

	A	B	C	D	∩	constraints (in E_k)	A	C	B	D	y_k and constraints (in E_k)
E_{17}	1	1	1	1	✓	$0 \in D^t_{\mathcal{G}}(a,b) \cap D^t_{\mathcal{G}}(c,d)$	1	1	1	1	$b = -a, d = -c, y_{17} = a$
E_{18}	1	1	1	0	∅	$0 = c \in D(a,b)$	1	1	1	0	impossible
E_{19}	1	1	0	1	∅	$0 = d \in D(a,b)$	1	0	1	1	impossible
E_{20}	1	1	0	0	✓	$t = 0; b = -a$	1	0	1	0	$y_{20} = a = -b$
E_{21}	1	0	1	1	∅	$0 = a \in D(c,d)$	1	1	0	1	impossible
E_{22}	1	0	1	0	∅	$0 = a = c$	1	1	0	0	impossible
E_{23}	1	0	0	1	∅	$0 = a = d$	1	0	0	1	impossible
E_{24}	1	0	0	0	∅	——	1	0	0	0	——
E_{25}	0	1	1	1	∅	$0 = b \in D(c,d)$	0	1	1	1	impossible
E_{26}	0	1	1	0	∅	$0 = b = c$	0	1	1	0	impossible
E_{27}	0	1	0	1	∅	$0 = b = d$	0	0	1	1	impossible
E_{28}	0	1	0	0	∅	——	0	0	1	0	——
E_{29}	0	0	1	1	✓	$t = 0; d = -c$	0	1	0	1	$y_{29} = -c = d$
E_{30}	0	0	1	0	∅	——	0	1	0	0	——
E_{31}	0	0	0	1	∅	——	0	0	0	1	——
E_{32}	0	0	0	0	✓	$t = 0$	0	0	0	0	$y_{32} = 1$

Since we are assuming that t is a witness for the antecedent of the implication corresponding to the exchange principle, and $W = \bigcup_{k=17}^{32} E_k$, Table 2 shows that except for E_{17}, E_{20}, E_{29} and E_{32}, all other E_k ($17 \leq k \leq 32$) must be empty. Thus, $W = E_{17} \cup E_{20} \cup E_{29} \cup E_{32}$ (disjoint union).

Notation as above, let $\mathfrak{q} = \{k \in \{1, \ldots, 32\} : E_k \neq \emptyset\}$; set $\Delta_k = \Sigma(E_k)$, for $k \in \mathfrak{q}$. By Fact 2.5, there is a natural isomorphism between K and $\prod_{k \in \mathfrak{q}} K/\Delta_k$. Hence, there is $v \in K$ so that $v/\Delta_k = y_k/\Delta_k$, for each $k \in \mathfrak{q}$.

Let $V = (A \cap C) \cup (B \cap D)$; we shall verify that

(#) $\quad z = v_{e_V} \in D_{\mathcal{G}}^t(ae_A, -ce_C) \cap D_{\mathcal{G}}^t(-be_B, de_D) = D_{\mathcal{G}}^t(\mathfrak{a}, -\mathfrak{c}) \cap D_{\mathcal{G}}^t(-\mathfrak{b}, \mathfrak{d})$,

establishing [RS 3'] and showing that $\mathcal{G}$ is a real semigroup.

We have $V = V \cap (\neg W \cup W) = (V \cap \neg W) \cup (V \cap W)$. We discuss the following cases:

I. In $A \cap C \cap \neg W$. If $\sigma \in A \cap C \cap \neg W$, then $\sigma(z) = 0$; the pertinent line in Table 1 is E_{11} ($E_{16} = E_{12} = E_{15} = \emptyset$).

• If $\sigma \in E_{11}$, then $\sigma(z) = 0 \in D_3^t(0,0) \cap D_3^t(\sigma(-b), \sigma(d)) = D_3^t(\sigma(-\mathfrak{b}), \sigma(\mathfrak{b}))$, as needed.

Similarly, one treats the case of each $\sigma \in B \cap D \cap \neg W$.

II. In $A \cap C \cap W$. The pertinent line here is E_{32} in Table 2. But then we have $\sigma(z) = 0 = \sigma(\mathfrak{a}) = \sigma(\mathfrak{b}) = \sigma(\mathfrak{c}) = \sigma(\mathfrak{d})$ and the desired conclusion is immediate. A similar argument applies of the case $\sigma \in B \cap D \cap W$.

If $\sigma \in X \setminus V$, then

$$\sigma \in (\neg A \cap \neg B) \cup (\neg A \cap \neg D) \cup (\neg C \cap \neg B) \cup (\neg C \cap \neg D),$$

which may be written as $(X \setminus V) \cap (\neg W \cup W)$.

III. In $\neg A \cap \neg B \cap \neg W$. The pertinent lines of Table 1 are $E_1 - E_3$.

III.1. If $\sigma \in E_1$, then

$$\sigma(z) = \sigma(v) = \sigma(y_1) \in D_3^t(\sigma(a), \sigma(-c)) \cap D_3^t(\sigma(-b), \sigma(d))$$
$$= D_3^t(\sigma(\mathfrak{a}), \sigma(-\mathfrak{c})) \cap D_3^t(\sigma(-\mathfrak{b}), \sigma(\mathfrak{d}));$$

III.2. If $\sigma \in E_2$, then

$$\sigma(z) = \sigma(y_2) = \sigma(-b) \in D_3^t(\sigma(a), \sigma(-c)) \cap D_3^t(\sigma(-b), 0)$$
$$= D_3^t(\sigma(\mathfrak{a}), \sigma(-\mathfrak{c})) \cap D_3^t(\sigma(-\mathfrak{b}), 0).$$

Similarly, one treats the cases in which $\sigma \in E_3$ and $\sigma \in \neg C \cap \neg D \cap \neg W$.

IV. In $\neg A \cap \neg D \cap \neg W$. The pertinent lines in Table 1 are E_1, E_3, E_5, E_7.

IV.1. If $\sigma \in E_1$, the argument is just as case III.1. above;

IV.2. If $\sigma \in E_3$, then $\sigma(z) = \sigma(y_3) = \sigma(a) \in D_3^t(\sigma(a), 0) \cap D_3^t(\sigma(-b), \sigma(d))$
$$= D_3^t(\sigma(\mathfrak{a}), 0) \cap D_3^t(\sigma(-\mathfrak{b}), \sigma(\mathfrak{d}));$$

IV.3 If $\sigma \in E_5$, then $\sigma(z) = \sigma(y_5) = \sigma(d) \in D_3^t(\sigma(a), \sigma(-c)) \cap D_3^t(0, \sigma(d))$
$$= D_3^t(\sigma(\mathfrak{a}), \sigma(-\mathfrak{c}) \cap D_3^t(0, \sigma(\mathfrak{d}));$$

IV.4. If $\sigma \in E_7$, $\sigma(z) = \sigma(y_7) = \sigma(a) = \sigma(d) \in D_3^t(\sigma(a), 0) \cap D_3^t(0, \sigma(d))$
$$= D_3^t(\sigma(\mathfrak{a}), 0) \cap D_3^t(0, \sigma(\mathfrak{d})),$$

as needed. The case in which $\sigma \in \neg B \cap \neg C \cap \neg W$ is handled similarly.

V. In $\neg A \cap \neg B \cap W$. The relevant lines in Table 2 are E_{17} and E_{20}.

- If $\sigma \in E_{17}$, then $\sigma(z) = \sigma(v) = \sigma(\mathfrak{a}) \in D_3^t(\sigma(\mathfrak{a}), \sigma(-c))$, because $\sigma(\mathfrak{a})$ is a unit in 3.
- If $\sigma \in E_{20}$, then $\sigma(z) = \sigma(v) = \sigma(y_{20}) = \sigma(\mathfrak{a}) = \sigma(-\mathfrak{b}) \in D_3^t(\sigma(\mathfrak{a}), 0) \cap D_3^t(\sigma(-\mathfrak{b}), 0)$, as needed. The case $\neg C \cap \neg D \cap W$ can be treated similarly.

VI. In $\neg A \cap \neg D \cap W$ and $\neg B \cap \neg C \cap W$. The relevant line is E_{17} in Table 2 and same argument used for $\sigma \in E_{17}$ in (V) above also applies here.

This completes the proof of (#) and that $\mathcal{G}$ is a real semigroup. ∎

We can now state, with notation as in 3.9 and recalling that $L(G^\times)$ is lattice-isomorphic to Id_G:

Theorem 3.15 (Structure Theorem for Boolean RS) *If G is a Boolean RS, there is a natural RS-isomorphism between G and $\mathcal{G}(G^\times, L(G^\times))$, given by the map $a \in G \xmapsto{f} \nabla a e_{L(a)} \in \mathcal{G}(G^\times, L(G^\times))$.*

Proof. To make matters clearer, we maintain a distinction between the homeomorphic spaces X_G and $X_{G^\times}$. Let $\eta : X_G \longrightarrow X_{G^\times}$, $\eta(\tau) = \tau \upharpoonright G^\times$, be the homeomorphism in Theorem 3.6.(3). By 3.9, for each $a \in G$, $L(a) = \eta[Z(a)]$ and so, $\eta^{-1}[\neg L(a)] = [\![a = 1]\!]_{X_G} \cup [\![a = -1]\!]_{X_G} = [\![a^2 = 1]\!]_{X_G}$. Hence, recalling 3.5.(d.2), for all $a \in G$ and $\tau \in X_G$,

(I) $\begin{cases} (1)\ \tau \in Z(a) \text{ iff } \tau \upharpoonright G^\times \in L(a); \\ (2)\ \tau \in [\![a^2 = 1]\!]_{X_G} = [\![\nabla(a^2) = 1]\!]_{X_G} \text{ iff } \tau(a) = \tau(\nabla a) \text{ iff } \tau \upharpoonright G^\times \in \neg L(a). \end{cases}$

We first note that $f(0) = \nabla 0 e_X = 0$, $f(1) = \nabla 1 e_\emptyset = 1$ and $f(-1) = \nabla(-1)e_\emptyset = -1$. Next, we show that f preserves products. For $a, b \in G$, we have

$$\begin{cases} f(ab) = \nabla(ab)e_{L(ab)} = \nabla(ab)e_{L(a) \cup L(b)} & \text{and} \\ f(a)f(b) = \nabla a \nabla b e_{L(a)} e_{L(b)} = \nabla a \nabla b e_{L(a) \cup L(b)}. \end{cases}$$

For $\sigma \in X_{G^\times}$:
- If $\sigma \in L(a) \cup L(b)$, then both $f(ab)$ and $f(a)f(b)$ are zero;
- If $\sigma \notin L(a) \cup L(b)$, let τ_s be the unique element of X_G so that $\sigma = \tau_s \upharpoonright G^\times$. Then, by (I).(2) we get $\tau_s \in [\![a^2 = 1]\!]_{X_G} \cap [\![b^2 = 1]\!]_{X_G} = [\![(ab)^2 = 1]\!]_{X_G}$ and so

(*) $\tau_s(a) = \tau_s(\nabla a)$, $\tau_s(b) = \tau_s(\nabla b)$ and $\tau_s(ab) = \tau_s(\nabla(ab)) = \tau_s(\nabla a)\tau_s(\nabla b)$.

Since $\nabla x \in G^\times$ for $x \in G$ and $\tau_s \upharpoonright G^\times = \sigma$, (*) entails $\sigma(\nabla(ab)) = \sigma(\nabla a)\sigma(\nabla b)$ and so, for all $\sigma \in X$, $\sigma(f(ab)) = \sigma(f(a)f(b))$, yielding $f(ab) = f(a)f(b)$ in $\mathcal{G}$, as needed.

f is injective. If, for $a, b \in G$, we have $\nabla a e_{L(a)} = \nabla b e_{L(b)}$, then 3.12.(a) entails $L(a) = L(b)$ and for all $\sigma \notin L(a) = L(b)$, the equalities $\sigma(\nabla a) = \sigma(\nabla b)$. Note that $L(a) = L(b)$ entails $Z(a) = Z(b)$ in X_G. We have $a = \nabla a \cdot a^2$ and $b = \nabla b \cdot b^2$ (3.5.(a.2)); for $\tau \in X_G$:

- If $\tau \in Z(a) = Z(b)$, then $\tau(a) = \tau(b) = 0$;
- If $\tau \in [\![a^2 = 1]\!]_{X_G} = [\![b^2 = 1]\!]_{X_G}$, then $\tau \restriction G^\times \in \neg L(a) = \neg L(b)$ and so $\tau(\nabla a) = \tau(\nabla b)$, entailing $\tau(a) = \tau(b)$.

Since τ is arbitrary in X_G, we obtain $a = b$, establishing the injectivity of f.

f is surjective. Let $t = ue_{L(a)} \in \mathcal{G}$, for some $u \in G^\times$ and $a \in G$. Let $b \in G$ be given by $b = ua^2 = \nabla b \cdot b^2$; then, $b^2 = a^2$ and so $L(a) = L(b)$. Thus, $t = ue_{L(b)}$; we show that $f(b) = \nabla b e_{L(b)} = ue_{L(b)} = t$. If $\sigma \in L(b)$, then $\sigma(f(b)) = 0 = \sigma(t)$. If $\sigma \in \neg L(b)$, let $\tau_s \in X_G$ satisfy $\tau_s \restriction G^\times = \sigma$; then, $\tau_s \in [\![b^2 = 1]\!]_{X_G}$ and so, by 3.5.(d.2), $\tau_s(b) = \tau_s(\nabla b) = \tau_s(u)$. Since $u, \nabla b \in G^\times$, we obtain $\sigma(u) = \sigma(\nabla b)$. Now, Lemma 3.12.(a) entails $f(b) = \nabla b e_{L(b)} = ue_{L(b)}$, as desired.

To finish the proof, observe that the arguments presented above show that for each $a \in G$ and all $\tau \in X_G$,

(**) $\quad\quad\quad \tau(a) \;=\; \tau \restriction G^\times (\nabla a e_{L(a)}) \;=\; \tau \restriction G^\times (f(a))$.

Since $\tau \longmapsto \tau \restriction G^\times$ and f are bijections, (**) implies that f must be an isomorphism, ending the proof. ∎

We end this section with two themes: the first is a characterization of semi-real rings (commutative and unitary) whose associated RS is Boolean; the second is to establish that the class of Boolean RSs is closed under a number of important constructions.

Proposition 3.16 *The real spectrum of a reduced semireal unitary commutative ring is Boolean iff its real closure is von Neumann regular.*

Proof. The equivalence is forthcoming from the following two well-known facts:

(1) The real spectrum of a reduced, semireal unitary commutative ring is homeomorphic to the Zariski spectrum of its real closure (cf. 13.6.3, p. 534, [Dickmann, Schwartz & Tressl, 2019]);

(2) The Zariski spectrum of a unitary commutative ring is Boolean iff it is von Neumann regular (cf. second paragraph, p. 71, [Dickmann, Schwartz & Tressl, 2019]). ∎

Remark 3.17 The characterization in 3.16 can, perhaps, be sharpened. To give an example, just consider the ring of integers $\mathbb{Z}$: its real spectra is Boolean, but it is very far from being von Neumann regular. It is an interesting – and seemingly hard – question to obtain a characterization in terms of the original ring. ∎

Theorem 3.18 *a) The class of Boolean real semigroups is closed under arbitrary*

(1) *Boolean extensions;* (2) *filtered colimits;* (3) *Products;*
(4) *RS-sums;* (5) *Reduced products; in particular, ultraproducts.*

b) *Let G, H be RSs and let $f : G \longrightarrow H$ be a surjective RS-morphism. If G is a Boolean RS, then the same is true of H. In particular, the class of Boolean RSs is closed under quotients by RS-congruences.*

Proof. a) (1) Let X be a Boolean space, G be a Boolean RS and let $T = \mathbb{C}(X, G)$ be the Boolean power of G by X, that is, the set of all locally constant G-valued maps on X. By Thm. 2.5 in [Dickmann, Miraglia & Petrovich, 2017], T is a real semigroup, whose space of RS-characters is $X \times X_G$ (with the product topology) and so T is also a Boolean RS.

(2) Since Boolean RSs are Horn-geometric axiomatizable, it follows from Prop. 3.2 in [Dickmann, Miraglia & Petrovich, 2017] that the class of Boolean RSs is closed under arbitrary directed colimits (or inductive limits).

(3) Again, the Horn-geometric axiomatizability of Boolean RSs and a classical result by Kiesler, Galvin and Shelah (Thm. 6.2.5', p. 366, [Chang & Keisler, 1990]), guarantees that the class of Boolean RSs is closed under arbitrary products.

(4) Let $\mathcal{R} = \{G_i : i \in I\}$ be a non-empty family of Boolean RSs. For the definition of the RS-sum of $\mathcal{R}$, $\bigoplus_{i \in I} G_i$, we refer the reader to Def. 4.2 in [Dickmann, Miraglia & Petrovich, 2017]. If I is finite, then, $\bigoplus_{i \in I} G_i$ is the product of the G_i (Prop. 4.3.(a), [Dickmann, Miraglia & Petrovich, 2017]), a case already covered by (3). Henceforth, we assume I is *infinite*. Let Fin(I) be the set of all non-empty finite subsets of I; for each $F \in$ Fin(I), define

$$G_F^\flat = \left(\prod_{i \in F} G_i\right) \times \mathbf{3},$$

with its natural product structure; note that $G_F^\flat$ is a Boolean RS (by (3)). Further, By Lemma 4.1.(b) of [Dickmann, Miraglia & Petrovich, 2017], if $K \subseteq J \in$ Fin(I), there are RS-morphisms (in fact, pure embeddings) $t_{KJ} : G_K^\flat \longrightarrow G_J^\flat$; moreover, it is shown in the proof of item (b) of Prop. 4.3 in [Dickmann, Miraglia & Petrovich, 2017], that $\bigoplus_{i \in I} G_i$ is the inductive limit of the $G_F^\flat$, with $F \in$ Fin(I), partially ordered under inclusion. The desired conclusion follows from item (2).

(5) It is well-known that reduced powers by a filter are inductive limits of products, and that this construction preserves Horn-geometric theories.

b) It is well-known (and straightforward to check) that positive $\forall \exists$ sentences are preserved by surjective L-morphisms, where L is any first-order language with equality. Hence, by the equivalence in item (a) of Theorem 3.6, the desired conclusion is immediately forthcoming. ∎

4 Morphisms of Boolean Real Semigroups

Since in this section we shall be dealing with several real semigroups, to ease presentation introduce the following

4.1 Notation If G_i is an RS, $K_i := G_i^\times$ is its RSG of units, $a \in G_i$, $u \in K_i$, $\varepsilon \in 3$ and $\mu \in \{1, -1\}$, write
$[\![a = \varepsilon]\!]_i = \{\tau \in X_{G_i} : \tau(a) = \varepsilon\}$ and $[\![u = \mu]\!]_i^\times = \{\sigma \in K_i : \sigma(u) = \mu\}$.
If $\eta_i : X_{G_i} \longrightarrow X_{K_i}$, $\tau \longmapsto \tau \restriction K_i$, is the homeomorphism in Theorem 3.6.(3) and $a \in G_i$, then, with the notation above and in 3.9, $L(a) = [\![\nabla a^2 = -1]\!]_i^\times$ = $\eta_i[Z(a^2)]$, i.e., $\eta^{-1}[L(a)] = Z(a)$. The reader should keep in mind that $L(a^2) = L(a)$ (cf. 3.9.(a)) and $\neg L(a^2) = \neg L(a)$. ∎

Let $F : G_1 \longrightarrow G_2$ be an RS-morphism and let $f = F \restriction K_1 : K_1 \longrightarrow K_2$ (a RSG-morphism). The map f yields, by composition, a continuous map, $f_* : X_{K_2} \longrightarrow X_{K_1}$, $f_*(\sigma) = \sigma \circ f$. We start with the following

Lemma 4.2 *Let e_i be idempotents in G_i, $i = 1, 2$. The following are equivalent:*

(1) (a) $f_*(L(e_2)) \subseteq L(e_1)$ and (b) $f_*(\neg L(e_2)) \subseteq \neg L(e_1)$;
(2) $\nabla e_2 = f(\nabla e_1)$.

Proof. (1) ⇒ (2): For $\sigma \in X_{K_2}$, suppose $\sigma(\nabla e_2) = -1$, i.e., $\sigma \in [\![\nabla e_2 = -1]\!]$; then, (1.(a)) entails $f_*(\sigma) = \sigma \circ f \in L(e_1) = [\![\nabla e_1 = -1]\!]$ and so $\sigma(f(\nabla e_1)) = -1$. A similar argument, employing (1.(b)) shows that $\sigma(\nabla e_2) = 1$ entails $\sigma(f(\nabla e_1)) = 1$, and the equality in (2) follows immediately.
(2) ⇒ (1): For $\sigma \in X_{K_2}$, we have two possibilities:
- If $\sigma \in L(e_2)$, then $\sigma(\nabla e_2) = -1 = \sigma(f(\nabla e_1))$ and so $f_*(\sigma) \in L(e_1)$, verifying (1.(a));
- If $\sigma \in \neg L(e_2)$, then $\sigma(\nabla e_2) = 1 = \sigma(f(\nabla e_1))$, and so $f_*(\sigma) \in [\![\nabla e_1 = 1]\!] = \neg L(e_1)$, proving (1.(b)), as needed. ∎

Remarks 4.3 a) An RS-morphism, $F : G_1 \longrightarrow G_2$, gives rise to two maps:
- A RSG morphism $f_F := F \restriction K_1 : K_1 \longrightarrow K_2$;

- A lattice morphism, $h_F : \text{Id}(G_1) \longrightarrow \text{Id}(G_2)$, given by $h_F(e) = F(e)$. To see h_F is indeed a lattice morphism note that for idempotents x, y in G_1, we have, recalling 1.27,
- $h_F(x \vee y) = h_F(xy) = F(xy) = F(x)F(y) = h_F(x) \vee h_F(y)$;
- $x \wedge y \in D^t_{\mathcal{G}_1}(x, y)$, and so $h_F(x \wedge y) = F(x \wedge y) \in D^t_{\mathcal{G}_2}(F(x), F(y))$, yielding $h_F(x \wedge y) = h_F(x) \wedge h_F(y)$.

b) F may be obtained back from the pair $\langle f_F, h_F \rangle$: for $a \in G_1$, 3.5.(a.2) yields $a = \nabla a \cdot a^2$ and so $F(a) = f_F(\nabla a)F(a^2) = f_F(\nabla a)h_F(a^2)$. ∎

Lemma 4.4 *Let $F : G_1 \longrightarrow G_2$ be an RS-morphism. To simplify exposition, write f for f_F.*
a) For each $e \in \text{Id}(G_1)$, we have $\nabla F(e) = f(\nabla e)$.
b) The pair $\langle f, h_F \rangle$ satisfies the conditions (1.(a)) and (1.(b)) in 4.2, i.e., for all $e \in \text{Id}(G_1)$,

(*)   (i) $f_*[L(h_F(e))] \subseteq L(e)$ and   (ii) $f_*[\neg L(h_F(e)] \subseteq \neg L(e)$.

Proof. a) By 3.5.(a.3), we have $e \in D^t_{\mathcal{G}_1}(1, \nabla e)$, whence $F(e) \in D^t_{\mathcal{G}_2}(1, f(\nabla e))$, which yields, by 3.5.(c), [4]

(I) $\nabla F(e) \in D_{K_2}(1, f(\nabla e))$.

Let $\tau \in X_{G_2}$; then:
- If $\tau(f(\nabla e)) = 1$, relation (I) implies $\tau(\nabla F(e)) = 1$;
- If $\tau \in [\![\nabla F(e) = 1]\!]$, then, by 3.5.(a.1), $\tau \in [\![F(e) = 1]\!]$, i.e., $\tau \circ F \in [\![e = 1]\!] = [\![\nabla e = 1]\!]$, whence $\tau(F(\nabla e)) = \tau(f(\nabla e)) = 1$.

Thus, for all $\tau \in X_{G_2}$, $\tau(\nabla F(e)) = 1$ iff $\tau(f(\nabla e)) = 1$; since both $\nabla F(e)$ and $f(\nabla e) \in K_2$, we conclude $\nabla F(e) = f(\nabla e)$, as desired.
Item (b) is immediate from (a) and 4.2: just take $e_2 = F(e) = h_F(e)$. ∎

Remark 4.5 Let G_1, G_2 be RSs, let $f : G_1^\times \longrightarrow G_2^\times$ be an RSG-morphism and $h: \text{Id}(G_1) \longrightarrow \text{Id}(G_2)$ be a lattice morphism, such that a pair $\langle f, h \rangle$ satisfies the conditions in (*) of 4.4.(b). Let $x, y \in \text{Id}(G_1)$ and suppose $\sigma \in L(h(x)) \cap L(h(y)) \subseteq X_{K_2}$. Then, (i) in (*) of 4.4.(b), yields

$\sigma \circ f \in L(x)$ and $\sigma \circ f \in L(y)$ and so $\sigma \circ f \in L(x) \cap L(y)$.

This applies to any Boolean combination of $L(h(x))$, $L(h(y))$, employing (i) and (ii) in (*) of 4.4.(b), and used below without comment. ∎

Definition 4.6 *Let G_i, $i = 1, 2$ be RSs. A pair $\langle f, h \rangle$, where $f : K_1 \longrightarrow K_2$ is an RSG-morphism and $h : \text{Id}(G_1) \longrightarrow \text{Id}(G_2)$ is a (bounded) lattice*

---
[4] Recall: if u is a unit in a Boolean RS, $\nabla u = u$ and 1, $f(\nabla e) \in K_2$.

morphism, is **compatible** *if they satisfy the conditions in* $(*)$ *of* 4.4.(b). *Write* $\mathcal{C}(G_1, G_2)$ *for the set of all compatible pairs between* G_1 *and* G_2.

By 4.4, every RS-morphism, $F : G_1 \longrightarrow G_2$ yields a compatible pair, while 4.3.(b) shows that F may be obtained back from its compatible pair. To establish a bijective correspondence between $Hom_{RS}(G_1, G_2)$ and $\mathcal{C}(G_1, G_2)$ it suffices to establish:

Theorem 4.7 *If* $\langle f, h \rangle \in \mathcal{C}(G_1, G_2)$, *the map* $F(f, h) : G_1 \longrightarrow G_2$, *given by* $a \in G_1 \longmapsto f(\nabla a)h(a^2)$ *is an RS-morphism. Moreover:*
a) $F(f,h) \upharpoonright K_1 = f$ *and* $F(f,h) \upharpoonright Id(G_1) = h$;
b) $G = G_1 = G_2$, *then* $F(f,h) = Id_G$ *iff* $f = Id_{K_{G^\times}}$ *and* $h = Id_{Id(G)}$.

Proof. We first verify (a) and (b). Once again, to simplify presentation, write F for $F(f, h)$. For $u \in K_1$, $F(u) = f(\nabla u)u^2 = f(u)$ and so $F \upharpoonright K_1 = f$; next, if $a \in G_1$, then $F(a^2)^2 = F(a^2) = (f(\nabla a^2))^2 h(a^2)^2 = h(a^2)$ (recall: $f(\nabla a^2) \in K_2$), completing the verification of (a). Item (b) is clear.

We now turn to the proof that F is an RS-morphism. Note that $F(1) = 1$, $F(-1) = -1$ and $F(0) = 0$. We must show that F preserves products and representation.

I. F preserves products. For $a = \nabla a \cdot a^2$ and $b = \nabla b \cdot b^2 \in G_1$, we have [5]:
(i) $F(ab) = f(\nabla(ab))h(a^2 b^2) = f(\nabla(ab))h(a^2 \vee b^2) = f(\nabla(ab))[h(a^2) \vee h(b^2)]$
$\qquad = f(\nabla(ab))h(a^2)h(b^2);$
(ii) $F(a)F(b) = f(\nabla a \nabla b)h(a^2)h(b^2)$
  For $\tau \in X_{G_2}$:
(1) If $\tau \in [\![h(a^2) = 0]\!]_2$ or $\tau \in [\![h(b^2) = 0]\!]_2$ then the both terms in (i) and (ii) are zero, as needed;
(2) If $\tau \in [\![h(a^2) = 1]\!]_2 \cap [\![h(b^2) = 1]\!]_2 = [\![h(a^2 b^2) = 1]\!]_2$, then $\tau \upharpoonright K_2 \in \neg L(h(a^2)) \cap \neg L(h(b^2))$; now compatibility entails $\tau \upharpoonright K_2 \circ f \in \neg L(a) \cap \neg L(b) = \neg L(ab)$. The equality in (σ) of 3.9.(b) obtains

$$\tau(F(ab)) = \tau(f(\nabla(ab))) = \tau(f(\nabla a)f(\nabla b)) = \tau(F(a)F(b)),$$

completing the proof of that F preserves products.

II. F preserves representation. For $a, b, c \in G_1$, suppose that $a \in D_{G_1}^t(b, c)$. Fix $\tau \in X_{G_2}$; set $\sigma := \tau \upharpoonright G_2^\times$ and let $\mu \in X_{G_1}$ be such that $\mu \upharpoonright G_1^\times = f \circ \sigma$.
  We discuss two cases:

---
[5] Recall that for idempotents x, y, we have $x \vee y = xy$.

II.1. $\tau \in [\![h(a^2) = 0]\!]_2 \cup [\![h(b^2) = 0]\!]_2 \cup [\![h(c^2) = 0]\!]_2$. Since $a \in D_{G_1}^t(b,c)$ iff $-b \in D_{G_1}^t(-a, c)$ iff $-c \in D_{G_1}^t(b, -a)$, it suffices to discuss the case $\tau \in [\![h(a^2) = 0]\!]_2$, for the others can be similarly treated.

So assume $\tau(h(a^2)) = 0$ (and so $\sigma \in L(h(a^2))$, hence, by compatibility, $\sigma \circ f \in L(a)$. We claim that in this case
$$\tag{\#} \tau(h(b^2)) = \tau(h(c^2)).$$
Indeed, suppose $\tau(h(b^2)) = 0$ and $\tau(h(c^2)) = 1$ (i.e., $\sigma \in L(h(b^2)) \cap \neg(L(h(c^2)))$). Then, compatibility yields
$$\sigma \circ f \in L(a) \cap L(b) \cap \neg L(c).$$
Since $\mu \restriction G_1^\times = \sigma \circ f$, we get $\mu(a) \in Z(a)$, $\mu(b) \in Z(b)$, while $\mu \in [\![\nabla c^2 = 1]\!]_1 = [\![c^2 = 1]\!]_1$, whence $\mu(c) \neq 0$. But then, $a \in D_{G_1}^t(b, c)$ implies $\mu(a) = 0 \in D_3^t(0, \mu(c))$, which is impossible since $\mu(c) \neq 0$. A similar argument shows that $\tau(h(b^2)) = 1$ and $\tau(h(c^2)) = 0$ to be untenable, establishing (#).
If $\tau(h(a^2)) = \tau(h(c^2)) = 0$, then $\tau(F(a)) = \tau(F(b)) = \tau(F(c)) = 0$, and we are done.

Henceforth, assume $\tau \in [\![h(b^2) = 1]\!]_2 \cap [\![h(c^2) = 1]\!]_2$, whence, $\sigma \in L(h(a^2))$ and $\sigma \in \neg L(h(b^2)) \cap \neg L(h(c^2))$. Compatibility yields $\sigma \circ f \in L(a)$ and $\sigma \circ f \in \neg L(b) \cap \neg L(c)$, hence $\mu \in Z(a)$ and $\mu \in [\![\nabla b^2 = 1]\!]_1 = [\![b^2 = 1]\!]_1$ and $\mu \in [\![\nabla c^2 = 1]\!]_1 = [\![c^2 = 1]\!]_1$. In particular, $\mu(b), \mu(c) \neq 0$. From $a \in D_{G_1}^t(b, c)$, we obtain $\mu(a) = 0 \in D_3^t(\mu(b), \mu(c))$, and so
$$\tag{*} \mu(c) = -\mu(b) \neq 0.$$
If $\mu \in [\![b = 1]\!]_1 \cap [\![c = -1]\!]_1$, 3.5.(a.1) yields $\mu(\nabla b) = 1$ and $\mu(\nabla c) = -1$ (recall: $\mu(c) \neq 0$); a similar argument applies in case $\mu(b) = -1$ and $\mu(c) = 1$, and so (*) entails $\mu(\nabla b \cdot \nabla c) = -1$, that is, $\mu \restriction G_1^\times(\nabla c) = \sigma \circ f(\nabla c) = -\mu \restriction G_1^\times(\nabla b) = -(\sigma \circ f(\nabla b))$. Unraveling notation yields $\tau(f(\nabla c)) = -\tau(f(\nabla b))$, and so (recall: $\tau(h(b^2)) = \tau(h(c^2)) = 1$), $\tau(F(a)) = 0 \in D_3^t(\tau(f(\nabla b)), -\tau(f(\nabla b)))$, ending the discussion of case II.1.

II.2. $\tau \in [\![h(a^2) = 1]\!]_2 \cap [\![h(b^2) = 1]\!]_2 = [\![h(a^2 b^2) = 1]\!]_2$ and $\tau \in [\![h(c^2) = 1]\!]_2$. For $x \in \{a, b, c\}$, $\tau(F(x)) = \tau(f(\nabla x))$. By 3.5.(c), we have $\nabla a \in D_{G_1}^t(\nabla b, \nabla c) \cap K_1 = D_{K_1}(\nabla b, \nabla c)$ and so $f(\nabla a) \in D_{G_2}^t(f(\nabla b), f(\nabla c))$, because f is a RSG-morphism. Hence, $\tau(F(a)) \in D_3^t(\tau(F(b)), \tau(F(c)))$.
Since τ is arbitrary in X_{G_2}, the proof is complete. ∎

We now have

Proposition 4.8 *Let G_i be Boolean RSs and let $K_i = G_i^\times$, $i = 1, 2, 3$.*

a) If $\langle f, h \rangle \in \mathcal{C}(G_1, G_2)$ and $\langle g, k \rangle \in \mathcal{C}(G_2, G_3)$, then
(1) $\langle g \circ f, k \circ h \rangle \in \mathcal{C}(G_1, G_3)$.

(2) $F(g \circ f, k \circ h) = F(g,k) \circ F(f,h)$.

b) For an RS-morphism $F : G_1 \longrightarrow G_2$, the following are equivalent,
(1) F is an RS-isomorphism;
(2) f_F is a RSG-isomorphism and $h_F : \mathrm{Id}(G_1) \longrightarrow \mathrm{Id}(G_2)$ is a lattice isomorphism (and $\langle f_F, h_F \rangle \in \mathcal{C}(G_1, G_2)$).

Proof. a) (1) Clearly, it suffices to show that $\langle g \circ f, h \circ k \rangle$ satisfy conditions (i) and (ii) in (*) of Lemma 4.4.(b). Note that $(g \circ f)_* = f_* \circ g_*$. If $e \in \mathrm{Id}(G_1)$, then $h(e) \in \mathrm{Id}(G_2)$ and so
$$f_*[g_*[L(k(h(e)))]] \subseteq f_*[L(h(e)] \subseteq L(e),$$
proving (i) in (*) of 4.4.(b). The same argument will yield (ii) in (*) of 4.4.(b), establishing (a.1). Item (a.2) is clear.

b) (1) $\Rightarrow$ (2). Clearly, if F is an RS-isomorphism, then $f_F : K_1 \longrightarrow K_2$ is an RSG-isomorphism and $h_F : \mathrm{Id}(G_1) \longrightarrow \mathrm{Id}(G_2)$ is a lattice isomorphism.

(2) $\Rightarrow$ (1). To simplify exposition, write f for f_F and h for h_F. Let $g : K_2 \longrightarrow K_1$ and $k : \mathrm{Id}(G_2) \longrightarrow \mathrm{Id}(G_1)$ be the inverse isomorphisms of f and h, respectively. Since f is a RS-isomorphism, its dual $f_* : X_{K_2} \longrightarrow X_{K_1}$ is a homeomorphism, whose inverse is g_*. Note that $\{L(e), \neg L(e)\}$ and $\{L(h(e), \neg L(h(e))\}$ are clopen partitions of X_{K_1} and X_{K_2} respectively; but then, because $\langle f, h \rangle$ satisfies conditions (i) and (ii) in (*) of Lemma 4.4.(b), we conclude that for all $e \in \mathrm{Id}(G_1)$,

(I) $\qquad f_*[L(h(e)] \;=\; L(e)$ and $f_*[\neg L(h(e)] \;=\; \neg L(e)$.

We now show that $\langle g, k \rangle \in \mathcal{C}(G_2, G_1)$. Indeed, if $e' \in \mathrm{Id}(G_2)$, then $k(e') \in \mathrm{Id}(G_1)$ and so the first equality in (I) obtains $f_*[L(h(k(e')))] = L(k(e'))$, whence (recall: g_* is the inverse of f_*) $g_*[L(k(e'))] = L(h(k(e'))) = L(e')$ as needed. A similar argument shows that $\langle g, k \rangle$ satisfies (ii) in (*) of 4.4.(b) and so $\langle f, k \rangle \in \mathcal{C}(G_2, G_1)$. Since both f, g and h, k are inverses to one another, items (a) and yield (recall: $F = F(f, h)$), $F(g, k) \circ F = Id_{G_1}$, $F \circ F(g, h) = Id_{G_2}$ and F is an RS-isomorphism, ending the proof. ∎

5 Quotients of Boolean Real Semigroups

Here we characterize RS-quotients of Boolean RSs, showing that if G is a Boolean RS, **any** RS-congruence on G is determined by a subsemigroup of G [6], generated by a saturated subgroup of $G^\times$ (Theorem 5.5) [7].

---

[6] Called subspaces in [Marshall, 1996].
[7] Conversely, by Theorem II.3.8, [Dickmann & Petrovich, 2016], any saturated subgroup of an RS H, rise to an RS-congruence on H.

5.1 Notation. a) In this section we fix:

- A **Boolean** real semigroup, G, and let $K := G^\times$ be the RSG of units in G (cf. 3.6.(3));
- An RS-congruence on G, $\equiv$, writing $G_\equiv$ for the quotient RS and $\pi : G \longrightarrow G_\equiv$ for the natural quotient RS-morphism. Let $K_\equiv := G_\equiv^\times$ be the RSG of units of the Boolean RS $G_\equiv$ (by 3.18.(b)) and let $X_\equiv$ be the space of RS-characters of $G_\equiv$.

b) Let $X_G \xrightarrow{\eta} X_K$ be the homeomorphism $\tau \in X_G \longmapsto \tau \restriction K \in X_K$ (3.6.(3)).

c) Let $\Sigma = \ker(\pi \restriction K)$, a (proper) saturated subgroup of K (recall: $\pi \restriction K$ is a RSG-morphism).

d) Let $\mathcal{H} = \{\sigma \circ \pi \in X_G : \sigma \in X_\equiv\}$. Thus, if $\pi_* : X_\equiv \longrightarrow X_G$ is the dual of π, induced by composition, we have $\mathcal{H} = \pi_*[X_\equiv]$ [8].

e) If $\tau \in X_G$, set $P(\tau) = \{x \in G : \tau(x) \in \{0, 1\}\}$. ∎

Remarks 5.2 a) By definition (cf. II.2.1 and I.1.25, [Dickmann & Petrovich, 2016]), an RS-congruence on G is also a *proper* congruence of the ternary semigroup underlying G, i.e., $\equiv$ is a proper subset of $G \times G$ and for all $x \in G$, $x \equiv -x$ implies $x = 0$. In particular,

(i) $\pi(0)$, $\pi(1)$ and $\pi(-1)$ are pairwise distinct;

(ii) For $a, b, c, d \in G$, $a \equiv b$ and $c \equiv d \Rightarrow ac \equiv bd$.

Item (ii) above will be frequently used below without explicit mention.

b) By items (ii) and (v) of Proposition II.2.8, [Dickmann & Petrovich, 2016], we have:

(1) $\mathcal{H}$ is a proconstructible subset of X_G, whence closed in X_G (X_G is Boolean).

(2) For all $a, b \in G$,

(#) $\qquad\qquad a \equiv b$ iff For all $p \in \mathcal{H}$, $p(a) = p(b)$.

c) If $e, f \in \mathrm{Id}(G)$, it follows straightforwardly from Lemma 3.5.(a.1) (or its item (d)) that $\nabla e = \nabla f$ iff $e = f$.

d) Recall (cf. Notation I.1.4 and Definition I.4.1, [Dickmann & Petrovich, 2016] or Definition 3.1, p. 112, [Dickmann & Petrovich, 2004]) that a subset A of G is:

- A **subsemigroup** of G if it is closed under products and contains 1;
- **Saturated** if for all $a, b \in G$, $a, b \in A$ implies $D_G(a, b) \subseteq A$. Similarly, one defines when A is **transversally saturated (or t-saturated)**, replacing

---

[8] In [Dickmann & Petrovich, 2016], (cf. II.2.7.(i)), π_* is written π^*, while $\mathcal{H}$ is sometimes written $\mathcal{H}_\equiv$.

D_G by D_G^t.

e) If $n \geq 1$ is an integer, write $n\langle 1\rangle$ for the n-form whose entries are all equal to 1. It follows easily from (D) in 1.17 and 1.20.(c) that $D_3(n\langle 1\rangle) = \{0, 1\}$. ∎

We also mention item (2.b) of Proposition I.4.6, [Dickmann & Petrovich, 2016], namely:

Fact 5.3 If A is a subset of G, closed under products and containing 1, then the saturated subsemigroup generated by A, $[A]$, is given by

$$[A] = \bigcup \{D_G(\varphi) : \varphi \text{ is a } n\text{-form with coefficients in } A, n \geq 1\}. \qquad \blacksquare$$

Proposition 5.4 With notation as above, and for $a, b \in G$

a) $a \equiv b \Leftrightarrow (i) \nabla a \equiv \nabla b$ and $(ii) a^2 \equiv b^2$.

b) If $u \in K$ and $u \equiv a$, then $u \equiv \nabla a$ and $a^2 \equiv 1$.

c) If $w \in K_\equiv$, then there is $v \in K$ so that $\pi(v) = w$. In particular, $\pi \upharpoonright K : K \longrightarrow K_\equiv$ is surjective. Moreover, for $u, v \in K$, $u \equiv v$ iff $uv \in \Sigma$ (cf. 5.1.(c)).

d) If $e \in \mathrm{Id}(G_\equiv)$, there is $x \in \mathrm{Id}(G)$ so that $\pi(x) = e$.

Proof. a) Since for all $x \in G$, $x = \nabla x \cdot x^2$ (3.5.(a.2)), only the implication $\Rightarrow$ needs proof. Assume $a \equiv b$; then, termwise multiplication of this relation by itself yields $a^2 \equiv b^2$, i.e., (ii). To establish (i), fix $p \in \mathcal{H}$, then $a \equiv b$, (#) in 5.2.(b.2) and (ii) entail

(I) $\qquad p(a) = p(\nabla a)p(a^2) = p(b) = p(\nabla b)p(b^2) = p(\nabla b)p(a^2)$.

If $p(a^2) = 0 = p(b^2)$, $p \in Z(a) \cap Z(b)$ and so 3.5.(a.1) yields $p(\nabla a) = p(\nabla b) = -1$. If, $p(a^2) = p(b^2) = 1$, then (I) entails $p(a) = p(\nabla a) = p(b) = p(\nabla b)$. Since p is arbitrary in $\mathcal{H}$, (#) in 5.2.(b.2) obtains $\nabla a \equiv \nabla b$, as needed.

Item (b) is immediate from (a), recalling that for all $u \in K$, $\nabla u = u$.

c) Let $w \in K_\equiv$; then, there is $a \in G$ so that $\pi(a) = w$. Thus, $\pi(a)^2 = \pi(a^2) = w^2 = 1$, and so $a^2 \equiv 1$. Thus, $a = \nabla a \cdot a^2 \equiv \nabla a \in K$, entailing $\pi(\nabla a) = w$. For the second statement in (c), since $u, v \in K$, we have $\pi(u) = \pi(v) \in K_\equiv$. Thus,

$$u \equiv v \quad \text{iff} \quad \pi(u) = \pi(v) \quad \text{iff} \quad \pi(uv) = 1 \quad \text{iff} \quad uv \in \ker(\pi \upharpoonright K) = \Sigma.$$

d) If $e \in \mathrm{Id}(G_\equiv)$, there is $x \in G$ so that $\pi(x) = e$. But then, $\pi(x)^2 = \pi(x^2) = e^2$, as needed. ∎

The main result in this section is the following

Theorem 5.5 With notation as in items (c) and (d) of 5.1, let $\Delta(\Sigma)$ be the saturated subsemigroup generated by Σ in G. Now define,

$$\begin{cases} \mathcal{H}_\Sigma & = \{\tau \in X_G : \Sigma \subseteq \ker(\tau \upharpoonright K)\}; \\ \mathcal{H}_{\Delta(\Sigma)} & = \{\tau \in X_G : \Delta(\Sigma) \subseteq P(\tau)\}. \end{cases}$$

Then,

a) For all $a, b \in G$, $a \equiv b \Rightarrow$ For each $\tau \in \mathcal{H}_\Sigma$, $\tau(a) = \tau(b)$.

b) $\mathcal{H} = \mathcal{H}_\Sigma = \mathcal{H}_{\Delta(\Sigma)}$. In particular, all of the these three sets of RS-characters induce the congruence $\equiv$ on G.

c) The map $\widehat{\pi} : K/\Sigma \longrightarrow K_\equiv$, given by $\widehat{\pi}(u/\Sigma) = \pi(u)$ is a RSG-isomorphism.

d) (1) For all $e, e' \in (\mathrm{Id}(G)$,

(∗) $\qquad\qquad e \equiv e'$ iff $\pi(e) = \pi(e')$ iff $\nabla e \cdot \nabla e' \in \Sigma$.

(2) The restriction of $\equiv$ to $\mathrm{Id}(G)$ is a lattice congruence and $\pi \upharpoonright \mathrm{Id}(G) : \mathrm{Id}(G) \longrightarrow \mathrm{Id}(G_\equiv)$ is the natural quotient map.

Proof. a) For $a, b \in G$, suppose $a \equiv b$; then 5.4.(a) entails (i) $\nabla a \equiv \nabla b$ and (ii) $a^2 \equiv b^2$. From (i) and 5.4.(c) we obtain $\nabla a \cdot \nabla b \in \Sigma$ and so $\tau(\nabla a) = \tau(\nabla b)$ (recall: $\Sigma \subseteq \ker(\tau \upharpoonright K)$). Now (ii) and 5.4 yield $\nabla a^2 \equiv \nabla b^2$, that just as above yields $\tau(\nabla(a^2)) = \tau(\nabla(b^2))$. Since τ is a RS-morphism, 4.4.(a) entails $\nabla \tau(a^2) = \tau(\nabla(a^2)) = \tau(\nabla(b^2)) = \nabla \tau(b^2)$, and so $\tau(a^2) = \tau(b^2)$ (by 5.2.(c)). But then,
$$\tau(a) = \tau(\nabla a \cdot a^2) = \tau(\nabla a)\tau(a^2) = \tau(\nabla b)\tau(b^2) = \tau(b),$$
as needed.

b) We first verify that $\mathcal{H} = \mathcal{H}_\Sigma$. If $p \in \mathcal{H}$, then since $K = \ker(\pi \upharpoonright K)$, it is clear that $\Sigma \subseteq p \upharpoonright K$, and $\mathcal{H} \subseteq \mathcal{H}_\Sigma$. For the reverse inclusion, suppose $\tau \in \mathcal{H}_\Sigma$; since $\equiv$ is an RS-congruence, item (a) implies that there is a unique RS-morphism, $\sigma : G_\equiv \longrightarrow 3$ (i.e., $\sigma \in X_\equiv$) making the diagram below left commutative:

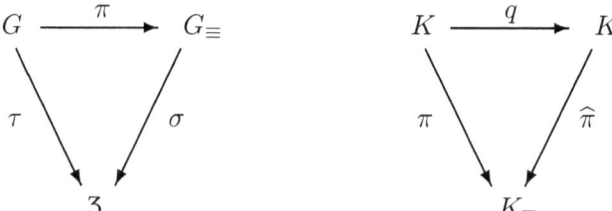

and so $\tau = \sigma \circ \pi \in \mathcal{H}$, establishing that $\mathcal{H} = \mathcal{H}_\Sigma$. If $\gamma \in \mathcal{H}_{\Delta(\Sigma)}$, since $\Sigma \subseteq \Delta_\Sigma$, then $\Sigma \subseteq P(\gamma)$; but then, because Σ consists of units, we obtain $\Sigma \subseteq \ker(\gamma \upharpoonright K)$, showing $\mathcal{H}_{\Delta(\Sigma)} \subseteq \mathcal{H}_\Sigma$. Now let $p \in \mathcal{H}$; we already know that $\Sigma \subseteq \ker(p \upharpoonright K)$. If $a \in \Delta_\Sigma$ (the subsemigroup generated by Σ in G), by 5.3, there is a n-form φ over Σ ($n \geq 1$) so that $a \in D_G(\varphi)$. Hence, $p(a) \in D_3(n\langle 1 \rangle)$ and so by 5.2.(d), $p(a) \in \{0, 1\}$, whence $p \in \mathcal{H}_{\Delta(\Sigma)}$. Therefore,

$$\mathcal{H} \subseteq \mathcal{H}_{\Delta(\Sigma)} \subseteq \mathcal{H}_{\Sigma},$$

and the first part of the proof establishes the desired equality.

c) Let $q : K \longrightarrow K/\Sigma$ be the natural RSG quotient morphism and let $\pi \upharpoonright K : K \longrightarrow K_\equiv$ be the induced RSG morphism, which is surjective by 5.4.(c). Since the maps q and $\pi \upharpoonright K$ have the same kernel (namely Σ), and q is an RSG-quotient morphism, there is, by Proposition 2.21.(b), p. 43, [Dickmann & Miraglia, 2000], a unique RSG-morphism, $\widehat{\pi} : K/\Delta \longrightarrow K_\equiv$, making the diagram above right commutative. To show $\widehat{\pi}$ is a RSG-isomorphism, we prove that $\pi \upharpoonright K$ is *regular* (Definition 2.22, p. 43, [Dickmann & Miraglia, 2000]). Since $\equiv$ is the equivalence relation corresponding to a subspace (by item (b)), we employ the characterization of D^t in the quotient $G_\equiv$, appearing in Theorem II.3.8.(c) of [Dickmann & Petrovich, 2016]. Let $u_i \in K$, $i = 1, 2, 3$ and assume $\pi(u_1) \in D_{K_\equiv}(\pi(u_2), \pi(u_3)) = D^t_{G_\equiv}(\pi(u_2), \pi(u_3)) \cap K_\equiv$. By item (c) in Theorem II.3.8, [Dickmann & Petrovich, 2016],

(I) $\begin{cases} \exists\ x_i \in G,\ i = 1, 2, 3,\ \text{so that} \quad (1)\ x_i \equiv u_i^2 = 1;\ \text{and} \\ \qquad\qquad\qquad\qquad\qquad\quad (2)\ u_1 x_1 \in D^t_G(u_2 x_2, u_3 x_3). \end{cases}$

Then (I.1) yields $u_i x_i \equiv u_i$, whence $\nabla(u_i x_i) \equiv \nabla u_i = u_i$, while (I.2) and 3.5.(c) imply $\nabla(u_1 x_1) \in D^t_G(\nabla(u_2 x_2), \nabla(u_3 x_3))$. Hence, setting $u'_i = \nabla(u_i x_i) \in K$, $i = 1, 2, 3$, obtains

$$u'_1 \in D^t_G(u'_2, u'_3) \cap K = D_K(u'_2, u'_3), \text{ with } u'_i \equiv u_i\ (i = 1, 2, 3),$$

establishing the regularity of $\pi \upharpoonright K$. By Proposition 2.22, p. 43, in [Dickmann & Miraglia, 2000], $\widehat{\pi}$ is an RSG-isomorphism, as needed.

d) (1) The first equivalence in is clear; for the second, $e \equiv e'$ entails, by 5.4.(a), $\nabla e \equiv \nabla e'$, and so since these terms are units, 5.4.(c) obtains $\nabla e \cdot \nabla e' \in \Sigma$. For the converse, if this latter condition is satisfied then, recalling 4.4.(a), we get $\nabla \pi(e) = \pi(\nabla e) = \pi(\nabla e') = \nabla \pi(e')$, and so 5.2.(c) implies $\pi(e) = \pi(e')$, as needed.

By 4.3.(a), the RS-morphism π yields a lattice morphism, $\mathrm{Id}(G) \xrightarrow{h_\pi} \mathrm{Id}(G_\equiv)$, given by $e \in \mathrm{Id}(G) \longmapsto h_\pi(e) = \pi(e) \in \mathrm{Id}(G_\equiv)$, that by item (d) is surjective. By (*) in item (d.1),

$$e \equiv e' \text{ iff } h_\pi(e) = \pi(e) = \pi(e') = h_\pi(e'),$$

and so the restriction of $\equiv$ to $\mathrm{Id}(G)$ is equal to the kernel of the lattice morphism h_π [9], a lattice congruence, whose natural quotient projection is $\pi \upharpoonright \mathrm{Id}(G) : \mathrm{Id}(G) \longrightarrow \mathrm{Id}(G_\equiv)$ [10]. ∎

---

[9] $\ker(h_\pi) = \{\langle x, y \rangle \in \mathrm{Id}(G) \times \mathrm{Id}(G) : h_\pi(x) = h_\pi(y)\}$.

[10] It is not true, in general, that the congruence of distributive lattices are given by either filters or ideals.

Remarks 5.6 a) With notation as in 5.1, by Theorem 5.5, any RS-congruence on a Boolean RS is induced by a subspace of X_G of a special kind: note that $\mathcal{H}_\Sigma = \bigcap \{[\![u = 1]\!]_{X_G} : u \in \Sigma\}$. In fact, items (c) and (d.1) of 5.5 show that such an RS-congruence is essentially determined by the saturated subgroup Σ of K.

b) If B is a Boolean algebra, any non-empty closed set, C, of the Stone space of B, $S(B)$, is of the form described in item (a). Indeed, if $U = S(B) \setminus C$, then there is a family of finite subsets of B, $\mathcal{C} = \{J \subseteq B : J \text{ is finite}\}$, so that $U = \bigcup_{J \in \mathcal{C}} \bigcap_{u \in J} [\![u = 1]\!]$. For $J \in \mathcal{C}$, set $u_J = \bigwedge_{u \in J} u$; then, $[\![u_J = 1]\!] = \bigcap_{u \in J} [\![u = 1]\!]$, whence $U = \bigcup_{J \in \mathcal{C}} [\![u_J = 1]\!]$ and so $C = \bigcap_{J \in \mathcal{C}} [\![-u_j = 1]\!]$.

In particular, if $P = \mathbb{C}(X, 3)$ is a Post algebra of order 3 and B is the Boolean algebra of units in P, then X_P is naturally homeomorphic to $S(B)$ and so any closed subset of X_P is a subspace of the kind described in (a).

c) It is not clear that *every* closed subset of the space of orders of a Boolean RS is a subspace, in particular of the form described in item 5.6.(a). In fact, this is not clear even for reduced special groups. ■

References

Andradas, C., Bröcker, L. & Ruiz, J. (1996), *Constructible Sets is Real Geometry*, Vol. 33 of *Modern Surveys in Math*, Springer-Verlag.

Atiyah, M. F. & Macdonald, I. G. (1969), *Introduction to Commutative Algebra*, Addison-Wesley Publ. Co.

Bochnak, J., Coste, M. & Roy, M.-F. (1998), *Real Algebraic Geometry*, Vol. 36 of *Ergeb. Math.*, Springer-Verlag.

Chang, C. C. & Keisler, H. J. (1990), *Model Theory*, North-Holland Publ. Co.

Dickmann, M. & Miraglia, F. (2000), *Special Groups : Boolean-Theoretic Methods in the Theory of Quadratic Forms*, Vol. 689 of *Memoirs of the AMS*, AMS.

Dickmann, M. & Miraglia, F. (2015), *Faithfully Quadratic Rings*, Vol. 1128 of *Memoirs of the AMS*, AMS.

Dickmann, M. & Petrovich, A. (2004), 'Real semigroups and abstract real spectra, I', *Contemporary Math.* **344**, 99–119.

Dickmann, M. & Petrovich, A. (2008), *The Three-Valued Logic of Quadratic Form Theory over Real Rings*, in **Andrezj Mostowski and**

Foundational Studies, A. Ehrenfeucht and V. W. Marek and M. Srebrny (eds.), IOS Press.

Dickmann, M. & Petrovich, A. (2012), 'Spectral Real Semigroups', *Annales de la Faculté des Sciences de Toulouse (Mathématiques)* **XXI**, 359–412.

Dickmann, M. & Petrovich, A. (2016), 'Real Semigroups, Real Spectra and Quadratic Forms over Rings', (in preparation); February 10th, 2016 version available at http://www.ime.usp.br/~miraglia/textos/RS-fev-16.pdf.

Dickmann, M., Marshall, M. & Miraglia, F. (2005), 'Lattice-ordered reduced special groups', *Annals of Pure and Appl. Logic* **132**, 27–49.

Dickmann, M., Miraglia, F. & Petrovich, A. (2017), 'Constructions in the category of real semigroups', *Contemporary Math.* **697**, 109–136.

Dickmann, M., Schwartz, N. & Tressl, M. (2019), *Spectral Spaces*, Vol. 35 of *New Mathematical Monographs*, Cambridge University Press.

Marshall, M. (1996), *Spaces of Orderings and Abstract Real Spectra*, Vol. 1636 of *Lecture Notes in Mathematics*, Springer-Verlag.

Kolgomorov-Veloso Problems and Dialectica Categories[‡]

Valeria de Paiva* Samuel G. da Silva [†]

* Topos Institute
Berkeley, CA
valeria.depaiva@gmail.com

[†] Departamento de Matemática
Instituto de Matemática e Estatística
Universidade Federal da Bahia
samuel@ufba.br

Dedicatory. This work is dedicated to Professor Paulo Veloso with our deepest appreciation for all he has done for research and teaching in Logic and Theoretical Computer Science in Brazil. Personally Professor Veloso has been a friend and mentor to both of us. He was also kind enough to accept to supervise me (Valeria de Paiva) knowing that I would be leaving the program, if my scholarship abroad was granted, which did happen. For this and many other acts of kindness, academic and personal, we are grateful and happy to celebrate his strong influence in our work.

1 Introduction

Blass' seminal paper [Blass, 1995] makes a surprising connection between de Paiva's Dialectica categories [Paiva, 1991], Vojtáš' methods to prove inequalities between cardinal characteristics of the continuum [Vojtáš, 1993]

[‡]The authors would like to thank Professor Andreas Blass for his work in on the interaction between Dialectica Categories and cardinal invariants of the continuum, especially for the seminal article [Blass, 1995] that was the original source of our joint research project [Silva & Paiva, 2017]. We would also like to thank Professor Wagner Sanz for, not only calling our attention to Veloso's work on problems, but also for providing us with the necessary means to compare our work to his own. His presentation [Sanz, 2020] at the online seminar "Logicians in Quarantine" (sponsored by the Brazilian Logic Society and by the Interest Group on Logic of the Brazilian Computing Society) during the COVID-19 pandemics was our starting point for our investigation of problems. Thus we would like to thank the organizers of these splendid seminars as well.

and the complexity theoretical notions of problems and reductions developed in [Impagliazzo & Levin, 1990]. Blass did not mention Kolmogorov's very abstract notion of *problem* ([Kolmogorov, 1991]), which is not related to specific complexity issues. Kolmogorov did investigate a notion of abstract problem, producing an alternative intuitive semantics for Propositional Intuitionistic Logic, part of the celebrated BHK interpretation of Intuitionistic Logic.

Kolmogorov's problems were cited as an inspiration by Veloso when he developed his own theory of abstract problems in the Eighties [Veloso, 1984], but the two frameworks were not formally connected. This note recalls Veloso's 'Teoria de Problemas' (Theory of Problems) and shows how it can be related to the Dialectica construction [Paiva, 1989b], via Kolmogorov's concepts. To establish this relationship, we use the modification of the Dialectica construction considered by the second author [Silva & Paiva, 2017; Silva, 2020] for other set-theoretical purposes. This modification corresponds to a condition of non-triviality of the collections of problems or solutions, first suggested by [Moore et al., 2004].

The categorical connection between Dialectica models, Kolmogorov's problems, Veloso's problems and Blass' problems shows that the use of categories really allows us to connect extremely different areas of mathematics, using simple methods. In the case of this paper, it also allow us to determine where exactly the foundational choices of axioms are important. We show that while Blass and Kolmogorov's notions of problem can be investigated using the set-theoretical framework of **ZF**, Veloso's problems commit us to a stronger set-theory, as discussed in the following sections. Whether this requirement of stronger foundations is a bug or a feature depends, perhaps, on personal taste and conviction. From our part we are happy to note that, like many other questions in Mathematics, as soon as one investigates them a little, the notion of 'problem' points out to grand challenges in the foundations of Mathematical Logic, to wit whether one wants to accept or not the Axiom of Choice within their chosen framework.

2 Kolmogorov Problems

Blass ([Blass, 1995]) noticed that de Paiva's Dialectica construction GC [Paiva, 1989b], when the base category C is the category **Sets**, is the dual of Vojtáš's category GT of generalized Galois-Tukey connections [Vojtáš, 1993]. Since the Dialectica construction has been generalized in many different directions, instead of writing G**Sets** we will write $\text{Dial}_2(\textbf{Sets})$ for this category, to make explicit the object 2 (the object of truth-values) where our relations map into, as well as the category C that is Sets, where objects 'live'. We recall the definition of the category below.

Definition 2.1 (Dialectica category [Paiva, 1989b]). *The* **dialectica category** *$Dial_2(\textbf{Sets})$ has as objects triples of the form $A = (U, X, \alpha)$, where U and X are sets and $\alpha \subseteq U \times X$ is a set-theoretical relation, which can be written, equivalently, as $\alpha : U \times X \to 2$. If $A = (U, X, \alpha)$ and $B = (V, Y, \beta)$ are objects of $Dial_2(\textbf{Sets})$, a morphism from A to B is a pair of functions in Sets, (f, F), $f : U \to V$, $F : Y \to X$ such that*

For all $u \in U$ and $y \in Y$, $u\alpha F(y)$ implies $f(u)\beta y$.

Blass noticed that the opposite, dual Dialectica Category $Dial_2(\textbf{Sets})^{op}$ can be taken to intuitively mean that objects represent *problems* and morphisms stand for reductions between problems. Thus a triple $P = (I, S, \sigma)$ of $Dial_2(\textbf{Sets})^{op}$, under this interpretation, represents a problem, whose instances are elements of I, the set of possible solutions is given by S and the relation σ can be read as "is solved by", that is, if z is an instance of the problem P and s is a possible solution then $z\sigma s$ states that "s solves z". Blass associates these problems with a concept of many-one reduction of search problems in complexity theory, a very restricted kind of problem, for which he refers to [Impagliazzo & Levin, 1990; Venkatesan & Levin, 1988].

Veloso's theory of problems ([Veloso, 1984]), following Pólya, suggests that in order to understand a problem one should consider the following initial questions:

1. What is the unknown?

2. What are the data?

3. What is the condition?

These questions correspond directly to the elements of the triples of information which characterize a problem, which will be referred to, in this work, as a *Kolgomorov problem*. They also correspond precisely to Blass' interpretation above.

Definition 2.2 (Kolgomorov problems). *A* **Kolgomorov problem** *is a triple $P = (I, S, \sigma)$, where I and S are sets and $\sigma \subseteq I \times S$ is a set theoretical relation. We say that:*

- *I is the set of* **instances** *of the problem P;*

- S is the set of **possible solutions** for the instances I; and

- σ is the **problem condition**, i.e. the relation σ holds between z and s, in symbols $z\,\sigma\,s$ if the solution s satisfies the problem condition σ for the instance z, or, more briefly, "s σ-solves z".

Kolgomorov's 1932 paper, republished in English in 1991 ([Kolmogorov, 1991]) has two parts. The first section, which introduces Kolgomorov's problems, says that "If the intuitionistic cognitive presuppositions are not accepted then one should take into account only the first section". In this section he introduces problems via mathematical examples, for instance "Find any four integers, x, y, z and n such that $x^n + y^n = z^n$, for $n > 2$". (Note that Kolmogorov states explicitly, in page 152, that "We never assume a problem to be solvable".)

We can identify the *objects* of the category $\mathrm{Dial}_2(\mathbf{Sets})$ with Kolgomorov problems. What would the *morphisms* represent in this case? To answer this, we consider the morphisms of the opposite of the Dialectica category, $\mathrm{Dial}_2(\mathbf{Sets})^{\mathrm{op}}$. In $\mathrm{Dial}_2(\mathbf{Sets})^{\mathrm{op}}$, a morphism from an object $P' = (I', S', \sigma')$ to an object $P = (I, S, \sigma)$ is a pair of functions (f, F), where $f\colon I \to I'$ and $F\colon S' \to S$ are such that the following condition holds

$$(\forall z \in I)\ (\forall t \in S')\ [f(z)\,\sigma'\,t \longrightarrow z\,\sigma\,F(t)].$$

So for all instances of problems z of P and all solutions of problems t of P', if the instance of problem $f(z)$ has σ'-solution t then the instance z has σ-solution $F(t)$.

If we regard P and P' as Kolgomorov problems, the existence of a morphism from P' to P ensures that there is a *reduction* of the problem P to the problem P' – because the act of solving an instance of P may be reduced to the act of solving an instance of P'. More precisely, if one wants to solve a particular instance z of the Kolgomorov problem P, it suffices to find a solution t for the instance $f(z)$ of P' – since $F(t)$ will provide a solution for the initial instance z of P.

Kolmogorov discusses a few number-theoretical and geometrical problems, as well as abstract, logical rules ones. In what follows, we present a number of examples from daily mathematical practice to show how they are coded as Kolgomorov problems. We also show reductions between those problems which can be seen as morphisms of $\mathrm{Dial}_2(\mathbf{Sets})^{\mathrm{op}}$. As a piece of notation, if X is a set and $n \in \mathbb{N}$ then $[X]^n$ denotes the family of all subsets of X which have precisely n elements and $[X]^{\leqslant n}$ means $\bigcup_{m \leqslant n} [X]^m$.

Example 2.3. *Analytical geometry is entirely based on the reduction of geometrical problems to equation solving problems.*

We present some practical examples to explain what we mean by the statement above.

Let π be any plane of the 3-dimensional Euclidean space $\mathbb{R}^3$ and let $F\colon \mathbb{R}^2 \to \pi$ be a coordinate system as usual (i.e., for every pair (u,v) of real numbers we associate the point $P = F(u,v)$ of the plane which has coordinates (u,v) – that is, $P = P_{(u,v)}$). Every line l of the plane π is then represented by an equation of the form $ax + by = c$, where a and b are real numbers such that $a \neq 0$ or $b \neq 0$ and $c \in \{0,1\}$. Let E be the family of all equations of the described canonical form and let $\mathcal{L}$ denotes the family of all lines of the plane π. The decision problem of "whether a given point lies on a given line" is $(\mathcal{L}, \pi, \ni)$, and the problem of "whether a given pair of real numbers satisfy a given equation" is $(E, \mathbb{R}^2, \zeta)$, on which an equation $ax + by = c$ is ζ-related to a pair (u,v) of real numbers if $au + bv = c$. Then we can reduce the geometrical problem $(\mathcal{L}, \pi, \ni)$ to the algebraic problem $(E, \mathbb{R}^2, \zeta)$ using the morphism (f, F), where $f \colon \mathcal{L} \to E$ is defined by putting $f(l) = \operatorname{eq}(l)$ (where $\operatorname{eq}(l)$ is the canonical equation which represents l) and $F\colon \mathbb{R}^2 \to \pi$ is the coordinate system. Indeed, if (u,v) satisfies the equation of a line l we know that the corresponding point $P_{(u,v)}$ lies in l.

A slight variation of what we have just done reduces the problem of finding the intersection point of two distinct lines to the problem of solving a linear system with two equations over two variables. The geometrical problem of finding the intersection point of two distinct lines is $([\mathcal{L}]^2, \pi, \xi)$, where $\{l_1, l_2\} \xi P$ means that $P \in l_1 \cap l_2$ for any distinct lines l_1, l_2 of π and for every point P in π. The algebraic problem of finding the solution for a linear system with two equations over two variables is $([E]^2, \mathbb{R}^2, \lambda)$, where the relation λ in $\{a_1 x + b_1 y = c_1, a_2 x + b_2 y = c_2\} \lambda (u,v)$ is the conjunction of "$a_1 x + b_1 y = c_1$"$\zeta(u,v)$ and "$a_2 x + b_2 y = c_2$"$\zeta(u,v)$. The morphism which gives the reduction is (g, F), where $g \colon [\mathcal{L}]^2 \to [E]^2$ is given by $g(\{l_1, l_2\}) = \{f(l_1), f(l_2)\} = \{\operatorname{eq}(l_1), \operatorname{eq}(l_2)\}$ for all distinct lines l_1, l_2 of π and $F\colon \mathbb{R}^2 \to \pi$ is still the coordinate system. Now, if l_1 and l_2 are distinct lines and (u,v) is a solution of the system $\{\operatorname{eq}(l_1), \operatorname{eq}(l_2)\}$ then the point $P_{(u,v)}$ lies in the intersection of the lines l_1 and l_2.

Example 2.4. *The problem of finding vectors in the intersection of the kernels of a finite family of linear functionals over $\mathbb{R}^n$ reduces to the problem of finding the solutions of a homogeneous system of linear equations.*

Let $n \geqslant 2$ and $\{e_1, e_2, \ldots, e_n\}$ denote the canonical basis of the n-dimensional Euclidean space $\mathbb{R}^n$, regarded as a vector space over $\mathbb{R}$. A *linear functional* over $\mathbb{R}^n$ is a linear transformation from $\mathbb{R}^n$ into $\mathbb{R}$. Let $(\mathbb{R}^n)^*$ (the *dual* of $\mathbb{R}^n$) denote the family of all linear functionals over $\mathbb{R}^n$. It is well-known that the dimension of the dual space is also n, and that $\{\mathbf{x_1}, \mathbf{x_2}, \ldots, \mathbf{x_n}\}$ is a basis of the dual space – where, for $1 \leqslant i \leqslant n$, $\mathbf{x_i} : \mathbb{R}^n \to \mathbb{R}$ is the linear functional which satisfies $\mathbf{x_i}(e_j) = \delta_{ij}$ (where $\delta_{ij} = 1$ if $i = j$ and it is zero otherwise) for $1 \leqslant j \leqslant n$ and is then extended to all linear functionals by linearity.

Let $\mathcal{E}$ denote the family of all linear equations of the form $a_1 x_1 + a_2 x_2 + \ldots + a_n x_n = 0$, where $a_1, \ldots, a_n$ are real numbers. A linear functional $\mathbf{T}$ in the dual space of $\mathbb{R}^n$ may be written, in a unique way, in the form $\mathbf{T} = \sum_{i=1}^{n} a_i \mathbf{x_i}$.

Thus we may define a translation function $g : (\mathbb{R}^n)^* \to \mathcal{E}$ in the obvious way, i.e. by putting

$$g(\mathbf{T}) = \text{``}a_1 x_1 + a_2 x_2 + \ldots + a_n x_n = 0\text{''}$$

if $\mathbf{T} = \sum_{i=1}^{n} a_i \mathbf{x_i}$.

The problem of finding vectors in the intersection of the kernel of finite families of linear functionals is $([(\mathbb{R}^n)^*]^{\leqslant n}, \mathbb{R}^n, \alpha)$, where $\mathcal{H} \alpha (c_1, c_2, \ldots, c_n)$ means that $(c_1, c_2, \ldots, c_n) \in \bigcap_{h \in \mathcal{H}} \text{Ker}(h)$ for any finite family $\mathcal{H}$ of m linear functionals with $m \leqslant n$ and any n-tuple $(c_1, c_2, \ldots, c_n) \in \mathbb{R}^n$ (recall that the dimension of the dual space is n as well, so we may only consider finite families with no more than n linear functionals). The problem of solving a homogeneous system of no more than n linear equations is $([\mathcal{E}]^{\leqslant n}, \beta, \mathbb{R}^n)$, where $\mathcal{G} \beta (c_1, c_2, \ldots, c_n)$ means that, for any $(c_1, c_2, \ldots, c_2) \in \mathbb{R}^n$ and any $\mathcal{G} \in [E]^m$ with $m \leqslant n$, $(c_1, c_2, \ldots, c_2)$ solves each one of the m linear equations of the finite family $\mathcal{G}$. A morphism which codes the usual and easily verifiable procedure of reducing the first problem to the second is now given by (f, Id), where $f : [(\mathbb{R}^n)^*]^{\leqslant n} \to [\mathcal{E}]^{\leqslant n}$ is given by $f(\mathcal{H}) = \{g(h) : h \in \mathcal{H}\}$ for any finite family $\mathcal{H}$ of no more than n linear functionals and Id is the identity function from the *set* $\mathbb{R}^n$ into the *vector space* $\mathbb{R}^n$ – i.e., we have reduced a problem stated in the context of the linear structure of $\mathbb{R}^n$ and its dual to a more naive, pedestrian problem of finding solutions of a homogeneous system of equations.

Veloso developed further notions of *viable problems*, *links* between problems (and *reduction links* between problems) in his theory of mathematical problems [Veloso, 1984]. However, as Veloso himself was aware of (see further discussion in the final section), these definitions make essential use of the Axiom of Choice ([Martin-Löf, 2006],[Moore, 1982]), and this use will be detailedly discussed in the next section.

3 Veloso Problems and the Axiom of Choice

In our previous work [Silva & Paiva, 2017] we described a modification of the category $\text{Dial}_2(\textbf{Sets})^{op}$, which Blass calls the category $\mathcal{PV}$ in [Blass, 1995]. Blass, describing the category $\mathcal{PV}$, explains that it has as objects problems, together with their instances and respective solutions. Moreover, morphisms of the category $\mathcal{PV}$ identify reductions of classes of (complexity) problems to others. The modification of the category in [Silva & Paiva, 2017], following the work of [Moore et al., 2004], insists on some conditions of non-triviality of the objects. These conditions can be paraphrased as 'there are no problems without a solution' and 'there are no solutions that solve all problems at once'. There are also some constraints on the cardinality of the sets of instances and solutions of $\mathcal{PV}$; all constituent sets of objects of $\mathcal{PV}$ are bounded above by the cardinality of the *continuum* (which is $\mathfrak{c} = 2^{\aleph_0} = |\mathbb{R}| = |\mathcal{P}(\mathbb{N})|$).

The mentioned non-triviality conditions are not found in the original Dialectica categorical construction ([Paiva, 1989b],[Paiva, 1991]), where trivial objects are actually required to provide truth-values (or units for the categorical operators) associated to the logical connectives of Girard's Linear Logic [Girard, 1987]. The non-triviality conditions are also not found in Kolmogorov's work. He describes *meaningless* problems as the ones that do not have a solution.

The non-triviality conditions will characterize, exactly, a class of problems we refer to as *Veloso problems*. We note, however, that there are no upper bounds for the cardinality of the sets of instances and solutions of the problems originally discussed by Veloso.

We first formally present the category $\mathcal{PV}$, as well as stratified versions of it which were introduced in [Silva, 2020]. The following definition should be considered within **ZFC**, since we refer to well-ordered cardinals in its first clause.

Definition 3.1 (Category $\mathcal{PV}$ [Silva & Paiva, 2017]). *The* **category** $\mathcal{PV}$ *is the subcategory of* $\text{Dial}_2(\textbf{Sets})^{op}$ *whose objects are the triples* $A = (U, X, \alpha)$ *satisfying the following three clauses, which will be referred to as the* **MHD** *conditions (MHD stands for Moore, Hrušák and Džamonja [Moore et al., 2004]):*

(1) The cardinalities of the (non-empty) constituent sets are bounded

above by the cardinality of the continuum – i.e. $0 < |U|, |X| \leqslant 2^{\aleph_0}$.

(2) Every problem has a solution – i.e.

$$(\forall u \in U)(\exists x \in X)[u \, \alpha \, x].$$

(3) There are no solutions that solve all the problems at once – i.e.

$$(\forall x \in X)(\exists u \in U)[\neg (u \, \alpha \, x)].$$

The morphisms between objects of $\mathcal{PV}$ are the same morphisms of $\mathrm{Dial}_2(\mathbf{Sets})^{op}$ – that is, a morphism from an object $B = (V, Y, \beta)$ to an object $A = (U, X, \alpha)$ is a pair of functions (f, F), where $f : U \to V$ and $F : Y \to X$ are such that

$$(\forall u \in U)\,(\forall y \in Y)\,[f(u) \, \beta \, y \longrightarrow u \, \alpha \, F(y)].$$

Morphisms of $\mathcal{PV}$ induce the **Galois-Tukey pre-order** introduced by Vojtáš, which is defined in the following way: if $A = (U, X, \alpha)$ and $B = (V, Y, \beta)$ are objects of $\mathcal{PV}$, then we have

$$A \leqslant_{GT} B \iff \text{There is a morphism from } B \text{ to } A.$$

The diagram below represents the situation where $A \leqslant_{GT} B$:

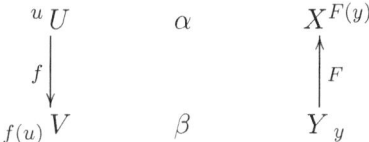

Given an object $A = (U, X, \alpha)$ of $\mathcal{PV}$, its *dual object* is given by $A^* = (X, U, \alpha^*)$, where $x\alpha^*u$ means that $\neg(u\alpha x)$. One can easily check (via a contrapositive argument) that:

$$\text{If } A \leqslant_{GT} B, \text{ then } B^* \leqslant_{GT} A^*.$$

In [Silva, 2020] parametrized, stratified versions of $\mathcal{PV}$ were introduced – the $\mathcal{PV}_X$ categories, for X an infinite set. The main goal of the proposed stratification was to generalize features of $\mathcal{PV}$ to sets of higher cardinalities, since the constituents of the objects of $\mathcal{PV}$ are bounded above by the cardinality of the continuum (which is the cardinality of the power set of the naturals). Notice that, indeed, in **ZFC** the categories $\mathcal{PV}$ and $\mathcal{PV}_\mathbb{N}$ coincide, and so the whole idea of the stratified versions was to generalize $\mathcal{PV}$ (in a choiceless context) to any other other infinite set X. These categories are introduced in the **ZF** setting, as their definitions use the notion of *domination* ($\preccurlyeq$) between sets instead of that of well-ordered cardinals (for a more detailed discussion on the subtleties of comparing sizes in the absence of **AC**, we refer to Section 2 of [Silva, 2020]). Recall that a set A is *dominated* by a set B, $A \preccurlyeq B$, if there is an injective function from A into B.

Definition 3.2 (Categories $\mathcal{PV}_X$). Let X be an infinite set in **ZF**. The category $\mathcal{PV}_X$ is the subcategory of $Dial_2(\mathbf{Sets})^{op}$ whose objects (U, X, α) are those which satisfy the following $\mathbf{MHD}_X$ conditions (since they are a restricted form of the **MHD** conditions):

(1) U, V are non-empty sets and $U, V \preccurlyeq \mathcal{P}(X)$, that is, sets U, V are dominated by $\mathcal{P}(X)$.

(2) $(\forall u \in U)(\exists x \in X)[u\,\alpha\,x]$

(3) $(\forall x \in X)(\exists u \in U)[\neg(u\,\alpha\,x)]$

The morphisms between objects of $\mathcal{PV}_X$ are the same morphisms of $Dial_2(\mathbf{Sets})^{op}$.

We also define, in the expected way, a Galois-Tukey ordering $\leqslant_{GT}$ on the objects of $\mathcal{PV}_X$.

The categories $\mathcal{PV}_X$ are used in [Silva, 2020] to provide, after a quantification over all infinite sets, equivalences of the Axiom of Choice **AC**. The following equivalences are proved in [Silva, 2020]:

Theorem 3.3 (da Silva [Silva, 2020]). *The Axiom of Choice is equivalent to the following statements:*

(*) *"For every infinite set X, $(\mathcal{P}(X), \mathcal{P}(X), =)$ is a maximum element in the Galois-Tukey ordering $\leqslant_{GT}$ on $\mathcal{PV}_X$."*

and

(**) *"For every infinite set X, $(\mathcal{P}(X), \mathcal{P}(X), \neq)$ is a minimum element in the Galois-Tukey ordering $\leqslant_{GT}$ on $\mathcal{PV}_X$."*

This theorem shows that stratifying objects and features of the Dialectica construction using the Galois-Tukey ordering is equivalent to accepting the Axiom of Choice. This may be a surprise to mathematicians not used to the "thinking about foundational issues" exercise.

In this work we are interested in the investigation of general problems, which means that, in particular, the sets of instances and solutions should have no upper bounds on their cardinalities. To do so in a categorical setting, we provide the following definition, which should be considered in the setting of **ZF**.

Definition 3.4 (Unbounded $\mathcal{PV}$ category). *The **unbounded** $\mathcal{PV}$ **category** $\mathcal{PV}_{unb}$ is the the subcategory of $Dial_2(\mathbf{Sets})^{op}$ whose objects (U, X, α) satisfy the **MHD** conditions (2) and (3) but where the **MHD** condition (1) is relaxed to allow sets of arbitrary large cardinality. Thus we require **MHD** conditions (2) and (3) together with*

(1)' U and X are non-empty sets

A triple (U, X, α) is an object of $\mathcal{PV}_{unb}$ if it is an object of $\mathcal{PV}_Y$ for some infinite set Y.

The morphisms between objects of $\mathcal{PV}_{unb}$ are the very same morphisms of $Dial_2(\mathbf{Sets})^{op}$.

The unbounded $\mathcal{PV}$ category captures, precisely Veloso problems – these are the Kolgomorov problems which are **viable** and **non-generic**, as we define now.

Definition 3.5 (Veloso problem). Let $P = (I, S, \sigma)$ be a Kolgomorov problem.

(i) P is said to be a **viable** problem if the domain of σ is the whole set I – or, equivalently, if for every instance z of I there is some possible solution s in S that solves the problem, so that the relation $z\,\sigma\,s$ holds.

(ii) A possible solution s will be said to be a **generic solution** for P if it solves all of its instances – that is, if for every instance z of I one has $z\,\sigma\,s$ for this particular solution s.

(iii) The problem P is said to be **non-generic** if it has no generic solutions – equivalently, for every possible solution $s \in S$ there is some instance $z \in I$ such that $\neg(z\sigma s)$.

(iv) P will be said to be a **Veloso problem** if it is viable and non-generic.

Notice that the fact that a problem $P = (I, S, \sigma)$ is non-generic is equivalent to the viability of the dual problem $P^* = (S, I, \sigma^*)$, where $s\,\sigma^*\,z$ means

$\neg(z\sigma s)$ for all $z \in I$ and $s \in S$.

It should be clear that viability and non-genericity are conditions asking for non-triviality of a given problem. For instance, it is well-known that the following statement is an Axiom of Incidence, in Hilbert's plane geometry:

(†) *"For every line l, there are at least two points which lie in the line l and at least one point which does not lie in the line l"*

Together with the axiom "For every pair of distinct points there is only one line which passes through those points", the above statement (†) corresponds, precisely, to the viability and non-genericity requirements for the problem of determining whether a given line contains a given pair of distinct points of the plane or $(L, P, \supseteq)$, where

L = the set of all lines of the plane; and

P = the family of all pairs of two distinct points of the plane.

Notice also that non-genericity establishes a dimension for the preceding problem: if all points were in the same line l we would not be studying the geometry of the plane, only the geometry of a line.

The following example shows that certain mathematical notions are equivalent to the viability of certain problems. Recall that a subset of a topological space is *dense* if its closure is equal to the whole space. It is a textbook easy exercise to show that dense sets are precisely those which intersect any non-empty open set. Thus, the following holds:

Example 3.6. *Let (X, τ) be topological space and D be a proper subset of X. Then D is a dense subspace of X if, and only if, the Kolgomorov problem $(\tau \setminus \{\emptyset\}, D, \ni)$ is viable.*

Alternatively, D is a dense set if, and only if, for every non-empty open set U of X the problem $(\{U\}, D, \ni)$ is viable. This observation is relevant in the following example, which is, accordingly to Veloso ([Veloso, 1984], page 24), a *problem to prove* in Pólya's terminology – in contrast to the so-called *problems to find*. Recall that a topological space is a *Baire space* if the intersection of any countable family of dense subsets of the space is also dense subset of the space.

Example 3.7. *To prove the Baire Theorem for Complete Metric Spaces (i.e., the theorem which asserts that any complete metric space is a Baire space), it suffices to verify that, for any non-empty open set O and*

for any countable family $\{U_n : n \in \mathbb{N}\}$ of dense open sets, the problem $(\{O\}, \bigcap_{n\in\mathbb{N}} U_n, \ni)$ is a viable problem.

Later on we will show that the above example is easier to handle if transformed into a countable chain of viable related problems.

The notion of *viability* of a problem, introduced by Veloso in [Veloso, 1984] (page 25), is related to the notion of *solvability* of the problem. However, a solution of a problem for Veloso is represented by a function, not by a relation, as it is the case in our work. We will use the terminology *solution function* to denote these objects that are, instead of solution relations, solution functions.

Definition 3.8. *Let $P = (I, S, \sigma)$ be any viable Kolgomorov problem. A* **solution function** *for P is a function $f : I \to S$ satisfying $f \subseteq \sigma$ – or, equivalently, f is a S-valued function with domain I such that for every instance z of P we have that $f(z)$ solves z.*

Veloso (op. cit) has defined a notion of solvability according to his definition of solution – that is, in our terms, a Kolgomorov problem P is **solvable** if P has a solution function. As one of the easiest equivalent statements of the Axiom of Choice is precisely *"Every relation contains a function with the same domain"* (see page 131 of [Moore, 1982]), it is straightforward to check that, within **ZFC** (i.e., assuming the Axiom of Choice), the following equivalences hold:

Proposition 3.9 (viable is solvable in **ZFC**). *Let $P = (I, S, \sigma)$ be any Kolgomorov problem. The following statements are equivalent:*

(i) *The Kolmogorov problem P is viable.*
(ii) *The problem P is solvable.*
(iii) *The problem P satisfies the formula $(\forall z \in I)(\exists s \in S)[z \, \sigma \, s]$.*

The preceding proposition was also stated by Veloso. The only non-obvious implication $((i) \longrightarrow (ii))$ follows easily from the equivalence of the Axiom of Choice mentioned above.

Still following Veloso's definitions from [Veloso, 1984], next we define the notions of *links* and *reduction links* between problems. We point out that the notion of "reduction link" will be defined in terms of "solution functions". Again, this context is intrinsically associated to the Axiom of Choice, as we will discuss later.

Definition 3.10 (Links and reduction links [Veloso, 1984]). Let $P = (I, X, \sigma)$ and $P' = (I', X', \sigma')$ be two Kolgomorov problems.

(i) A **link** from P to P' is a pair of functions (f, F), where

$f : I \to I'$ is said to be a **translation function**; and
$F : S' \to S$ is said to be a **recovery function**.

(ii) A link (f, F) from P to P' is called a **reduction link** of P to P' if it lifts solution functions from P' to P, i.e. for every solution function g of P' the composite function $F \circ g \circ f$ is a solution function for P.

From now on we identify the class of all Veloso problems with the objects of the category $\mathcal{PV}_{unb}$. Next we show that morphisms of this category correspond to reduction links between the corresponding problems.

Proposition 3.11. Let $P = (I, X, \sigma)$ and $P' = (I', X', \sigma')$ be objects of $\mathcal{PV}_{unb}$, considered as Kolgomorov problems. If (f, F) is a morphism witnessing the order $P \leqslant_{GT} P'$, then (f, F) is a reduction link of P to P'.

Proof:. Let g be an arbitrary solution function for the problem P'. As we know that (f, F) is a morphism from P' to P, we know that for all $z \in I$ and $t \in S'$ the following implication holds:

$$f(z)\,\sigma'\,t \longrightarrow z\,\sigma\,F(t)$$

Fix an arbitrary $z \in I$. As $g \subseteq \sigma'$, for $t = g(f(z))$ we have that $f(z)\,\sigma'\,t$, and therefore

$$z \sigma F(t) = F(g(f(z)))$$

and thus $(z, F(g(f(z))) \in \sigma$. By the arbitrariness of z, we conclude that $F \circ g \circ f \subseteq \sigma$ and so the composite function $F \circ g \circ f$ is a solution function for P, as desired. ∎

So, given Kolgomorov problems P and P' with a Dialectica morphism (f, F) witnessing the inequality $P \leqslant_{GT} P'$ in the Galois-Tukey ordering, we are allowed to interpret such inequality as a measure of complexity, since a solution of P may be reduced to a solution of P' – thus under $P \leqslant_{GT} P'$ we can say that P is as easy to solve as P', or that P is not more complicated to be solved than P'.

We have shown that the existence of solution functions for viable problems is ensured by the Axiom of Choice (Proposition 3.9). As Veloso's reduction links were defined in terms of solution functions, it is clear that this approach presupposes a choice principle. In the following results, we will show more, as it will be established that assuming that Veloso's approach holds in full generality gives rise to a number of equivalents of the Axiom of Choice (or of weak related statements). For the rest of this section, our base theory is **ZF**.

Theorem 3.12. **AC** *is equivalent to the following statement:*

"Every Veloso problem has a solution function".

Proof:. ($\Rightarrow$) Let $P = (I, S, \sigma)$ be a Veloso problem. By the **MHD** condition (1), P is viable and so $dom(\sigma) = I$. As in the proof of Proposition 3.9, the existence of a solution function is guaranteed by the equivalent of **AC** given by the statement "Every relation contains a function with same domain".
($\Leftarrow$) It was remarked in [Silva, 2020] that **AC** is equivalent to the following statement:

"For every infinite set X, there is a choice function defined on $\mathcal{P}(X) \setminus \{\emptyset\}$ – that is, there is $f : \mathcal{P}(X) \setminus \{\emptyset\} \to X$ such that $f(Y) \in Y$ for all $\emptyset \neq Y \subseteq X$."

So, let X be any infinite set. The problem

$$(\mathcal{P}(X) \setminus \{\emptyset\}, X, \ni)$$

is clearly viable (since non-empty subsets have elements, by definition) and non-generic (given $x \in X$ one has that $Y = X \setminus \{x\}$ is not solved by x). A solution function for this problem is, clearly, a choice function for $\mathcal{P}(X) \setminus \{\emptyset\}$. As X was taken arbitrarily, we have established **AC** by the above remark. ■

It follows from the previous theorem that in the absence of the Axiom of Choice there will be Veloso problems in $\mathcal{PV}_{unb}$ without any solution function.

Notice that, if P' is a Veloso problem without a solution function, it is vacuously true that *any link* from P to P' (for *any* given P) is a reduction link. Motivated by this, we introduce the following definition:

Definition 3.13. *Let P, P' be problems in $\mathcal{PV}_{unb}$. A morphism (f, F) from P' to P is said to be **realized as a reduction link** if the statement "(f, F) is a reduction link from P to P'" holds non-vacuously.*

If all morphisms have to be realized as reduction links, then the Axiom of Choice must be present, as the following theorem shows.

Theorem 3.14. AC *is equivalent to the statement*

"Every morphism of $\mathcal{PV}_{unb}$ *is realized as a reduction link".*

Proof:. ($\Rightarrow$) Given a morphism from P' to P, **AC** implies that P' has solution functions (by 3.12), and the rest follows from Proposition 3.11 and from the definition of "realized as a reduction link".
($\Leftarrow$) Given any infinite set X, we are able to consider the object of $\mathcal{PV}_{unb}$ given by
$$P = (\mathcal{P}(X) \setminus \{\emptyset\}, X, \ni).$$
and the corresponding identity morphism (id, id) from P to P. If we assume that (id, id) is realized as a reduction link then there is a solution link for P, which will be a choice function for
$$\mathcal{P}(X) \setminus \{\emptyset\}.$$
As X was taken arbitrarily, we obtain **AC** in the same way as in Theorem 3.12 ∎

There are several set-theoretical statements which are referred to, in the literature, as *weak choice principles*. Weak choice principles are implied by the Axiom of Choice **AC**, but they are not equivalent to it, and are often regarded as "fragments" or "partial cases" of **AC**. To identify the precise amount of choice which is needed for a particular argument/result is a fruitful and current line of research (akin to Reverse Mathematics) within Set Theory. This may be seen in the standard reference [Howard & Rubin, 1998] and in the dozens of papers which have cited it over the last twenty years.

The Axiom of Countable Choice (usually denoted by $\mathbf{AC}_\omega$) is one of the most celebrated weak choice principles. It corresponds to the restriction of the Axiom of Choice to countable families of non-empty sets, that is, it asserts that "Every countable family of non-empty sets has a choice function", or if $\{X_n : n \in \mathbb{N}\}$ if a countable family of non-empty sets then there is a function
$$f : \{X_n : n \in \mathbb{N}\} \to \bigcup_{n \in \mathbb{N}} X_n$$
such that $f(X_n) \in X_n$ for all $n \in \mathbb{N}$.

In the following theorem, we prove an equivalent of countable choice $\mathbf{AC}_\omega$ in terms of Veloso problems – more specifically, in terms of Veloso problems whose instance sets are countable.

Theorem 3.15. *The Axiom of Countable Choice* $\mathbf{AC}_\omega$ *is equivalent to the following statement:*

"Every Veloso problem whose set of instances is countable has a solution function".

Proof:. ($\Rightarrow$) Let $P = (I, S, \sigma)$ be a Veloso problem, on which the instance set I is a countable set. Enumerate $I = \{z_n : n \in \mathbb{N}\}$. As P is viable, we know that
$$(\forall n \in \mathbb{N})(\exists s \in S)[z_n \, \sigma \, s]$$
and so for every $n \in \mathbb{N}$ the set
$$F_n = \{s \in S : z_n \, \sigma \, s\}$$
is a non-empty set. Applying $\mathbf{AC}_\omega$ to the countable family of non-empty sets given by
$$\mathcal{F} = \{F_n : n \in \mathbb{N}\},$$
we get a choice function $g : \mathcal{F} \to \bigcup \mathcal{F}$ with $g \subseteq \sigma$ – so that $z_n \, \sigma \, g(F_n)$ for all $n \in \mathbb{N}$. A solution function f for the problem $P = (I, S, \sigma)$ is now easily defined by putting, for every $z \in I$,
$$f(z) = g(F_m) \iff z = z_m.$$

($\Leftarrow$) Let $\mathcal{F} = \{X_n : n \in \mathbb{N}\}$ be a countable family of non-empty sets. We have to show that there is a choice function for such family, i.e. we have to exhibit a function $f : \mathcal{F} \to \bigcup \mathcal{F}$ such that $f(X_n) \in X_n$ for all $n \in \mathbb{N}$.

Consider the set X given by
$$X = \bigcup_{n \in \mathbb{N}} (\{n\} \times X_n)$$
and let $\sigma \subseteq \mathbb{N} \times X$ be the binary relation defined in the following way:
$$n \, \sigma \, x \iff \Pi_1(x) = n,$$
where Π_1 denotes the projection on the first coordinate. In other words, for all $(m, z) \in X$ we have that $n \, \sigma \, (m, z)$ if, and only if, $n = m$.

Consider the problem $(\mathbb{N}, X, \sigma)$. As $\mathcal{F}$ is supposed to be a family of non-empty sets, such problem is viable, and (as $\mathbb{N}$ is infinite) it is easy to check that it is also non-generic. So, $(\mathbb{N}, X, \sigma)$ is a Veloso problem. By hypothesis, there is a solution function of $(\mathbb{N}, X, \sigma)$, say $g : \mathbb{N} \to X$. Now we define
$$f : \mathcal{F} \to \bigcup_{n \in \mathbb{N}} X_n$$

by putting
$$f(F_n) = \Pi_2(g(n))$$
for all $n \in \mathbb{N}$. It should be clear that f is a choice function for $\mathcal{F}$. ∎

The reader may check that we could have proved versions of the three preceding theorems stated only in terms of viable problems. In fact, Veloso himself has observed that the equivalence between **AC** and the statement "Every viable problem has a solution" holds ([Veloso, 1984], page 35). However, as the problem $(\mathcal{P}(X) \setminus \{\emptyset\}, X, \ni)$ is a Veloso problem (i.e., viable and non-generic) for any infinite set X – and it is, basically, the problem which appears in several parts of the proofs -, we have preferred to point out that the class of Veloso problems is enough, in each case, to provide an equivalence with the corresponding choice principle.

The final result of this section will be stated in terms of viable problems only. First, we will introduce the notion of ω-chain of viable problems.

Definition 3.16 (ω-chain of viable problems). *Let $\{I_n : n \in \mathbb{N}\}$ be a family of non-empty sets and $\{\sigma_n : n \in \mathbb{N}\}$ be a family of binary relations such that, for all $n \in \mathbb{N}$,*
$$\sigma_n \subseteq I_n \times I_{n+1}.$$
*Let $S_n = I_{n+1}$ and $P_n = (I_n, S_n, \sigma_n)$ for all $n \in \mathbb{N}$. We say that $\langle P_n : n \in \mathbb{N}\rangle$ is an ω-**chain of viable problems** if all problems P_n are viable, i.e. for all $n \in \mathbb{N}$ one has*
$$\forall z \in I_n \ \exists s \in S_n \ [z \, \sigma_n \, s].$$

The preceding definition aims to capture the very common situation in Mathematics on which one has to solve a $\mathbb{N}$-sequence of chained problems, where the solution of the n-th problem is an instance of the $(n+1)$-problem. In such situation one wishes, of course, to produce a sequence of solutions.

Definition 3.17. *Let $\langle P_n : n \in \mathbb{N}\rangle$ be an ω-chain of viable problems, where $P_n = (I_n, S_n, \sigma_n)$ and $S_n = I_{n+1}$ for all $n \in \mathbb{N}$. A **solution sequence** for $\langle P_n : n \in \mathbb{N}\rangle$ is a sequence $\langle z_n : n \in \mathbb{N}\rangle$ such that $z_n \in I_n$ and $z_n \sigma_n z_{n+1}$ for all $n \in \mathbb{N}$.*

We present below a mathematical example where the above concepts play an important role.

Example 3.18. *The usual proof of the Baire Category Theorem for complete metric spaces given in the textbooks may be encoded by a ω-chain of viable problems.*

Let us see why. Let $\{U_n : n \in \mathbb{N}\}$ be a countable family of dense open sets of a complete metric space M and let O be an arbitrary non-empty open set (as in Example 3.7). We have to show that the intersection of all the U_n's meets O. We let I_n be the family of all non-empty open sets whose diameter is less than $\frac{1}{n+1}$ and whose closures are included in $U_n \cap O$, and let σ_n be the reverse inclusion. It is easy to check that all of the problems $\langle I_n, S_n, \sigma_n \rangle$ (where $S_n = I_{n+1}$) are viable problems. If $\langle V_n : n \in \mathbb{N} \rangle$ is a solution sequence for such ω-chain of problems, then $\left(\bigcap_{n \in \mathbb{N}} U_n \right) \cap O$ is ensured to be non-empty, since $\{\overline{V_n} : n \in \mathbb{N}\}$ is a decreasing sequence of non-empty closed sets whose diameters converge to zero, so it has non-empty intersection in the complete metric space M – and, by the definition of the I_n's, any point of such intersection testifies that $\left(\bigcap_{n \in \mathbb{N}} U_n \right) \cap O \neq \emptyset$.

What is usually not said in the textbooks about the preceding example is that a solution sequence for the ω-chain of viable problems is given by the *Principle of Dependent Choices*, denoted by **DC**, which is another celebrated weak choice principle[1]. The Principle of Dependent Choices states:

"If δ is a binary relation on a non-empty set A (i.e. $\delta \subseteq A \times A$) satisfying

$$(\forall x \in A)(\exists y \in A)[x \, \delta \, y],$$

then there is a sequence $\langle x_n : n \in \mathbb{N} \rangle$ of elements of A such that $x_n \, \delta \, x_{n+1}$ for all $n \in \mathbb{N}$".

The principle **DC** is regarded as the precise amount of choice needed to make a countable number of consecutive arbitrary choices. It is well-known that the Axiom of Choice is stronger than the Principle of Dependent Choice which is stronger than the Axiom of Countable Choice, i.e.

$$\mathbf{AC} \Rightarrow \mathbf{DC} \Rightarrow \mathbf{AC}_\omega,$$

and that none of those implications is reversible. More information on these and many other weak choice principles may be found in [Howard & Rubin, 1998].

In what follows, we present equivalents of **DC** in terms of ω-chains of viable problems.

Theorem 3.19. *The following statements are equivalent:*

*(i) The Principle of Dependent Choices, **DC**;*

---

[1] In fact, the Principle of Dependent Choices is *equivalent* to the Baire Theorem for complete metric spaces, as shown in [Blair, 1977].

(ii) For any ω-chain of viable problems $\langle P_n : n \in \mathbb{N} \rangle$ (where, for all $n \in \mathbb{N}$, $P_n = (I_n, S_n, \sigma_n)$ and $S_n = I_{n+1}$) and for every instance $z \in I_0$, there is a solution sequence $\langle z_n : n \in \mathbb{N} \rangle$ for this ω-chain with $z_0 = z$; and

(iii) Every ω-chain of viable problems has a solution sequence.

Proof:.(i) $\Rightarrow$ (ii): Let $\langle P_n : n \in \mathbb{N} \rangle$ be as in the statement. We define a set A whose elements are, precisely, all $(k+1)$-tuples

$$(z_0, z_1, \ldots, z_k),$$

where k ranges over all natural numbers, $z_0 = z$, $z_i \in I_i$ for all $0 \leqslant i \leqslant k$ and also $z_i \sigma_i z_{i+1}$ for all $0 \leqslant i \leqslant k - 1$ if $k > 0$. We define, over the tuples of A, a relation δ such that

$$(w_0, w_1, \ldots, w_k) \, \delta \, (t_0, t_1, \ldots, t_j)$$

if $k < j$ and $w_i = t_i$ for all $0 \leqslant i \leqslant k$ – that is, δ is the usual strict prefix order over the tuples.

As all of the problems of the ω-chain are viable by definition, it is easy to check that the set A and the relation δ satisfy the requirements one needs to apply the Principle of Dependent Choices. So, by **DC**, there is a sequence $\langle s_n : n \in \mathbb{N} \rangle$ of tuples such that $s_n \, \delta \, s_{n+1}$ for all $n \in \mathbb{N}$. If we identify each tuple with the corresponding finite sequence, we get a compatible family of functions (in fact, an increasing chain of compatible finite functions), and so the union of such family is a function. It is easy to check that the obtained sequence $z = \bigcup_{n \in \mathbb{N}} s_n$ is a solution sequence for the ω-chain with $z_0 = z$, as desired.

(ii) $\Rightarrow$ (iii): Obvious.

(iii) $\Rightarrow$ (i): If A and δ are under the hypothesis of **DC**, then A is a non-empty set and (A, A, δ) is a viable problem. Then we just have to consider the ω-chain $\langle P_n : n \in \mathbb{N} \rangle$ where $I_n = S_n = A$ and $\sigma_n = \delta$ for all $n \in \mathbb{N}$. By (iii) we may assume that there is a solution sequence for this ω-chain, say $\langle z_n : n \in \mathbb{N} \rangle$. So, this sequence satisfies $z_n \, \delta \, z_{n+1}$ for all $n \in \mathbb{N}$. As A and δ were taken arbitrarily, we have just established **DC** – and the proof is then finished. ∎

We have shown that the Axiom of Countable Choice, the Principle of Dependent Choices and the fully-fledged Axiom of Choice correspond to natural (sub-)classes of problems, as described by Kolmogorov and Veloso.

4 Tight Coupled Problems

In this short section we discuss how the notion of a "tight coupled reduction link" between problems can be seen as a previously known variant of the Dialectica construction we have discussed so far.

The way in which the notion of *reduction link* of P to P' was defined (item (ii) of Definition 3.10) seems to suggest that there is no tight connection or "coupling", in general, between problems P and P' – as pointed out by Veloso in the page 28 of [Veloso, 1984]. Veloso argues that a reduction link of P to P' is, in most cases, *uncoupled* in the following sense: after applying the translation function f on an instance z of P, we may forget completely about P and only care about solving the instance $f(z)$ of P' – and, after that, any solution $t \in S'$ of the instance $f(z)$ of P' will generate a solution of the instance z of P by applying the recovery function F. However, in certain situations (see example below), it is interesting to allow the use of some additional information during the recovery process.

For this case, Veloso defines *tightly coupled links* from P to P' in the following way: the translation function is still some $f : I \to I'$ but the recovery function is a function $F \colon I \times S' \to S$ so that the information on the original instance z of P may be used at the time of recovery – that is, the pair (f, F) needs to satisfy the following condition: for any $z \in I$ and any $t \in S'$,

$$f(z)\,\sigma'\,t \longrightarrow z\,\sigma\,F(z,t).$$

This requirement on the tightly coupled links corresponds, precisely, to the morphisms of the dual of Dial(**Sets**) (the simplest case of the "original" Dialectica category, inspired by Gödel's Dialectica Interpretation and introduced by the first author in [Paiva, 1989a]).

Example 4.1 (Reduction with tight coupled link). *Using tight coupled links, one can formally prove that the problem of finding a normal line of a surface through a certain point is not more complicated than the problem of finding orthogonal planes to a certain plane.*

In this example, we use *surface* for a two-dimensional differential manifold $S \subseteq \mathbb{R}^3$, and therefore given a point $x \in S$ there is a tangent plane $T_x(S)$ centred at x. Recall that if x is a point of a plane $\pi \subseteq \mathbb{R}^3$ then a line l through x is the *normal line* of the plane π (through x) if l is perpendicular to all lines of π which go through x, and if x is a point of a surface S then the normal line of S through x is the normal line of the tangent plane $T_x(S)$ through x. Two intersecting planes π and ρ are said to be *orthogonal* if for every point of the intersection line the normal lines of π are contained in ρ and vice-versa. Notice that if l is the normal line of a plane π through x then l is included in every plane ρ which is orthogonal to π and passes through x – in other words, all planes which are orthogonal to a given plane π and passes through a given point share the normal line of π through this very same point.

Let $\mathcal{L}$ be the family of all lines of the 3-dimensional Euclidean space $\mathbb{R}^3$ and $\mathcal{P}$ the family of all planes of $\mathbb{R}^3$. For a given surface S, the problem of finding the normal lines through each point of the surface is $(S, \mathcal{L}, \sigma)$, where $x \sigma l$ means "l is the normal line of S through x" for every point $x \in S$ and every line $l \in \mathcal{L}$, and the problem of finding orthogonal planes is $(\mathcal{P}, \mathcal{P}, \xi)$, where $\pi \xi \rho$ means "π and ρ are orthogonal planes" for all planes π and ρ in $\mathbb{R}^3$.

A tight coupled link from $(S, \mathcal{L}, \sigma)$ to $(\mathcal{P}, \mathcal{P}, \xi)$ is given by the pair (f, F), where $f : S \to \mathcal{P}$ is defined by putting $f(x) = T_x(S)$ for all $x \in S$ and $F \colon S \times \mathcal{P} \to \mathcal{L}$ is given by $F(x, \rho) = \varphi(T_x(S), \rho, x)$, where $\varphi : \mathcal{P} \times \mathcal{P} \times \mathbb{R}^3 \to \mathcal{L}$ is defined by putting $\varphi(\pi, \rho, x) =$ the line l contained in the plane $t(x, \rho)$ (where $t(x, \rho)$ is ρ itself if $x \in \rho$ or is the unique plane parallel to ρ passing through x otherwise) which is perpendicular to the intersecting line of π and $t(x, \rho)$ through x, if π and ρ are orthogonal planes; and any previously fixed line otherwise. In view of the fact remarked at the end of the previous paragraph, it is easy to see that $F(x, \rho)$ is the normal line of $T_x(S)$ through x (and thus, of S) whenever ρ is a plane orthogonal to $f(x) = T_x(S)$. Notice that the described reduction link formalizes the following mental procedure: "if I know how to produce an orthogonal plane for any given plane, then I know how to produce the normal line of a surface S at a point x: we take any plane ρ that is orthogonal to $T_x(S)$, translate it to x – via parallel translation – and then consider the perpendicular line (through x and contained in the translated plane) of the intersection line of $T_x(S)$ and the translated plane".

This shows one case where the information on the original problem instance (the original point x) is required for the recovery function F of the tight coupled link (f, F) that reduces the original problem of "finding normal lines to a given point in the surface" to the new problem of "finding orthogonal planes". This problem is particularly nice, as it seems to connect to approaches to automatic differentiation under development using categorical machinery in [Elliott, 2018]. More research work is required here.

5 Conclusions and Further Work

We have shown that the work of Kolmogorov can be regarded as a bridge between the abstract problems Veloso discussed in his Theory of Mathematical Problems and the complexity problems Blass discussed in [Blass, 1995]. This bridge can be seen by means of the categorical Dialectica constructions GC and DC introduced in [Paiva, 1991].

Veloso's *Critical Retrospect* in [Veloso, 1984] already observed that his approach on viable Kolgomorov problems relies heavily on the Axiom of Choice and asks whether such approach (together with some suggested techniques on decomposition of problems) was "tainted" by **AC** – since the mathematical entities whose existence depend on the Axiom of Choice (and statements about those entities) are the usual examples of non-constructive notions in Mathematics, and he aimed his approach to incorporate constructivity at some level. In this paper we have proceed with this line of research and we introduced a restricted class of Kolgomorov problems, which we call *Veloso Problems*, and we have shown that if we restrict ourselves to this class and assume that the presented machinery holds in full generality then we get, again, equivalences of the Axiom of Choice. We have also shown that some related/restricted notions on problems are intrinsically associated to weak choice principles such as the Axiom of Countable Choice and the Principle of Dependent Choices. To determine precisely the deductive strength of assertions relating to Kolgomorov and Veloso problems (positioning them in the hierarchy of weak choice principles) seems to deserve further research, and the same could be said about the following question:

Question 5.1 (implicit in [Veloso, 1984]). *How to deal with the above mentioned non-constructive aspects of Kolgomorov-Veloso theory of problems within constructive environments?*

The work in this paper shows that some answers to this question are obtained by focusing on *relations*, instead of *functions*. In fact, equivalences with choice principles arise from assuming either the existence of solution functions (Theorems 3.12, 3.14 and 3.15) or of solution sequences (Theorem 3.19).

Veloso (op.cit.) presents some arguments against the use of relations instead of functions in the representation of solutions of problems. He argues that if the solution of a problem can be represented as a relation, then the problem condition itself, σ, would be a solution of the problem, and so nothing more would need to be done; to know the specification of the problem would solve it automatically. Second, he argues that, if one wants to assume that some instance $z \in I$ could be solved by more than one element of the set of possible solutions S then, even in this case, it is possible to work with a representation using functions, but in this case one should work with the so-called *multifunctions*, or multivalued functions. If I is the set of instances and S is the set of possible solutions of a viable problem P, a multifunction solution f would be, formally, a function with domain I and codomain $\mathcal{P}(S)$ (i.e, the set of all subsets of S) such that, for every $z \in I$, $f(z)$ is the *subset* of S given by $\{s \in S : z \sigma s\}$ – that is, the use of multivalued solution functions consists in associating each instance $z \in I$ to the set of *all* of its particular solutions.

Replying to the first argument, in practical applications, to know the *definition* of the condition problem σ does not give a solution automatically (even less the set of all solutions) for an instance $z \in I$. In simple/naive cases, the more realistic approach would be, probably, to proceed with some decision problem/search problem considering, for each $z \in I$, the set $\{z\} \times S$ as a domain and with $(\{z\} \times S) \cap \sigma$ as the set of yes-instances for this problem. So, to know the problem condition does not solve the problem. For instance, one may consider the viable problem $(\{\zeta\}, \mathbb{C}, \sigma)$, on which ζ is Riemann's zeta function and σ is the relation given by $\zeta \sigma c \iff \zeta(c) = 0$ – that is, σ is the restriction to $\{\zeta\}$ of the very general problem of finding zeros of analytic functions. Despite σ being perfectly (and easily) defined, we still do not know (after more than 150 years, see [Bombieri, 2008]) whether there are complex numbers (apart from the even negative numbers) with real part distinct from $\frac{1}{2}$ which solve the problem.

For the second objection against relations as solutions, we would like to mention two points. The first is that, if we decide to work with the powerset $\mathcal{P}(S)$ instead of S, then we are dramatically increasing the cardinality of the sets we are dealing with. For problems where any instance has only a *finite* set of solutions there is perhaps no major issue but we want to work with theories where problems have instances with possible infinite solutions. (Actually the set-theory applications do insist on infinite sets.) As some features of the theory may depend on the cardinalities of the constituents of the triples of the corresponding categories (as in the definitions of $\mathcal{PV}$ and $\mathcal{PV}_X$ for a given infinite set X, see Definitions 3.1 and 3.2), it would not be desirable such a dramatic increase in cardinality. The second argument is that if we assume $f(z)$ to be the set of all solutions (for f a multivalued solution function and z some instance of the problem) then we have all solutions for z "locked inside a box"; these solutions may become individually inaccessible and this precludes a qualitative analysis of them. We prefer to have the possibility of comparing distinct solutions for any fixed instance of the problem we are solving. Thus we insist that relations are a better modelling tool than functions, one that allows us to move on to functions, when and if we feel **AC** is adequate.

A second main conclusion for us is that the notions of reduction between problems, considered by Kolmogorov, Veloso and Blass, seen to be well modelled by the categorical morphisms in either $\text{Dial}_2(\textbf{Sets})$, its dual or the original dialectica construction. A skeptical reader may complain that the categorical language used is not buying us much. We beg to differ: the possibility of relating formally these, to begin with, quite 'woolly' notions of problems and solutions, seems a serious step forward in the hard task of detecting unwarranted foundational assumptions that tend to 'sneak' into mathematics. We still need to investigate whether the traditional tools of category theory, e.g. products, coproducts, exponentials, (co-)limits, etc. can be leveraged to our advantage. And, apart of such traditional tools, we are also interested in the investigation of the possible interactions between the Dialectica categorical modelling of problems and the so-called *lenses* ([Hedges, 2016], [Spivak, 2019]). Lenses are constructions used in situations where some structure is converted to different forms – through actions and observations between environments and agents – in such a way that all changes made can be reflected as updates to the original structure. Such constructions have attracted the attention of several researchers over the past ten years (see also [nLab authors, 2020], and references therein).

A possible avenue for further work from this point on would be to connect the categorical semantics meaning of $\text{Dial}_2(\textbf{Sets})$ (logically speaking $\text{Dial}_2(\textbf{Sets})$ models Linear Logic, together with Intuitionistic Propositional Logic) to, yet to be conceived, models of Ecumenical Propositional Logic. Ecumenical Propositional Logic [Pimentel et al., 2019] is Prawitz' recent suggestion of how to consider under the same umbrella both intuitionsitic and classical principles, as used by mathematicians. Since both Kolmogorov and Veloso mentioned their intentions of being understood by both classical and intuitionistic mathematicians, it would be extremely nice if the categorical models discussed here could help with modelling ecumenical logic. However new insights will be required to deal with the traditional issues of modelling categorically classical logic.

References

Blair, C. E. (1977), 'The Baire category theorem implies the principle of dependent choices.', *Bull. Acad. Pol. Sci., Sér. Sci. Math. Astron. Phys.* **25**, 933–934. ISSN 0001-4117.

Blass, A. (1995), 'Questions and answers—a category arising in linear logic, complexity theory, and set theory', *Advances in linear logic* **222**, 61–81.

Bombieri, E. (2008), The Riemann Hypothesis: Official Problem Description, Clay Mathematics Institute, 2008, pp.1–5.

Elliott, C. (2018), 'The simple essence of automatic differentiation', *Proceedings of the ACM on Programming Languages* **2**(ICFP), 1–29.

Girard, J.-Y. (1987), 'Linear logic', *Theoretical computer science* **50**(1), 1–101.

Hedges, J. (2016), 'Lenses for philosophers', Blog post. Available at. URL https://julesh.com/2018/08/16/lenses-for-philosophers.

Howard, P. & Rubin, J. E. (1998), *Consequences of the axiom of choice.*, Vol. 59, Providence, RI: American Mathematical Society.

Impagliazzo, R. & Levin, L. A. (1990), No better ways to generate hard np instances than picking uniformly at random, *in* 'Proceedings [1990] 31st Annual Symposium on Foundations of Computer Science', pp. 812–821 vol.2.

Kolmogorov, A. (1991), 'On the interpretation of intuitionistic logic', *Selected Works of A.N. Kolmogorov, Volume 1, Mathematics and Mechanics, ed. V. M. Tikhomirov*.

Martin-Löf, P. (2006), '100 years of Zermelo's axiom of choice: what was the problem with it?', *The Computer Journal* **49**(3), 345–350. DOI 10.1093/comjnl/bxh162. ISSN 0010-4620. URL https://doi.org/10.1093/comjnl/bxh162.

Moore, G. H. (1982), 'Zermelo's axiom of choice. Its origins, development and influence.', Studies in the History of Mathematics and Physical Sciences, 8. New York - Heidelberg - Berlin: Springer-Verlag. XIV, 410 pp.

Moore, J., Hrušák, M. & Džamonja, M. (2004), 'Parametrized ◊ principles', *Transactions of the American Mathematical Society* **356**(6), 2281–2306.

nLab authors (2020), 'lens (in computer science)', Blog post, https://ncatlab.org/nlab/show/lens+(in+computer+science). Revision 11.

Paiva, V. C. d. (1989a), The dialectica categories, *in* 'Categories in Computer Science and Logic, Proceedings of a Summer Research Conference, held June 14–20, 1987 (eds J. Gray and A Scedrov)', American Mathematical Society, pp. 23–47.

Paiva, V. d. (1989b), A dialectica-like model of linear logic, *in* D. Pitt, D. Rydeheard, P. Dybjer, A. Pitts & A. Poigne, eds, 'Category Theory and Computer Science', Springer, pp. 341–356.

Paiva, V. d. (1991), *The Dialectica Categories*, Computer Laboratory, University of Cambridge.

Pimentel, E., Pereira, L. C. & de Paiva, V. (2019), 'An ecumenical notion of entailment', *Synthese* pp. 1–23. https://doi.org/10.1007/s11229-019-02226-5.

Sanz, W. (2020), Kolgomorov and the general theory of problems, To appear in a Volume dedicated to Prof. Dr. P. Schröeder-Heister (Tübingen).

Silva, S. G. d. (2020), 'The Axiom of Choice and the Partition Principle from Dialectica Categories', *Logic Journal of the IGPL*. DOI 10.1093/jigpal/jzaa023. ISSN 1367-0751. URL https://doi.org/10.1093/jigpal/jzaa023.

Silva, S. G. d. & Paiva, V. C. V. d. (2017), 'Dialectica categories, cardinalities of the continuum and combinatorics of ideals', *Logic Journal of the IGPL* **25**(4), 585–603. DOI 10.1093/jigpal/jzx016. ISSN 1367-0751. URL https://doi.org/10.1093/jigpal/jzx016.

Spivak, D. I. (2019), 'Lenses: applications and generalizations', Talk at the Special Session on Applied Category Theory, AMS Western Sectional Meeting, Riverside.

Veloso, P. (1984), 'Aspectos de uma teoria geral de problemas', *Cadernos de História e Filosofia da Ciência* **7**, 21–42.

Venkatesan, R. & Levin, L. A. (1988), Random instances of a graph coloring problem are hard, *in* 'STOC '88: Proceedings of the twentieth annual ACM symposium on Theory of computing', pp. 217–222.

Vojtáš, P. (1993), Generalized Galois-Tukey-connections between explicit relations on classical objects of real analysis., *in* 'Set theory of the reals. Proceedings of a winter institute on set theory of the reals held at Bar-Ilan University, Ramat-Gan (Israel), January 1991', Providence, RI: American Mathematical Society (Distrib.); Ramat-Gan: Bar-Ilan University, pp. 619–643.

On the construction of explosive relation algebras

Carlos G. Lopez Pombo[*]   Marcelo F. Frias[†]

Thomas S.E. Maibaum[‡]

[*] Universidad de Buenos Aires, Facultad de Ciencias Exactas y Naturales, Departamento de Computación and Instituto de Investigación en Ciencias de la Computación, CONICET–Universidad de Buenos Aires (ICC).
clpombo@dc.uba.ar

[†] Department of Software Engineering, Buenos Aires Institute of Technology (ITBA) and Consejo Nacional de Investigaciones Científicas y Técnicas (CONICET).
mfrias@itba.edu.ar

[‡] Emeritus Professor, Department of Computing and Software, McMaster University
tom@maibaum.org

Abstract

Fork algebras are an extension of relation algebras obtained by extending the set of logical symbols with a binary operator called *fork*. This class of algebras was introduced by Haeberer and Veloso in the early 90's aiming at enriching relation algebra, an already successful language for program specification, with the capability of expressing some form of parallel computation.

The further study of this class of algebras led to many meaningful results linked to interesting properties of relation algebras such as representability and finite axiomatizability, among others. Also in the 90's, Veloso introduced a subclass of relation algebras that are expansible to fork algebras, admitting a large number of non-isomorphic expansions, referred to as *explosive relation algebras*.

In this work we discuss some general techniques for constructing algebras of this type.

1 introduction

A *relation algebra* is an algebraic structure formed by three relational constants understood as the empty, universal and identity relations, typically represented by the symbols "0", "1" and "1'", respectively; two unary operators playing the role of complement of a relation (with respect to the universal relation) and transposition (or converse) of a relation, typically represented by the symbols "‾" and "⌣", respectively; and binary operators for product, co-product and relative product (also commonly referred to as composition) of relations, typically represented by symbols "·", "+" and ";", respectively.

In [Tarski, 1941], Tarski noted that the calculus of (binary) relations "[...] has had a strange and rather capricious line of historical development.", but leaving historical discussions aside, it is fair to consider that it was proposed, still not properly presented and formalised, by him in the previously cited work. There, Tarski committed himself to the development of the *calculus of relations* (CR). In the first place, Tarski introduces the *elementary theory of binary relations* (ETBR), as a logical formalisation of the algebras of binary relations in a kind of definitional extension of first-order logic (see [Tarski, 1941, Axs. 1.–12.]); then, the calculus of relations can be obtained from the elementary theory of binary relations by restricting the language to sentences without individual variables (see [Tarski, 1941, Thms. 1–15][1]). While the modern equational formulation of the calculus of relations is not explicit in Tarski's work, it is hinted at after the proof that [Tarski, 1941, Thm. 32] (originally proved by Shröder in [Schöder, 1895, 1), pp. 150–153]) follows from Axioms 1 to 15, by proving a general metalogical result stating that any sentence of the calculus of relations can be transformed into an equivalent sentence of the form $R = S$.

At the end Tarski states five questions related to the calculus of relations, its class of models and the *algebras of binary relations*[2], with three of them of particular interest to this work:

- Is every model of the calculus of relations isomorphic to an algebra of binary relations?

---

[1]Theorems I–VII, originally due to Huntington [Huntington, 1904, §1], provide a characterisation of the meaning of the absolute constants (i.e. the Boolean fragment of the logic), and Thms. 8–15 express the fundamental properties of the relative ones.

[2]Tarski refers to these structures as "a class of binary relations which contains 1, 0, 1', 0' and is closed under all the operations considered in the calculus [of relations]", without providing a proper name for such an intended class of models.

- Is it true that every formula that is valid in every algebra of binary relations is provable in the calculus of relations?

- Is it true that every formula of the elementary theory of binary relations can be transformed into an equivalent formula of the calculus of relations?

It was Lyndon in [Lyndon, 1950] who gave a negative answer to the first two questions by exhibiting a finite, non-simple and non-trivial algebra of relations that is not representable as an algebra of binary relations. After that, it was Monk in [Monk, 1964] who proved that the class of the algebras of binary relations cannot be finitely axiomatised. The third question was answered negatively by Tarski (hinted at in op. cit., pp. 88–89) by making more precise the existence of *uncondensable*[3] formulae proved by Korselt (and published in [Löwenheim, 1915, Thm. 1]). Tarski's detailed proof of the equipolence of the calculus of relations and the three variable fragment of the dyadic first-order predicate logic appeared for the first time in a book manuscript [Tarski, 1943–1945], and later was published in [Tarski & Givant, 1987, §3.9]. By mid-50's Tarski had already adopted a completely equational presentation for relation algebras (see [Tarski, 1955, pp. 60] and [Tarski, 1956, §3]).

The *fork algebras* (FA) were introduced by Armando Haeberer and Paulo A. S. Veloso in [Haeberer & Veloso, 1991] as extensions of relation algebras, obtained by adding a new operator called *fork* (typically represented as "∇"). They arose in the search for a formalism suitable for software specification and verification. In [Frias, 2002, Chap. 3, pp. 20] Frias gave a detailed discussion of the evolution of fork algebras, focussing the reader's attention to the concepts behind such specific direction. The interpretation of ∇ is defined by the following first-order formula: given relations R and S,

$$R \nabla S = \{ \langle x, y \rangle \mid (\exists y_1, y_2 \in U)(\langle x, y_1 \rangle \in R \land \langle x, y_2 \rangle \in S \land y = y_1 \star y_2) \}$$

where $\star : U^2 \to U$ is an injective function acting as an encoding of pairs of elements U, the set over which the relations are defined.

The class of fork algebras have some particularly attractive features:

- every fork algebra is isomorphic to an algebra whose domain is a set of binary relations (Frias et al. in [Frias et al., 1997a] and, independently, Gyuris in [Gyuris, 1997]),

---

[3]To *condense* a formula, as used by Löwenheim following Schröder's terminology [Schöder, 1895, pp. 550], is to transform a formula of ETBR into another one in which no quantifiers or individual variables appear.

- it has a finite equational calculus (Frias et al. in [Frias et al., 1997b]),
- it has expressive power capable of providing an interpretation language for many logics. Given a logic $\mathcal{L}$, an interpretation is a relational algebraization of $\mathcal{L}$. This is done by resorting to a semantics-preserving mapping $T_\mathcal{L} : Formulas_\mathcal{L} \to RelDes(X)$ for some set of relational variables X, translating $\mathcal{L}$-formulas to relational terms. Some known interpretability results are: first-order predicate logic (FOL) in fork algebra [Frias, 2002], PDL in fork algebra [Frias & Orlowska, 1998], first-order dynamic logic (FODL) in fork algebra [Frias et al., 2002], LTL, TL [Frias & Lopez Pombo, 2003] and their first-order versions in fork algebra [Frias & Lopez Pombo, 2006], and propositional dynamic linear temporal logic (DLTL) in fork algebra [Frias et al., 2005], among others.

The existence of a representability theorem and a finetely axiomatizable complete calculus for the class of fork algebras motivates the study of the subclass of relation algebras obtained by taking the relational reduct of the fork algebras, resulting also in a subclass of the proper relation algebras.

In this paper we present some techniques for constructing relation algebras admitting a large amount of non-isomorphic expansions to a fork algebra. This class of algebras was introduced by Veloso in [Veloso, 1996a,b] and because they possess this property, they are called explosive. The definitions and results in this work are strongly inspired by the reports mentioned above and joint technical discussion with Paulo A.S. Veloso.

2 Preliminaries

In this section we fix notation and present definitions and results used in the rest of the paper. In general, we adopt the algebraic notation used in [Burris & Sankappanavar, 1981], but resorting to the symbols proposed by Tarski in [Tarski, 1941] and used in [Jónsson & Tarski, 1951, 1952]. In general, axioms, deduction rules and proofs will follow the notation used in [Enderton, 1972].

As we mentioned before, Tarski's development of relation algebras started by introducing the elementary theory of binary relations as a definitional extension of first-order logic with the relational operators proposed by Schröder in [Schöder, 1895] and then moved on to the calculus of relations by restricting the language to formulae stating properties of relational terms exclusively. The following definitions introduce the *calculus of relations*.

Definition 1 (Formulae of the calculus of relations). *Let $\mathcal{R}$ be a set of relation variables, then the set of relation designations is the smallest set $RelDes(\mathcal{R})$ such that:*

- $\mathcal{R} \cup \{1, 0, 1'\} \subseteq RelDes(\mathcal{R})$,
- *If $r, s \in RelDes(\mathcal{R})$, then $\{r+s, r\cdot s, \overline{r}, r;s, \breve{r}\} \subseteq RelDes(\mathcal{R})$.*

Then the set of formulae of CR is the smallest set $CRForm(\mathcal{R})$ such that:

- *If $r, s \in RelDes(\mathcal{R})$, then $r = s \in CRForm(\mathcal{R})$,*
- *If $f, g \in CRForm(\mathcal{R})$, then $\{\neg f, f \vee g\} \subseteq CRForm(\mathcal{R})$.*

The remaining propositional connectives can be defined as usual in terms of negation ($\neg$) and disjunction ($\vee$).

Definition 2 (The calculus of relations, [Tarski, 1941], pp. 76–77). *Let $\mathcal{R}$ be a set of relation variables, then CR is defined for the formulae in $CRForm(\mathcal{R})$ by[4]:*

- *the axioms for the Boolean operators, and*
- *the following axioms for the relational operators:*

$$(r = s \wedge r = t) \Longrightarrow s = t ,$$
$$r = s \Longrightarrow (r+t = s+t \wedge r\cdot t = s\cdot t) ,$$
$$r+s = s+r \wedge r\cdot s = s\cdot r ,$$
$$r+(s\cdot t) = (r+s)\cdot(r+t) \wedge r\cdot(s+t) = (r\cdot s)+(r\cdot t) ,$$
$$r+0 = r \wedge r\cdot 1 = r ,$$
$$r+\overline{r} = 1 \wedge r\cdot\overline{r} = 0 ,$$
$$\overline{1} = 0 ,$$
$$\breve{\breve{r}} = r ,$$
$$(r;s)\breve{} = \breve{s};\breve{r} ,$$
$$r;(s;t) = (r;s);t ,$$
$$r;1' = r ,$$
$$r;1 = 1 \vee 1;\overline{r} = 1 ,$$
$$(r;s)\cdot\breve{t} = 0 \Longrightarrow (s;t)\cdot\breve{r} = 0 .$$

---

[4]Tarski presented CR incorporating axioms in order to characterize the relative addition ($\dag$) and the diversity ($0'$) operators. Tarski's axioms for these two operators are:
$$r \dag s = \overline{\overline{r};\overline{s}}$$
$$0' = \overline{1'}$$

While Tarski did not commit to any set of inference rules for structuring deduction, one can assume any appropriate set for the boolean operators ("¬" and "∨"), and the equality ("=").

As we mentioned in the introduction, in Tarski's presentation of the representation problem [Tarski, 1941, pp. 88] there is only an implicit definition of the intended models of the calculus of relations. For the purpose of the present work, we adopt the formal definition given by Jónsson and Tarski in [Jónsson & Tarski, 1952].

Definition 3 (Proper relation algebras, [Jónsson & Tarski, 1952], Def. 4.23).
A proper relation algebra is an algebraic structure $\langle A, \cup, \cap, ^-, \emptyset, E, \circ, \smile, Id \rangle$ in which A is a set of binary relations on a set U, $\cup$, $\cap$ and $\circ$ are binary operations, $^-$ and $\smile$ are unary operations and $\emptyset$, E and Id are distinguished elements of A satisfying:

- *A is closed under $\cup$ (i.e. set union),*
- *A is closed under $\cap$ (i.e. set intersection),*
- *A is closed under $^-$ (i.e. set complement with respect to E),*
- *$\emptyset \in A$ is the empty relation on the set U,*
- *$E \in A$ and $\bigcup_{r \in A} r \subseteq E$,*
- *A is closed under $\circ$, defined as follows*

$$x \circ y = \{\, \langle a, b \rangle \in U \times U \mid (\exists c)(\langle a, c \rangle \in x \land \langle c, b \rangle \in y)\,\} ,$$

- *A is closed under $\smile$, defined as follows*

$$\breve{x} = \{\, \langle a, b \rangle \in U \times U \mid \langle b, a \rangle \in x \,\} ,$$

- *$Id \in A$ is the identity relation on the set U.*

The class of proper fork algebras will be denoted as PRA.

In [Jónsson & Tarski, 1952], Jónsson and Tarski proved that the axiom $r;1 = 1 \lor 1;\bar{r} = 1$ forces the models to be simple, a property that is not necessarily satisfied by the proper relation algebras, so that is why their equational presentation of CR does not include it.

Definition 4 (Equational formulae of the relational calculus).
Let $\mathcal{R}$ be a set of relation variables, then the set of formulas of CR is the set $\{\, r = s \mid r, s \in CRForm(\mathcal{R}) \,\}$.

Definition 5 (The equational calculus of relations, [Jónsson & Tarski, 1952], Def. 4.1). *Let $\mathcal{R}$ be a set of relation variables, then* CR *is defined for the formulae in $CRForm(\mathcal{R})$ by:*

- *the axioms for the Boolean operators, and*
- *the following axioms for the relational operators[5]: for all $r, s, t \in A$*

$$r;(s;t) = (r;s);t$$
$$(r+s);t = (r;t)+(s;t)$$
$$(r+s)^{\vee} = \breve{r}+\breve{s}$$
$$\breve{\breve{r}} = r$$
$$r;1' = r$$
$$(r;s)^{\vee} = \breve{s};\breve{r}$$
$$(r;s)\cdot t \leq (r\cdot(t;\breve{s}));(s\cdot(\breve{r};t))$$

As in the case of Def. 2, one can adopt any appropriate set of inference rules for the equality to structure proofs in this calculus.

Definition 6 (Relation algebras). *The class of relation algebras (*RA *for short) is the class of algebraic structures $\langle A, +, \cdot, ^-, 0, 1, ;, ^{\vee}, 1'\rangle$ satisfying the axioms in* CR.

Tarski's first question about the relation between the class of models of the calculus of relations, RA, and the class of concrete algebras of binary relations, PRA, is of utmost importance in the context of computer science. Let us formulate it in more formal terms.

Definition 7. *Given an algebra $\mathcal{A}$ and a class of algebras* K, $\mathcal{A}$ *is representable in* K *if there exists $\mathcal{B} \in$* K *such that $\mathcal{A}$ is isomorphic to $\mathcal{B}$. This notion generalises as follows: a class of algebras* K_1 *is representable in a class of algebras* K_2 *if every member of* K_1 *is representable in* K_2.

Then, Tarski's first question explores whether the class RA is representable in the class PRA. Lyndon's negative answer is devastating in practice. Consider the classical problem of formal verification in software engineering, stated as follows: given a specification of a software artefact written as a set of formulae $\Gamma \subseteq CRForm(\mathcal{R})$ (see Def. 4) and a desired property of such an artefact, formalised as a formula $\alpha \in CRForm(\mathcal{R})$, does $\Gamma \vdash^{CR} \alpha$? Then, we are interested in either constructing a proof for the previous judgement, or finding $\mathcal{A}$ such that $\mathcal{A} \models^{CR} \Gamma$ and $\mathcal{A} \not\models^{CR} \alpha$.

---

[5]The last axiom, known as the Dedekind formula, is equivalent to $r;s\cdot t = 0$ iff $t;\breve{s}\cdot r = 0$ iff $\breve{r};t\cdot s = 0$ known as *cycle rule*.

In general, we would like $\mathcal{A}$ to be a concrete model (i.e. $\mathcal{A} \in$ PRA) as it provides a natural interpretation of relations in set theoretical terms. Lyndon's answer can be summarised as follows: It might happen that $\Gamma \not\models^{CR} \alpha$ and for every $\mathcal{A} \in$ RA such that $\mathcal{A} \models^{CR} \Gamma$ and $\mathcal{A} \not\models^{CR} \alpha$, there is no $\mathcal{B} \in$ PRA such that $\mathcal{A} \simeq \mathcal{B}$ (i.e. every counterexample witnessing that $\Gamma \not\models^{CR} \alpha$ is a non-representable relation algebra and thus, a model of no interest in this context).

The following definitions and properties will be of interest in further sections of the paper.

Definition 8 (Full proper relation algebras). *An algebraic structure $\langle A, \cup, \cap, ^-, \emptyset, E, \circ, \smile, Id \rangle$ is said to be a full proper algebra of relations over a set S if:*

- $A = \{\, a \mid a \subseteq S \times S \,\}$ *(equivalently $A = \wp(S^2)$), and*
- $E = \{\, \langle a, b \rangle \mid a, b \in S \,\}$ *(equivalently $E = S \times S$ or $E = S^2$).*

Definition 9 (Ideal elements). *Let $\mathcal{A} \in$ RA with domain A, then $x \in A$ is an* ideal element *if and only if $x = 1; x; 1$.*

A formal definition of *simple algebra* can be found in [Burris & Sankappanavar, 1981, Chap. 2, Sec. 8], but for all practical purposes we will use the following theorem proved by Tarski.

Theorem 1 ([Jónsson & Tarski, 1952], Thm. 4.10). *Let $\mathcal{A} \in$ RA, then $\mathcal{A}$ is* simple *if and only if $\mathcal{A}$ has, at most, two ideal elements (i.e. 0 and 1).*

Notice that if a proper relation algebra is full, then it is simple. The proof of this property can be found in [Jónsson & Tarski, 1952, Thm. 4.10].

Definition 10. **1** *and* **2** *are the only relation algebras with 1 and 2 elements in their domain, respectively.*

To ease the reader's understanding, when we interpret **1** as a proper relation algebra we obtain $\langle \{\emptyset\}, \cup, \cap, ^-, \emptyset, \emptyset, \circ, \smile, \emptyset \rangle$, with $\emptyset$ being the empty relation. On the other hand, if $U = \{\bullet\}$, **2** is interpreted as a proper relation algebra over U as $\langle \{\emptyset, \{\langle \bullet, \bullet \rangle\}\}, \cup, \cap, ^-, \emptyset, \{\langle \bullet, \bullet \rangle\}, \circ, \smile, \{\langle \bullet, \bullet \rangle\} \rangle$, also with $\emptyset$ being the empty relation.

Definition 11 (Trivial and prime relation algebras). *Let $\mathcal{A} \in$ RA, then*

- $\mathcal{A}$ *is* trivial *if it is isomorphic to* **1** *(noted as $\mathcal{A} \simeq \mathbf{1}$) or to* **2**, *and*
- $\mathcal{A}$ *is* prime *if it is simple and non-trivial.*

In [Frias, 2002, Sec. 3], Frias pinpoints the motivations behind the introduction of *Fork algebra*. In [Haeberer & Veloso, 1991], Haeberer and Veloso started the study of this class of algebras in the search for a calculus suitable for program construction, derivation and verification.

If we recall Def. 3, a *proper fork algebra* is obtained by extending a proper relation algebra with a new operation called *fork* and usually symbolised with "∇". The introduction of this new operator induces a structure on the set over which the relations are defined. This is done by considering binary relations over the domain of a structure $\langle U, \star \rangle$ where $\star : U \times U \to U$ is injective. Then, given $r, s \in \wp(U \times U)$,

$$r \underline{\nabla} s = \{ \langle a, b \rangle \in U \times U \mid \exists x, y \in U \mid b = x \star y \land a\, r\, x \land a\, s\, y \} \quad (1)$$

Fork algebras evolved around the definition of the function $\star$. In [Haeberer & Veloso, 1991] proper fork algebras were presented on a domain of binary relations on the set of finite trees built up from applications of $\star$; in that sense $\star$ acts as a set theoretical pairing function. In [Veloso & Haeberer, 1991] Veloso and Haeberer moved to a definition where the domain is built from binary relations on finite strings; an immediate consequence of this decision is that $\star$ acts as string concatenation. Later on, in [Veloso et al., 1992] the base set is once again made from finite trees. In all the previously mentioned articles, no axiomatization is given. Mikulás et. al. proved, in [Mikulás et al., 2015, Thm. 3.4], that an extension of a proper relation algebra with projection operators, like the ones induced by the operator $\underline{\nabla}$ as defined in Eq. 1, is not finitely axiomatizable. Such a result necessarily excludes the previous, more intuitive, interpretations of $\star$ as being binary tree constructor, string concatenation, set-theoretical pair formation, etc.

If U is the base set of a fork algebra and $x \in U$, x is said to be a *urelement* if there are no $y, z \in A$ such that $x = y \star z$. Intuitively, a urelement is a non-splitting element of A. It is easy to prove that having urelements is equivalent to having a non-surjective function $\star$.

Before introducing proper fork algebras we introduce *star proper fork algebras* as follows.

Definition 12 (Star proper fork algebra). *A star proper fork algebra is a two-sorted algebraic structure $\langle A, U, \cup, \cap, ^-, \emptyset, E, \circ, ^\smile, Id, \underline{\nabla}, \star \rangle$ in which A is a set of binary relations on U; $\cup, \cap, \circ$ and $\underline{\nabla}$ are binary operations on A; $^-$ and $^\smile$ are unary operations on A; $\emptyset, E$ and Id are distinguished elements of A; and $\star$ is a binary operation on U satisfying:*

- $\langle A, \cup, \cap, ^-, \emptyset, E, \circ, ^\smile, Id \rangle$ *is a proper relation algebra on U,*

- $\star : U \times U \to U$ is injective on the restriction of its domain to E and
- $\mathcal{A}$ is closed under $\underline{\nabla}$ of binary relations, defined as follows:

$$r \underline{\nabla} s = \{\, \langle a, x \star y \rangle \in U \times U \mid a\, r\, x \wedge a\, s\, y \,\}\ .$$

If in addition, if $\star$ is required to be non-surjective, the algebra is referred to as a star proper fork algebra with urelements. The class of star proper fork algebras (resp. star proper fork algebras with urelements) will be denoted as ⋆PFA (resp. ⋆PFAU).

A graphical interpretation of the fork of binary relations is presented in Fig. 1.

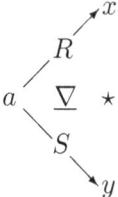

Figure 1: Graphical representation of "fork".

When dealing with algebras, the function $\mathbf{Rd}_T$ takes reducts to the type T; to take reduct of a class of algebras of type $\langle \mathcal{A}, \mathcal{F} \rangle$, to the type $\langle \mathcal{A}', \mathcal{F}' \rangle$[6] means to forget all the domains of $\mathcal{A}$ not mentioned in $\mathcal{A}'$ and the functions of $\mathcal{F}$ not mentioned in $\mathcal{F}'$. Notice that this definition requires $\mathcal{A}'$ to be included in $\mathcal{A}$, and $\mathcal{F}'$ to be included in $\mathcal{F}$.

Definition 13 (Proper fork algebras). *The class of* proper fork algebras *(denoted as* PFA*) is obtained from* ⋆PFA *as* $\mathbf{Rd}_T$ ⋆PFA*, where T is the similarity type* $\langle A, \cup, \cap, \bar{}, \emptyset, E, \circ, \breve{}, Id, \underline{\nabla} \rangle$ *and* $\mathbf{Rd}_T$ *is the algebraic operator for taking the reduct of an algebraic structure to the similarity type T.*

Analogously, the class of proper fork algebras with urelements *(denoted as* PFAU*) is obtained in the same way but applying the operator* $\mathbf{Rd}_T$ *on the class* ⋆PFAU*.*

Definition 14 (Full proper fork algebras).
A proper fork algebra $\langle A, \cup, \cap, \bar{}, \emptyset, E, \circ, \breve{}, Id, \underline{\nabla} \rangle$ *is said to be* full *if its relational reduct (which is a proper relation algebra)* $\langle A, \cup, \cap, \bar{}, \emptyset, E, \circ, \breve{}, Id \rangle$ *is full.*

---

[6]For a formal definition of the type of an algebraic language see [Burris & Sankappanavar, 1981, Defs. 1.1–1.3].

The class of *full proper fork algebras* (resp. *full proper fork algebras with urelements*) will be denoted as fPFA (fPFAU).

In the same way RA is the class of models of CR, and PRA its class of concrete models, one can formalise a calculus of fork algebras (resp. calculus of fork algebras with urelements) and establish the formal relationship between its class of abstract models and PFA. As Frias points out in [Frias, 2002], the current, most accepted, axiomatisation for the fork algebras is the one due to Haeberer et al. [Haeberer et al., 1993b,a].

Definition 15. *Let $\mathcal{R}$ be a set of relation variables, then the set of relation designations is the smallest set $RelDes(\mathcal{R})$ such that:*

- $\mathcal{R} \cup \{1, 0, 1'\} \subseteq RelDes(\mathcal{R})$,

- *If* $r, s \in RelDes(\mathcal{R})$, *then* $\{r+s, r \cdot s, \overline{r}, r; s, \breve{r}, r \nabla s\} \subseteq RelDes(\mathcal{R})$.

Then, the set of formulae is the set $\{r = s \mid r, s \in CFAForm(\mathcal{R})\}$.

Definition 16 (The calculus of fork algebras). *Let $\mathcal{R}$ be a set of relation variables, then the* calculus of fork algebras *(CFA for short) is defined for the formulae in $CFAForm(\mathcal{R})$ by:*

- *the axioms for the Boolean and the relational operators of Def. 5, and*

- *the following axioms for the fork operator: for all $r, s, t, u \in A$*

$$r \nabla s = (r;(1'\nabla 1)) \cdot (s;(1 \nabla 1'))$$
$$(r \nabla s);(t \nabla u)\breve{} = (r;\breve{t}) \cdot (s;\breve{u})$$
$$(1'\nabla 1)\breve{} \nabla (1 \nabla 1')\breve{} \leq 1'$$

Additionally, the calculus of fork algebras with urelements *(CFAU for short) is obtained by adding the axiom:*

$$1;(\overline{1 \nabla 1} \cdot 1');1 = 1$$

Once again, the calculus is completed by adopting any appropriate set of inference rules for the equality.

Definition 17. *The class of* fork algebras, *denoted as* FA *for short, (resp.* fork algebras with urelements, *denoted as* FAU*) is the class of algebraic structures $\langle A, +, \cdot, ^-, 0, 1, ;, \breve{}, 1', \nabla \rangle$ satisfying the axioms in CFA (resp. CFAU).*

The term $\overline{1\nabla 1}\cdot 1'$, appearing in the last axiom of Def. 16, characterises the partial identity on urelements. This term will be denoted by $1'_U$. Terms $(1'\nabla 1)^{\smile}$ and $(1\nabla 1')^{\smile}$, when interpreted in a proper fork algebra, act as projections of the first and second coordinates, respectively, of an element obtained by application of $\star$. These two terms will be denoted by π and ρ, respectively. Figs. 2a and 2b show a graphical representation of projections π and ρ.

(a) Graphical representation of π. (b) Graphical representation of ρ.

Figure 2: Graphical representation of projections.

These definitions allow us to rewrite the first, third and fourth axiom of Def 16 as follows:
$$r\nabla s = (r;\breve{\pi})\cdot(s;\breve{\rho})$$
$$\pi\nabla\rho \leq 1'$$
$$1;1'_U;1 = 1$$

By resorting to the identity between urelements we define $_U 1_U = 1'_U;1;1'_U$. Relation $_U 1_U$ relates every pair of urelements.

Checking that proper fork algebras (resp. proper fork algebras with urelements) are fork algebras (resp. fork algebras with urelements) is simple as it only requires to check that the structures defined in Def. 13 satisfy the axioms given in Def. 16. In [Frias et al., 1995a], Frias et al. proved the converse result by showing that FA is representable in PFA, but resorting to a non-equational axiom. Later on, in [Frias et al., 1995b], the same representability result was proved but only resorting to those equational axioms appearing in [Haeberer et al., 1993b].

Theorem 2 (Representability of FA in PFA, [Frias et al., 1995b], Thm. 3.7[7]). FA = I PFA[8] ∎

---
[7]The same representability result was obtained independently by Gyuris and presented in [Gyuris, 1997].

[8]The operator **I** closes a class of algebras of the same similarity type under isomorphisms.

The proof of Thm. 2, published also in [Frias, 2002, Sec. 4.1], can be easily adapted to a proof of the representability of FAU in PFAU.

Corollary 1.
FAU = **I** PFAU. ∎

The relational reduct of a fork algebra satisfies many specific properties, for example, they are representable in PRA. In [Veloso, 1996b], and briefly revisited in [Veloso, 1996c], Veloso explores the expansibility of a relation algebra to a fork algebra and, to that purpose, he presents the following definitions and results.

Definition 18. *Given* $\mathcal{F} \in$ FA, $\mathbf{Rd}_{\langle A,+,\cdot,\bar{\ },0,1,;,\check{\ },1'\rangle} \mathcal{F}$ *(called its relational reduct) will be denoted as* $\mathcal{F}_{RA}$.

Definition 19 (Fork index). *Let* $\mathcal{A} \in$ RA, *the fork index* φ *of* $\mathcal{A}$ *is defined as* $\varphi(\mathcal{A}) = |\{\mathcal{F} \in FA \mid \mathcal{F}_{RA} = \mathcal{A}\}|$.

Proposition 1 ([Veloso, 1996b], Sec. 5.1). *Let* $\mathcal{A} \in$ RA *with domain* A, *Then* $\varphi(\mathcal{A}) \leq |A|^2$. ∎

With this result, Veloso introduces the following classification of relation algebras.

Definition 20. *Let* $\mathcal{A} \in$ RA *with domain* A,
- $\mathcal{A}$ *is called* explosive *if* $\varphi(\mathcal{A}) = |A|^2$,
- $\mathcal{A}$ *is called* non-expansible *if* $\varphi(\mathcal{A}) = 0$,
- $\mathcal{A}$ *is called* rigid *if* $\varphi(\mathcal{A}) = 1$,
- $\mathcal{A}$ *is called* elastic *if* $\varphi(\mathcal{A}) = \infty$.

The aim of the present work is to contribute to the study of the first category of relation algebras defined above, i.e. the explosive relation algebras. Following [Veloso, 1996a,b], the class of explosive relation algebras will be denoted as EXP and, given $\kappa \geq \aleph_0$, EXP[κ] will denote the class of explosive relation algebras whose underlying set is of cardinality κ.

3 On the construction of explosive relation algebras

In [Veloso, 1996b, Sec. 6.2] Veloso presents the following three results. The first one, to be discussed more extensively in Sec. 3.1, constitutes the main motivation behind this work.

Proposition 2 (Existence of prime, big and explosive proper relation algebras, [Veloso, 1996b], Sec. 6.2). *Let $\kappa \geq \aleph_0$, there exists $\mathcal{R}_\kappa \in$ PRA prime and explosive such that $|\mathcal{R}_\kappa| = \kappa$ (i.e. $\mathcal{R}_\kappa$ has κ non-isomorphic expansions to fork algebras $\{\mathcal{F}_\gamma\}_{\gamma<\kappa}$).* ∎

The next property is a direct consequence of the fact that proper relation algebras having a different number of ideals cannot be isomorphic (the interested reader is pointed to [Jónsson & Tarski, 1952] for a discussion about the relation between homomorphisms and ideal elements). Then, to control the amount of ideal elements, we can combine prime algebras (see Prop. 11) with powers of 2 in a direct product.

Proposition 3 (Non-isomorphic combinations of prime, big and explosive proper relation algebras, [Veloso, 1996b], Sec. 6.2). *Let $\kappa \geq \aleph_0$ and $\mathcal{R}_\kappa \in$ PRA prime and explosive such that $|\mathcal{R}_\kappa| = \kappa$; then for each cardinal $\zeta < \kappa$, $2^\zeta \times \mathcal{R}_\kappa \in$ PRA is representable, explosive, $|2^\zeta \times \mathcal{R}_\kappa| = \kappa$ and has $2^{\zeta+1}$ ideal elements.* ∎

We will not discuss why direct product (used in the previous proposition) does not modify the number of posible expansions of a proper relation algebra to a fork algebra; the reader interested in the details of such a phenomenon is pointed to [Veloso, 1996a,b].

Theorem 3 (Many prime, big and explosive proper relation algebras, [Veloso, 1996b], Sec. 6.2). *Let $\kappa \geq \aleph_0$, then there exists κ non-isomorphic proper relation algebras of cardinality κ (i.e $|\text{EXP}[\kappa]| = \kappa$)* ∎

3.1 A prime, big and explosive proper relation algebra

In this section we review the construction of a prime, big and explosive proper relation algebra, presented by Veloso in [Veloso, 1996a] and used in the proof of Prop. 2.

Definition 21 ($\underline{2}$). *Let $\mathcal{F} = \langle A, \cup, \cap, ^-, \emptyset, E, \circ, ^\smile, Id, \underline{\nabla}\rangle \in$ PFA, then $\underline{2} = Id \underline{\nabla} Id$.*

Definition 22 (Subidentities of $\underline{2}$). *Let $\mathcal{F} = \langle A, \cup, \cap, ^-, \emptyset, E, \circ, ^\smile, Id, \underline{\nabla}\rangle \in$ PFA, then $Si_{\underline{2}}(\mathcal{F}) = \{\, a \in A \mid a \subseteq \underline{2} \cap Id\,\}$.*

Proposition 4. *Let $\mathcal{F}, \mathcal{G} \in$ PFA, if $\phi : \mathcal{F} \to \mathcal{G}$ is an isomorphism, then ϕ induces a bijection between $Si_{\underline{2}}(\mathcal{F})$ and $Si_{\underline{2}}(\mathcal{G})$.* ∎

Definition 23 (Fixpoints of $\star$). *Let $\star : U^2 \to U$, the fixpoints of $\star$ are defined as $fix(\star) = \{\, u \in U \mid u \star u = u \,\}$.*

This set can also be presented as a relation contained in the identity relation, as follows:

$$Id_{fix(\star)} = \left\{ \langle u, u \rangle \in U^2 \;\middle|\; u \in fix(\star) \right\} \qquad (2)$$

for which it is possible to prove the following properties.

Proposition 5. *Let* $\mathcal{F} = \langle A, \cup, \cap, ^-, \emptyset, E, \circ, ^\smile, Id, \underline{\nabla} \rangle \in$ PFA, *then* $\underline{2} \cap Id = Id_{fix(\star)}$. ∎

Proposition 6. *Let* $\mathcal{F} = \langle A, \cup, \cap, ^-, \emptyset, E, \circ, ^\smile, Id, \underline{\nabla} \rangle \in$ PFA *simple, such that* $\underline{\nabla}$ *is induced by* $\star : U^2 \to U$, *then* $Si_{\underline{2}}(\mathcal{F}) = \wp(Id_{fix(\star)}) \cap A$. ∎

Proposition 7 ([Veloso, 1996a], Sec. 6). *Let U be a set such that $|U| = \kappa$ and $\aleph_0 \leq \kappa$ then, for all $S \subseteq U$ such that $|S| < |U|$, there exists $\star_S : U^2 \to U$ bijective such that $fix(\star_S) = S$.*

Proof. If $|U| = \kappa$, then $|U^2| = \kappa$, $|Id| = \kappa$ and $|\overline{Id}| = \kappa$. Then, let $S \subseteq U$ such that $|S| < \kappa$, we know that the cardinality of the complement of S with respect to U (denoted as $\overline{S}$ when no ambiguity arises) is κ (denoted as $|\overline{S}| = \kappa$) and, therefore, it is possible to take $\overline{S} = A \cup \bigcup_{i \in \mathbb{N}} B_i$ such that:

- for all $i \in \mathbb{N}$, $A \cap B_i = \emptyset$,
- for all $i, j \in \mathbb{N}$, such that $i \neq j$, $B_i \cap B_j = \emptyset$,
- $|A| = \kappa$ and for all $i \in \mathbb{N}$, $|B_i| = \kappa$.

Then, there exists $g : \overline{S} \to \bigcup_{i \in \mathbb{N}} B_i$ bijective and without fixpoints defined as the union of the bijective functions $g_A : A \to B_0$ and $\{g_i : B_i \to B_{i+1}\}_{i \in \mathbb{N}}$. Notice that such union is disjoint as the domains and codomains of all of the functions are disjoint.

On the other hand, there exists $f : \overline{Id} \to A$, as a consequence of analysing the cardinality of $\overline{Id}$ and A and, consequently, it is possible to define $\star_S : U^2 \to U$ in the following way:

$$u \star_S v = \begin{cases} u & \text{, if } u \in S \text{ and } v = u. \\ g(u) & \text{, if } u \in \overline{S} \text{ and } v = u. \\ f(\langle u, v \rangle) & \text{, if } v \neq u. \end{cases}$$

From this we know that $\star_S : U^2 \to U$ is a bijective function as it is the union of bijective functions whose domains and codomains are pairwise disjoint such that $fix(\star_S) = S$. ∎

Now, it is possible to prove Prop. 2 by resorting to the construction of a proper relation algebra with the corresponding properties.

Proposition 2 (Existence of prime, big and explosive proper relation algebras, [Veloso, 1996b], Sec. 6.2). *Let $\kappa \geq \aleph_0$, there exists $\mathcal{R}_\kappa \in$ PRA prime and explosive such that $|\mathcal{R}_\kappa| = \kappa$ (i.e. $\mathcal{R}_\kappa$ has κ non-isomorphic expansions to fork algebras $\{\mathcal{F}_\gamma\}_{\gamma < \kappa}$).*

Proof. Let U be an infinite set such that $|U| = \kappa$ then, for all $\phi < \kappa$, there exists $S \subseteq U$ such that $|S| = \phi$. By Prop. 7, there exists $\star_S : U^2 \to U$ bijective inducing $\underline{\nabla}_S : (\wp(U^2))^2 \to \wp(U^2)$ and the corresponding projections π_S and ρ_S.

Let $\{S_\phi\}_{\aleph_0 \leq \phi \leq \kappa}$ such that $S_\phi \subseteq U$, for all $\aleph_0 \leq \phi \leq \kappa$ then, we define the set H in the following way[9]:

$$H = \wp_{fin}\left(U^2\right) \cup \bigcup_{\phi < \kappa} \left(\{\pi_{S_\phi}, \rho_{S_\phi}\} \cup \wp\left(Id_{fix(\star_{S_\phi})}\right)\right)$$

From the previous definition we know that $|H| = \kappa$. Therefore, it is enough to consider $\mathcal{R}_H$ as the subalgebra generated by H of the full proper relation algebra generated by U which, by [Burris & Sankappanavar, 1981, Sec. 3] has cardinality κ. Finally, by [Jónsson & Tarski, 1952, Thm. 4.11], as $\mathcal{R}_H$ is a subalgebra of a simple algebra, it is simple, and for each $\aleph_0 \leq \phi \leq \kappa$, S_ϕ induces $\underline{\nabla}_{S_\phi}$ determining a possible extension of $\mathcal{R}_H$ to a fork algebra. Notice that all of these possible extensions are pairwise non-isomorphic as they differ in the cardinality of the set of fixpoints for $\star$. ∎

Next, we reproduce the proof of Prop. 3 and Thm. 3.

Proposition 3 (Non-isomorphic combinations of prime, big and explosive proper relation algebras, [Veloso, 1996b], Sec. 6.2). *Let $\kappa \geq \aleph_0$ and $\mathcal{R}_\kappa \in$ PRA prime and explosive such that $|\mathcal{R}_\kappa| = \kappa$; then for all cardinal $\zeta < \kappa$, $2^\zeta \times \mathcal{R}_\kappa \in$ PRA is representable, explosive, $|2^\zeta \times \mathcal{R}_\kappa| = \kappa$ and has $2^{\zeta+1}$ ideal elements.*

Proof. Since $2^\zeta \leq \kappa$, the direct product $2^\zeta \times \mathcal{R}_\kappa$ is a representable relation algebra such that $|2^\zeta \times \mathcal{R}_\kappa| = \kappa$; it's prime factors are the rigid **2** and the explosive $\mathcal{R}_\kappa$, so $\phi(2^\zeta \times \mathcal{R}_\kappa) = \kappa$ leading to the fact that $2^\zeta \times \mathcal{R}_\kappa$ has $2^\zeta \cdot 2$ ideal elements. ∎

---
[9] $\wp_{fin}(A)$ is interpreted as $\{a \mid a \subseteq A \wedge |a| \in \mathbb{N}\}$, the finite powerset of A.

Theorem 3 (Many prime, big and explosive proper relation algebras, [Veloso, 1996b], Sec. 6.2). *Let $\kappa \geq \aleph_0$, then there exists κ non-isomorphic proper relation algebras of cardinality κ (i.e $|\mathsf{EXP}[\kappa]| = \kappa$)*

Proof. This theorem is a direct consequence of Prop. 3, as algebras with different cardinality of ideal elements cannot be isomorphic. ∎

3.2 Generalising the control of the fixpoints of $\star$

In the previous section we have shown the construction of a prime, big and explosive proper relation algebra where explosiveness is guaranteed by controlling the cardinality of the set of fixpoints of $\star$, each of which leads to a non-isomorphic fork algebra. In this section we propose a generalisation of the controlling technique of such fixpoints.

Let us first introduce some useful definitions.

Definition 24 (Binary trees). *Binary trees are the elements of BT, the smallest set of terms produced by the grammar* bt ::= nil | bin bt bt.

Definition 25. *The predicates $\bullet = \bullet \subseteq \mathsf{BT} \times \mathsf{BT}$ and $\bullet < \bullet \subseteq \mathsf{BT} \times \mathsf{BT}$ are defined as follows:*

$$
\begin{aligned}
\text{nil} &= \text{nil} \\
\text{bin } i\, d &= \text{bin } i'\, d' \quad \text{iff} \quad i = i' \text{ and } d = d' \\
\text{nil} &< \text{bin } i\, d \\
\text{bin } i\, d &< \text{bin } i'\, d' \quad \text{iff} \quad \text{bin } i\, d = i' \text{ or bin } i\, d < i' \text{ or} \\
&\qquad\qquad\qquad\qquad \text{bin } i\, d = d' \text{ or bin } i\, d < d'
\end{aligned}
$$

Definition 26 (Map). *Let U be a set, we define $map : \mathsf{BT} \times [U^2 \to U] \times U \to U$ as follows: let $f : U^2 \to U$ a binary function over U and $u \in U$*

$$
\begin{aligned}
map \text{ nil } f\, u &= u \\
map \text{ (bin } ab_1\, ab_2) \, f\, u &= f((map\ ab_1\ f\ u), (map\ ab_2\ f\ u))
\end{aligned}
$$

Definition 27 ($\underline{t}$). *Let $\mathcal{F} = \langle A, \cup, \cap, \bar{\ }, \emptyset, E, \circ, \smile, Id, \underline{\nabla} \rangle \in \mathsf{PFA}$ and $t \in \mathsf{BT}$, then $\underline{t} = map(t, \underline{\nabla}, Id)$.*

Definition 28 (Subidentities of $\underline{t}$). *Let $\mathcal{F} = \langle A, \cup, \cap, \bar{\ }, \emptyset, E, \circ, \smile, Id, \underline{\nabla} \rangle \in \mathsf{PFA}$ and $t \in \mathsf{BT}$, then $Si_{\underline{t}}(\mathcal{F}) = \{\, a \in A \mid a \subseteq \underline{t} \cap Id \,\}$.*

Proposition 8. *Let $\mathcal{F} = \langle F, \cup^{\mathcal{F}}, \cap^{\mathcal{F}}, \bar{\ }^{\mathcal{F}}, \emptyset^{\mathcal{F}}, E^{\mathcal{F}}, \circ^{\mathcal{F}}, \smile^{\mathcal{F}}, Id^{\mathcal{F}}, \underline{\nabla}^{\mathcal{F}} \rangle$ and $\mathcal{G} = \langle G, \cup^{\mathcal{G}}, \cap^{\mathcal{G}}, \bar{\ }^{\mathcal{G}}, \emptyset^{\mathcal{G}}, E^{\mathcal{G}}, \circ^{\mathcal{G}}, \smile^{\mathcal{G}}, Id^{\mathcal{G}}, \underline{\nabla}^{\mathcal{G}} \rangle$ be PFA and $t \in \mathsf{BT}$, if $\phi : \mathcal{F} \to \mathcal{G}$ is an isomorphism, then ϕ induces a bijection between $Si_{\underline{t}^{\mathcal{F}}}(F)$ and $Si_{\underline{t}^{\mathcal{G}}}(G)$.*

Proof. Let $\phi : F \to G$ be an isomorphism between $\mathcal{F}$ and $\mathcal{G}$, and $r \in F$ such that $r \in Si_{\underline{t}^{\mathcal{F}}}(F)$,

$$
\begin{array}{rll}
r \in Si_{\underline{t}^{\mathcal{F}}}(F) & \text{iff} & r \subseteq^{\mathcal{F}} \underline{t}^{\mathcal{F}} \cap^{\mathcal{F}} Id^{\mathcal{F}} \quad \text{[by Def. 28.]} \\
& \text{iff} & r \cup^{\mathcal{F}} \left(\underline{t}^{\mathcal{F}} \cap^{\mathcal{F}} Id^{\mathcal{F}}\right) = \underline{t}^{\mathcal{F}} \cap^{\mathcal{F}} Id^{\mathcal{F}} \quad \text{[by Def. of } \subseteq.\text{]} \\
& \text{iff} & \phi\left(r \cup^{\mathcal{F}} \left(\underline{t}^{\mathcal{F}} \cap^{\mathcal{F}} Id^{\mathcal{F}}\right)\right) = \phi\left(\underline{t}^{\mathcal{F}} \cap^{\mathcal{F}} Id^{\mathcal{F}}\right) \\
& & \quad \text{[because } \phi \text{ is an isomorphism.]} \\
& \text{iff} & \phi(r) \cup^{\mathcal{G}} \left(\phi\left(\underline{t}^{\mathcal{F}}\right) \cap^{\mathcal{G}} Id^{\mathcal{G}}\right) = \phi\left(\underline{t}^{\mathcal{F}}\right) \cap^{\mathcal{G}} Id^{\mathcal{G}} \\
& & \quad \text{[because } \phi \text{ is an isomorphism.]} \\
& \text{iff} & \phi(r) \cup^{\mathcal{G}} \left(\underline{t}^{\mathcal{G}} \cap^{\mathcal{G}} Id^{\mathcal{G}}\right) = \underline{t}^{\mathcal{G}} \cap^{\mathcal{G}} Id^{\mathcal{G}} \\
& & \quad \text{[by [Lopez Pombo et al., 2020, Lemma 1].]} \\
& \text{iff} & \phi(r) \subseteq^{\mathcal{G}} \underline{t}^{\mathcal{G}} \cap^{\mathcal{G}} Id^{\mathcal{G}} \quad \text{[by Def. of } \subseteq.\text{]} \\
& \text{iff} & \phi(r) \in Si_{\underline{t}^{\mathcal{G}}}(G) \quad \text{[by Def. 28.]}
\end{array}
$$

Thus, finishing the proof. ∎

Definition 29 (*t*-controlled fixpoints of $\star$). *Let $\star : U^2 \to U$ and $t \in \mathsf{BT}$, the t-controlled fixpoints of $\star$ are defined as $\mathit{fix}_t(\star) = \{\, u \in U \mid \mathit{map}\ t\ \star\ u = u \,\}$.*

Recalling the definition of the set of fixpoints of $\star$ given in Def. 23 we can present the t-controlled fixpoints of $\star$ as a partial identity as follows.

$$Id_{\mathit{fix}_t(\star)} = \left\{\, \langle u, u \rangle \in U^2 \,\middle|\, u \in \mathit{fix}_t(\star) \,\right\} \qquad (3)$$

for which it is possible to derive the following properties.

Proposition 9. *Let $\mathcal{F} = \langle F, \cup, \cap, ^-, \emptyset, E, \circ, \smile, Id, \underline{\nabla} \rangle \in \mathsf{PFA}$ with $\underline{\nabla}$ induced by $\star : U^2 \to U$ and $t \in \mathsf{BT}$ such that $t \neq \mathtt{nil}$, then $\underline{t} \cap Id = Id_{\mathit{fix}_t(\star)}$.*

Proof. Let $\langle u, v \rangle \in U^2$, then

$$
\begin{array}{rll}
\langle u, v \rangle \in \underline{t} \cap Id & \text{iff} & \langle u, v \rangle \in \mathit{map}\ t\ \underline{\nabla}\ Id \text{ and } u = v \quad \text{[by Defs. 27 and 3.]} \\
& \text{iff} & \langle u, u \rangle \in \mathit{map}\ t\ \underline{\nabla}\ Id \text{ and } u = v \\
& \text{iff} & u = \mathit{map}\ t\ \star\ u \text{ and } u = v \\
& & \quad \text{[by [Lopez Pombo et al., 2020, Lemma 2].]} \\
& \text{iff} & u \in \mathit{fix}_t(\star) \text{ and } u = v \quad \text{[by Def. 29.]} \\
& \text{iff} & \langle u, u \rangle \in Id_{\mathit{fix}_t(\star)} \text{ and } u = v \quad \text{[by Eq. 3.]} \\
& \text{iff} & \langle u, v \rangle \in Id_{\mathit{fix}_t(\star)}
\end{array}
$$

Thus, finishing the proof. ∎

Proposition 10. *Let $\mathcal{F} = \langle F, \cup, \cap, ^-, \emptyset, E, \circ, \smile, Id, \underline{\nabla} \rangle \in \mathsf{PFA}$ simple with $\underline{\nabla}$ induced by $\star : U^2 \to U$ and $t \in \mathsf{BT}$ such that $t \neq \mathtt{nil}$, then $Si_{\underline{t}}(\mathcal{F}) = \wp\left(Id_{\mathit{fix}_t(\star)}\right) \cap F$.*

Proof.

$$\begin{aligned}
Si_{\underline{t}}(\mathcal{F}) &= \{r \in F \mid r \subseteq t \cap Id\} & \text{[by Defs. 28.]}\\
&= \{r \in F \mid r \subseteq Id_{fix_t(\star)}\} & \text{[by Prop. 9.]}\\
&= \{r \in F \mid r \in \wp\left(Id_{fix_t(\star)}\right)\} & \text{[by Def. $\subseteq$.]}\\
&= \{r \in \wp(U^2) \mid r \in \wp\left(Id_{fix_t(\star)}\right) \text{ and } r \in F\}\\
&= \{r \in \wp(U^2) \mid r \in \wp\left(Id_{fix_t(\star)} \cap F\right)\}\\
&= \wp\left(Id_{fix_t(\star)} \cap F\right)
\end{aligned}$$

Thus, finishing the proof. ∎

The following property is analogous to Prop. 7 but relaxes the conditions over $\star$.

Proposition 11. *Let U be an infinite set $|U| = \kappa$ and $\aleph_0 \leq \kappa$ then for all $S \subseteq U$ such that $|S| < |U|$, there exists $\star_S : U^2 \to U$ injective such that given $t \in \mathsf{BT}$ where $t \neq \mathtt{nil}$, $|fix_t(\star_S)| = |S|$.*

Proof. If $|U| = \kappa$, then $|U^2| = \kappa$, $|Id| = \kappa$ and $|\overline{Id}| = \kappa$. Then, let $\{S_{t'}\}_{t' < t}$ be a finite family of sets such that:

- for all $t', t'' < t$, if $t' \neq t''$ then $S_{t'} \cap S_{t''} = \emptyset$, and
- for all $t' < t$, $|S_{t'}| = |S|$.

Analogous to what Veloso points out in the proof of Prop. 7, we know that $|\overline{S \cup \bigcup_{t' < t} S_{t'}}| = \kappa$ and, therefore, it is possible to take $\overline{S \cup \bigcup_{t' < t} S_{t'}} = \bigcup_{i \in \mathbb{N}} B_i$ such that:

- for all $i, j \in \mathbb{N}$, such that $i \neq j$, $B_i \cap B_j = \emptyset$, and
- $|A| = \kappa$ and for all $i \in \mathbb{N}$, $|B_i| = \kappa$.

Then, there exists $g : \bigcup_{i \in \mathbb{N}} B_i \to \bigcup_{i \in \mathbb{N}} B_i$ bijective and without fixpoints defined as the union of the bijective functions $\{g_i : B_i \to B_{i+1}\}_{i \in \mathbb{N}}$. Notice that such a union is disjoint as the domains and codomains of all of the functions are disjoint.

On the other hand, there exists an injective funtion $f : S^2 \to B_0$ and a finite family of bijective functions $\{h_{t'} : S \to S_{t'}\}_{t' < t}$, such that $h_{\mathtt{nil}} = id_S$. Then, it is possible to define $\star_S : U^2 \to U$ according to Table 1.

Table 1 shows how $\star_S : U^2 \to U$ is defined as the union of injective functions with disjoint domains and codomains. Therefore, $\star_S : U^2 \to U$ is injective.

$u \star_S v$	$S_{t''}$	B_i
$S_{t'}$	$\begin{cases} h_{\mathtt{bin}\ t'\ t''}\left(h_{t'}^{-1}(u)\right) \\ \quad ;\ \text{if}\ h_{t'}^{-1}(u) = h_{t''}^{-1}(v)\ \text{and} \\ \quad\ \mathtt{bin}\ t'\ t'' < t. \\ h_{t'}^{-1}(u) \\ \quad ;\ \text{if}\ h_{t'}^{-1}(u) = h_{t''}^{-1}(v)\ \text{and} \\ \quad\ \mathtt{bin}\ t'\ t'' = t. \\ f(h_{t'}^{-1}(u), h_{t''}^{-1}(v)) \\ \quad ;\ \text{otherwise}. \end{cases}$	$g(v)$
B_j	$g(u)$	$\begin{cases} g(u) & ;\ \text{if}\ i \geq j. \\ g(v) & ;\ \text{otherwise}. \end{cases}$

Table 1: Definition of $\star_S : U^2 \to U$ controlled by a binary tree.

On the one hand, by [Lopez Pombo et al., 2020, Lemma 3] we obtain that if $s \in S$, $s = map\ t \star_S s$ and, by Def. 29, that $S \subseteq \mathit{fix}_t (\star_S)$ and, consequently, that $|S| \leq |\mathit{fix}_t (\star_S)|$. On the other hand, for all $s \in \mathit{fix}_t (\star_S)$, $s \in S \cup \bigcup_{t' < t} S_{t'}$ and, consequently, $\mathit{fix}_t (\mathit{star}_S) \subseteq S \cup \bigcup_{t' < t} S_{t'}$ because, by the way in which $\star_S : U^2 \to U$ was constructed, $\overline{S \cup \bigcup_{t' < t} S_{t'}}$ does not contain fixpoints. Then, we obtain that $|\mathit{fix}_t (\mathit{star}_S)| \leq |S \cup \bigcup_{t' < t} S_{t'}| = |S|$. Jointly, these two results prove that $|\mathit{fix}_t (\mathit{star}_S)| = |S|$. ∎

From the previous result, it is possible to reproduce the result of Prop. 2, but with a minor modification because constructing the algebra requires the use of the set S_ϕ, instead of the set $\mathit{fix}_t (\star_{S_\phi})$, as there might be fixpoints outside S_ϕ[10]. Thereafter, Thm. 3 can be applied in order to guarantee the existence of infinitely many prime, big and explosive relation algebras obtained by controlling the fixpoints of $\star : U^2 \to U$ resorting to a term from BT.

Next theorem shows that Prop. 7 is a special case of Prop. 11.

Theorem 4. Let $\mathcal{F} = \langle F, \cup, \cap, ^-, \emptyset, E, \circ, {}^\smile, Id, \underline{\nabla} \rangle \in \mathsf{PFA}$, $Si_2(\mathcal{F}) = Si_{\mathtt{bin\ nil\ nil}}(\mathcal{F})$.

Proof. The proof is direct by using the definitions of $Si_2(\mathcal{F})$, 2, `bin nil nil` and $Si_{\mathtt{bin\ nil\ nil}}(\mathcal{F})$. ∎

Observing the generalised controlling technique of the fixpoints of $\star$ presented above, it is possible to establish certain relations between them.

---

[10] The reader should note the fact that while on the one hand, $S \subseteq \mathit{fix}_t (\star_S)$, on the other $\mathit{fix}_t (\mathit{star}_S) \subseteq S \cup \bigcup_{t' < t} S_{t'}$. Such asymmetry only allows us to guarantee that S_ϕ is a set of fixpoints but regarding as possible the existence of fixpoints of $\mathit{fix}_t (\star_S)$, which lay outside S_ϕ.

Let us first introduce some definitions. The next definition generalises BT by introducing *Binary tree contexts* which, like in rewriting systems such as λ-calculus, are defined to be binary tree terms with some holes in it, denoted as "[ ]".

Definition 30 (Binary tree contexts). *Binary tree contexts are defined to be* BTC, *the smallest set of terms produced by the following grammar* btc ::= `nil` | [ ] | `bin` btc btc.

Definition 31 (Substitution).   *Let $t \in$ BTC, we define the function* $\bullet[\bullet]:$ BTC $\times$ BTC $\to$ BTC *as follows:*

$$\begin{aligned}
[\,][t] &= t \\
\mathtt{nil}[t] &= \mathtt{nil} \\
(\mathtt{bin}\ i\ d)\,[t] &= \mathtt{bin}\ (i[t])\ (d[t])
\end{aligned}$$

Proposition 12. *Let $t \in$ BTC and $t' \in$ BT, then $t[t'] \in$ BT.*

Proof. The proof follows easily by induction on t. ∎

Definition 32 (Variants).   *Let $t \in$ BT, we define the variants of t as* $V_t = \{\, t' \in \mathsf{BTC} \mid t = t'[\mathtt{nil}]\,\}$.

Theorem 5. *Let $\star : U^2 \to U$ and $t, t' \in$ BT such that $t \neq \mathtt{nil}$ and $t' \neq \mathtt{nil}$, then $(\forall t'' \in V_{t'})\,\bigl(\mathit{fix}_t\,(\star) \cap \mathit{fix}_{t'}\,(\star) \subseteq \mathit{fix}_{t''[t]}\bigr)$.*

Proof. Let $u \in U$ such that $u \in \mathit{fix}_t\,(\star)$ and $u \in \mathit{fix}_{t'}\,(\star)$ then, we know that $\mathit{map}\ t\ \star\ u = u$ and $\mathit{map}\ t'\ \star\ u = u$. Let $t'' \in V_{t'}$, then $t''[t]$ is structurally equal to t' with the exception that some of its leaves (those that were [ ] in t'') were replaced by t and in t' are u. Then, using that $\mathit{map}\ t\ \star\ u = u$ and $\mathit{map}\ t'\ \star\ u = u$, we obtain that $\mathit{map}\ t''[t]\ \star\ u = u$ and, therefore, $u \in \mathit{fix}_{t''[t]}\,(\star)$. ∎

3.3 Controlling the fixpoints of $\star$ through π and ρ

In the previous section we presented the generalisation of the technique used by Veloso in [Veloso, 1996a,b] in the construction of a prime, big and explosive proper relation algebra where explosiveness is guaranteed by controlling the cardinality of the set of fixpoints of $\star$, each of which leads to a non-isomorphic fork algebra. In this section we show a similar construction but relying on the quasi-projections π and ρ.

In the forthcoming paragraph we focus on the use of π but it can be reproduced by means of analogous definitions and results for ρ.

Definition 33 ($\underline{\pi}$). Let $\mathcal{F} = \langle A, \cup, \cap, \bar{}, \emptyset, E, \circ, \smile, Id, \underline{\nabla} \rangle \in$ PFA, then $\underline{\pi} = (Id \underline{\nabla} E)^{\smile}$.

Definition 34 (Subidentities of $\underline{\pi}$). Let $\mathcal{F} = \langle A, \cup, \cap, \bar{}, \emptyset, E, \circ, \smile, Id, \underline{\nabla} \rangle \in$ PFA, then $Si_{\underline{\pi}}(\mathcal{F}) = \{\, a \in A \mid a \subseteq \underline{\pi} \cap Id \,\}$.

Proposition 13. Let $\mathcal{F}, \mathcal{G} \in$ PFA, if $\phi : \mathcal{F} \to \mathcal{G}$ is an isomorphism, then ϕ induces a bijection between $Si_{\underline{\pi}}(\mathcal{F})$ and $Si_{\underline{\pi}}(\mathcal{G})$.

Proof. The proof of this proposition is analogous to that of Prop. 8. ∎

Definition 35 (π-controlled fixpoints of $\star$). Let $\star : U^2 \to U$, the fixpoints of $\star$ are defined as $\mathit{fix}_\pi(\star) = \{\, u \in U \mid (\exists v \in U)(u \star v = u) \,\}$.

This set can also be presented as a relation contained in the identity relation, as follows:

$$Id_{\mathit{fix}_\pi(\star)} = \left\{\, \langle u, u \rangle \in U^2 \,\middle|\, u \in \mathit{fix}_\pi(\star) \,\right\} \tag{4}$$

for which it is possible to prove the following properties.

Proposition 14. Let $\mathcal{F} = \langle A, \cup, \cap, \bar{}, \emptyset, E, \circ, \smile, Id, \underline{\nabla} \rangle \in$ PFA, then $\underline{\pi} \cap Id = Id_{\mathit{fix}_\pi(\star)}$.

Proof.

$\langle u, v \rangle \in \underline{\pi} \cap Id$ iff $\langle u, v \rangle \in (Id \underline{\nabla} E)^{\smile}$ and $u = v$
[by Defs. 33 and 3 - Id.]
iff $\langle v, u \rangle \in Id \underline{\nabla} E$ and $u = v$
[by Def. 3 - $\smile$.]
iff there exist $r, s \in U$ such that $u = r \star s$, $\langle v, r \rangle \in Id$, $\langle v, s \rangle \in E$ and $u = v$
[by Def. 13 - $\underline{\nabla}$.]
iff there exist $r, s \in U$ such that $u = r \star s$, $v = r$ and $u = v$
[by Def. 3 - Id and E.]
iff there exists $s \in U$ such that $u = v \star s$ and $u = v$
iff there exists $s \in U$ such that $u = u \star s$ and $u = v$
iff $u \in \{\, u' \in U \mid (\exists s \in U)(u' = u' \star s) \,\}$ and $u = v$
iff $u \in \mathit{fix}_\pi(\star)$ and $u = v$
[by Def. 35.]
iff $\langle u, u \rangle \in Id_{\mathit{fix}_\pi(\star)}$ and $u = v$
[by Eq. 4.]
iff $\langle u, v \rangle \in Id_{\mathit{fix}_\pi(\star)}$

Thus, finishing the proof. ∎

Proposition 15. Let $\mathcal{F} = \langle A, \cup, \cap, ^-, \emptyset, E, \circ, \smile, Id, \underline{\nabla}\rangle \in$ PFA simple, such that $\underline{\nabla}$ is induced by $\star : U^2 \to U$, then $Si_{\underline{2}}(\mathcal{F}) = \wp(Id_{fix(\star)}) \cap A$.

Proof. The proof of this proposition is analogous to that of Prop. 10. ∎

The following property is analogous to Prop. 7 but resorting to π as a determinant means for controlling the fixpoints of $\star$.

Proposition 16. Let U be an infinite set such that $|U| = \kappa$ and $\aleph_0 \leq \kappa$ then, for all $S \subseteq U$ such that $|S| < |U|$, there exists $\star_S : U^2 \to U$ injective such that $fix_\pi(\star_S) = S$.

Proof. If $|U| = \kappa$, then $|U^2| = \kappa$, $|Id| = \kappa$ and $|\overline{Id}| = \kappa$. Then, let $S \subseteq U$ such that $|S| < \kappa$, we know that $|\overline{S}| = \kappa$ and, therefore, it is possible to take $\overline{S} = A \cup \bigcup_{i \in \mathbb{N}} B_i$ such that:

- for all $i \in \mathbb{N}, A \cap B_i = \emptyset$,
- for all $i, j \in \mathbb{N}$, such that $i \neq j$, $B_i \cap B_j = \emptyset$,
- $|A| = \kappa$ and for all $i \in \mathbb{N}$, $|B_i| = \kappa$.

Then, there exists $g : \overline{S} \to \bigcup_{i \in \mathbb{N}} B_i$ bijective and without fixpoints defined as the union of the bijective functions $g_A : A \to B_0$ and $\{g_i : B_i \to B_{i+1}\}_{i \in \mathbb{N}}$. Notice that such union is disjoint as the domains and codomains of all of the functions are disjoint.

Let P be a set such that $|P| = |S|$ and $l : S \to P$ be a bijective function. Then, there exists $f : (S \cup P)^2 \to A$ injective and, consequently, it is possible to define $\star_S : U^2 \to U$ according to Table 2

$u \star_S v$	S	P	A	B_j
S	$f(u.v)$	$\begin{cases} u & \text{; if } v = l(u). \\ f(u,v) & \text{; otherwise.} \end{cases}$	$g(v)$	$g(v)$
P	$f(u.v)$	$f(u.v)$	$g(v)$	$g(v)$
A	$g(u)$	$g(u)$	$g(u)$	$g(v)$
B_j	$g(u)$	$g(u)$	$g(u)$	$\begin{cases} g(u) & \text{; if } i \geq j. \\ g(v) & \text{; otherwise.} \end{cases}$

Table 2: Definition of $\star_S : U^2 \to U$ controlled by π.

Table 2 shows how $\star_S : U^2 \to U$ is defined as the union of injective functions with disjoint domains and codomains. Therefore, $\star_S : U^2 \to U$ is injective an injective function such that $fix(\star_S) = S$. ∎

Once again, it is possible to apply Prop. 2 and Thm. 3 in order to prove the existence of many prime, big and explosive proper relation algebras.

As we mentioned at the beginning of this section, analogous definitions and results can be developed for controlling the fixpoints of $\star$ but resorting to ρ.

3.4 Generalising the control of the fixpoints of $\star$ through the projections π and ρ

The generalisation of the controlling technique of the fixpoints of $\star$ through π and ρ presented in the previous section is somehow similar to what was presented in the previous section.

Let us first consider the following data type formalising non-empty sequences of relations "π" and "ρ".

Definition 36 (Sequences). *Sequences are the elements of* Sec, *the smallest set of terms produced by the following grammar* sec ::= elem $*$ | cons $*$ s, *where* $* \in \{\pi, \rho\}$ *and* $s \in$ Sec.

Definition 37. *The functions* $long :$ Sec $\to \mathbb{N}$, $\bullet[\bullet] :$ Sec $\times \mathbb{N} \to \{\pi, \rho\}$ *and* $\bullet|\bullet :$ Sec $\times \mathbb{N} \to$ Sec[11] *are defined as follows: let* $* \in \{\pi, \rho\}$ *and* $s \in$ Sec,

$$long(\texttt{elem } *) = 1$$
$$long(\texttt{cons } * \ s) = 1 + long(s)$$

$$(\texttt{elem } *)[1] = \texttt{elem } *$$
$$(\texttt{cons } * \ s)[i] = \begin{cases} * & ; \text{if } i = 1. \\ s[i-1] & ; \text{otherwise.} \end{cases}$$

$$(\texttt{elem } *)|1 = *$$
$$(\texttt{cons } * \ s)|i = \begin{cases} (\texttt{cons } * \ s) & ; \text{if } i = long(s) + 1. \\ s|i & ; \text{otherwise.} \end{cases}$$

Definition 38. *Let* $\mathcal{F} = \langle A, \cup, \cap, ^-, \emptyset, E, \circ, \smile, Id, \underline{\nabla} \rangle \in$ PFA, $* \in \{\pi, \rho\}$ *and* $s \in$ Sec, *then* $\underline{\texttt{elem } *} = *$ *and* $\underline{\texttt{cons } * \ s} = \underline{*} \circ \underline{s}$.

Definition 39 (Subidentities of $\underline{s}$). *Let* $\mathcal{F} = \langle A, \cup, \cap, ^-, \emptyset, E, \circ, \smile, Id, \underline{\nabla} \rangle \in$ PFA *and* $s \in$ Sec, *then* $Si_{\underline{s}}(\mathcal{F}) = \{ a \in A \mid a \subseteq \underline{s} \cap Id \}$.

---
[11] Note that the last two functions are partial and are only defined on those elements $n \in \mathbb{N}$ and $s \in$ Sec such that $1 \leq n \leq long(s)$.

Proposition 17. Let $\mathcal{F} = \langle F, \cup^\mathcal{F}, \cap^\mathcal{F}, ^{-\mathcal{F}}, \emptyset^\mathcal{F}, E^\mathcal{F}, \circ^\mathcal{F}, ^{\smile \mathcal{F}}, Id^\mathcal{F}, \underline{\nabla}^\mathcal{F} \rangle$ and $\mathcal{G} = \langle G, \cup^\mathcal{G}, \cap^\mathcal{G}, ^{-\mathcal{G}}, \emptyset^\mathcal{G}, E^\mathcal{G}, \circ^\mathcal{G}, ^{\smile \mathcal{G}}, Id^\mathcal{G}, \underline{\nabla}^\mathcal{G} \rangle$ be PFA and $s \in$ Sec, if $\phi: \mathcal{F} \to \mathcal{G}$ is an isomorphism, then ϕ induces a bijection between $Si_{\underline{s}^\mathcal{F}}(F)$ and $Si_{\underline{s}^\mathcal{G}}(G)$.

Proof. Let $\phi: \mathcal{F} \to \mathcal{G}$ be an isomorphism between $\mathcal{F}$ and $\mathcal{G}$, and $r \in F$ such that $r \in Si_{\underline{s}^\mathcal{F}}(F)$,

$$
\begin{array}{rll}
r \in Si_{\underline{s}^\mathcal{F}}(F) & \text{iff} \quad r \subseteq^\mathcal{F} \underline{s}^\mathcal{F} \cap^\mathcal{F} Id^\mathcal{F} & \text{[by Def. 39.]} \\
& \text{iff} \quad r \cup^\mathcal{F} \left(\underline{s}^\mathcal{F} \cap^\mathcal{F} Id^\mathcal{F} \right) = \underline{s}^\mathcal{F} \cap^\mathcal{F} Id^\mathcal{F} & \text{[by Def. of } \subseteq \text{.]} \\
& \text{iff} \quad \phi \left(r \cup^\mathcal{F} \left(\underline{s}^\mathcal{F} \cap^\mathcal{F} Id^\mathcal{F} \right) \right) = \phi \left(\underline{s}^\mathcal{F} \cap^\mathcal{F} Id^\mathcal{F} \right) & \\
& \multicolumn{2}{l}{\text{[because } \phi \text{ is an isomorphism.]}} \\
& \text{iff} \quad \phi(r) \cup^\mathcal{G} \left(\phi\left(\underline{s}^\mathcal{F}\right) \cap^\mathcal{G} Id^\mathcal{G} \right) = \phi\left(\underline{s}^\mathcal{F}\right) \cap^\mathcal{G} Id^\mathcal{G} & \\
& \multicolumn{2}{l}{\text{[because } \phi \text{ is an isomorphism.]}} \\
& \text{iff} \quad \phi(r) \cup^\mathcal{G} \left(\underline{s}^\mathcal{G} \cap^\mathcal{G} Id^\mathcal{G} \right) = \underline{s}^\mathcal{G} \cap^\mathcal{G} Id^\mathcal{G} & \\
& \multicolumn{2}{l}{\text{[by [Lopez Pombo et al., 2020, Lemma 5].]}} \\
& \text{iff} \quad \phi(r) \subseteq^\mathcal{G} \underline{s}^\mathcal{G} \cap^\mathcal{G} Id^\mathcal{G} & \text{[by Def. of } \subseteq \text{.]} \\
& \text{iff} \quad \phi(r) \in Si_{\underline{s}^\mathcal{G}}(G) & \text{[by Def. 39.]}
\end{array}
$$

Thus, finishing the proof. ∎

Definition 40 (*s-controlled fixpoints of* $\star$). Let $\star : U^2 \to U$ and $s \in$ Sec, the *s-controlled fixpoints of* $\star$ are defined as $fix_s(\star) = \{ u \in U \mid \langle u, u \rangle \in \underline{s} \}$.

Recalling the definition of the set of fixpoints of $\star$ given in Def. 40 we can present the s-controlled fixpoints of $\star$ as a partial identity as follows.

$$Id_{fix_s(\star)} = \left\{ \langle u, u \rangle \in U^2 \,\middle|\, u \in fix_s(\star) \right\} \tag{5}$$

for which it is possible to derive the following properties.

Proposition 18. Let $\mathcal{F} = \langle F, \cup, \cap, ^-, \emptyset, E, \circ, ^\smile, Id, \underline{\nabla} \rangle \in$ PFA with $\underline{\nabla}$ induced by $\star : U^2 \to U$ and $s \in$ Sec, then $\underline{s} \cap Id = Id_{fix_s(\star)}$.

Proof. The proof is analogous to that of Prop. 9. ∎

Proposition 19. Let $\mathcal{F} = \langle F, \cup, \cap, ^-, \emptyset, E, \circ, ^\smile, Id, \underline{\nabla} \rangle \in$ PFA simple with $\underline{\nabla}$ induced by $\star : U^2 \to U$ and $s \in$ Sec, then $Si_{\underline{s}}(\mathcal{F}) = \wp\left(Id_{fix_s(\star)} \right) \cap F$.

Proof. The proof is analogous to that of Prop. 10. ∎

Proposition 20. Let U be an infinite set $|U| = \kappa$ and $\aleph_0 \leq \kappa$ then for all $S \subseteq U$ such that $|S| < |U|$, there exists $\star_S : U^2 \to U$ injective such that given $s \in$ Sec, $|fix_s(\star_S)| = |S|$.

Proof. If $|U| = \kappa$, then $|U^2| = \kappa$, $|Id| = \kappa$ and $|\overline{Id}| = \kappa$. Then, let $\{S_i\}_{1 \leq i < long(s)}$ be a finite family of sets such that:

- for all $1 \leq i, j < long(s)$, such that $i \neq j$, $B_i \cap B_j = \emptyset$, and
- for all $1 \leq i < long(s)$, $|B_i| = |S|$.

Analogous to previous results, we know that $|S \cup \bigcup_{i=1}^{long(s)} S_i| = \kappa$ and, therefore, it is possible to take $\overline{S \cup \bigcup_{i=1}^{long(s)} S_i} = \bigcup_{i \in \mathbb{N}} B_i$ such that:

- for all $i, j \in \mathbb{N}$, such that $i \neq j$, $B_i \cap B_j = \emptyset$,
- for all $i \in \mathbb{N}$, $|B_i| = \kappa$.

Then, there exists $g : \bigcup_{i \in \mathbb{N}} B_i \to \bigcup_{i \in \mathbb{N}} B_i$ bijective and without fixpoints defined as the union of the bijective functions $\{g_i : B_i \to B_{i+1}\}_{i \in \mathbb{N}}$. Notice that such union is disjoint as the domains and codomains of all of the functions are disjoint.

Let P be a set such that $|P| = |S|$ and $l : S \to P$ be a bijective function. Then, there exists $f : (S \cup P)^2 \to B_0$ injective and a finite family of bijective functions $\{h_i^s : S \to S_i\}_{1 \leq i < long(s)}$, such that $h_0^s = id_S$. Then, it is possible to define $\star_S : U^2 \to U$ according to Table 3

Table 3 shows how $\star_S : U^2 \to U$ is defined as the union of injective functions with disjoint domains and codomains. Therefore, $\star_S : U^2 \to U$ is injective.

On the one hand, by [Lopez Pombo et al., 2020, Lemma 6] we obtain that if $u \in S$, $\langle u, u \rangle \in \underline{s}$ and, by Def. 40, that $S \subseteq \mathit{fix}_{\underline{s}}(\star_S)$ and, consequently, that $|S| \leq |\mathit{fix}_{\underline{s}}(\star_S)|$. On the other hand, for all $u \in \mathit{fix}_{\underline{s}}(\star_S)$, $u \in S \cup \bigcup_{i=1}^{long(s)} S_i$ and, consequently, $\mathit{fix}_{\underline{s}}(\mathit{star}_S) \subseteq S \cup \bigcup_{i=1}^{long(s)} S_i$ because, by the way in which $\star_S : U^2 \to U$ was constructed, $\overline{S \cup \bigcup_{i=1}^{long(s)} S_i}$ does not contain fixpoints. Then, we obtain that $|\mathit{fix}_{\underline{s}}(\mathit{star}_S)| \leq |S \cup \bigcup_{i=1}^{long(s)} S_i| = |S|$. Jointly, these two results prove that $|\mathit{fix}_{\underline{s}}(\mathit{star}_S)| = |S|$. ∎

Once again, from the previous result, it is possible to reproduce the result of Prop. 2. Analogous to what we did in the proof of Prop. 11, we must consider the use of the set S_ϕ, instead of the set $\mathit{fix}_t\left(\star_{S_\phi}\right)$, for constructing the algebra, as there might be fixpoints outside S_ϕ. Thereafter, Thm. 3 can be applied in order to guarantee the existence of infinitely many prime, big and explosive relation algebras obtained by controlling the fixpoints of $\star : U^2 \to U$ resorting to a sequence from Sec.

$u \star_S v$	S_j	P
S_i	$f(u, v)$	$\begin{cases} h^s_{i+1}\left(h^{s-1}_i(u)\right) & \text{; if } 1 \leq i < long(s) \\ & \text{and } s[long(s)-i] = \pi. \\ & \text{and } v = l\left(h^{s-1}_i(u)\right). \\ h^{s-1}_i(u) & \text{; if } i = long(s) - 1 \\ & \text{and } s[1] = \pi. \\ & \text{and } v = l\left(h^{s-1}_i(u)\right). \\ f\left(h^{s-1}_i(u), v\right) & \text{; otherwise.} \end{cases}$
P	$\begin{cases} h^s_{i+1}\left(h^{s-1}_i(v)\right) & \text{; if } 1 \leq i < long(s) \\ & \text{and } s[long(s)-i] = \rho. \\ & \text{and } u = l\left(h^{s-1}_i(v)\right). \\ h^{s-1}_i(v) & \text{; if } i = long(s) - 1 \\ & \text{and } s[1] = \rho. \\ & \text{and } u = l\left(h^{s-1}_i(v)\right). \\ f\left(u, h^{s-1}_i(v)\right) & \text{; otherwise.} \end{cases}$	$f(u, v)$
B_m	$g(u)$	$g(u)$

Table 3: Definition of $\star_S : U^2 \to U$ controlled by a non empty sequence.

Once again, from observing the generalised controlling technique of the fixpoints of $\star$ presented above, it is possible to establish certain relations between them.

Definition 41. *The functions* $\bullet +\!\!+ \bullet : \text{Sec}^2 \to \text{Sec}$ *is defined as follows: let* $* \in \{\pi, \rho\}$ *and* $s, s' \in \text{Sec}$

$$(\text{elem } *) +\!\!+ s = \text{cons} * s$$
$$(\text{cons} * s') +\!\!+ s = \text{cons} * (s' +\!\!+ s)$$

Theorem 6. *Let* $\star : U^2 \to U$ *and* $s, s' \in \text{Sec}$, $\text{fix}_s(\star) \cap \text{fix}_{s'}(\star) \subseteq \text{fix}_{s +\!\!+ s'}$.

Proof.

$$\begin{array}{lll}
u \in \text{fix}_s(\star) \cap \text{fix}_{s'}(\star) & \text{iff} & u \in \text{fix}_s(\star) \text{ and } \text{fix}_{s'}(\star) \\
& \text{iff} & \langle u, u \rangle \in \underline{s} \text{ and } \langle u, u \rangle \in \underline{s'} \quad \text{[by Def. 40.]} \\
& \text{implies} & \langle u, u \rangle \in \underline{s} \circ \underline{s'} \quad \text{[by Def. 3 - } \circ.] \\
& \text{iff} & \langle u, u \rangle \in \underline{s +\!\!+ s'} \\
& & \text{[by [Lopez Pombo et al., 2020, Lemma 8].]} \\
& \text{iff} & u \in \text{fix}_{s +\!\!+ s'} \quad \text{[by Def. 40.]}
\end{array}$$

Thus, finishing the proof. ∎

Finally, it is possible to connect t-controled fixpoints and s-controlled fixpoints of $\star$ by considering properties like the next one.

Definition 42. *The predicates* $\bullet = \bullet \subseteq \text{BT} \times \text{BT}$ *and* $\bullet < \bullet \subseteq \text{BT} \times \text{BT}$ *are defined as follows:*

$$\begin{array}{rl}
s \ll t \text{ iff} & (s = \text{elem } \pi \text{ and } t = \text{bin nil } t') \text{ or} \\
& (s = \text{elem } \rho \text{ and } t = \text{bin } t' \text{ nil}) \text{ or} \\
& (s = \text{cons } \pi \ s' \text{ and } t = \text{bin } t' \ t'' \text{ and } s' \ll t') \text{ or} \\
& (s = \text{cons } \rho \ s' \text{ and } t = \text{bin } t' \ t'' \text{ and } s' \ll t'')
\end{array}$$

Theorem 7. *Let* U *be an infinite set* $|U| = \kappa$ *with* $\aleph_0 \leq \kappa$, $S \subseteq U$ *such that* $|S| < |U|$, $\star_S : U^2 \to U$ *injective and* $t \in \text{BT}$ *then, for all* $s \in \text{Sec}$, $s \ll t$ *implies* $\text{fix}_t(\star_S) \subseteq \text{fix}_s(\star_S)$.

Proof. The proof of this theorem follows from Def. 29, 40 and 42, and applying [Lopez Pombo et al., 2020, Lemma 9]. ∎

4 Conclusions

As we mentioned at the beginning of this work, binary relations are ubiquitous in computer science as they provide the concept perfectly fitted for formalising programs by rationalising them as the connection between its inputs and its outputs. In this context, the language of relation algebras is expected to provide the reasoning tool for program verification, derivation and refinement. The mismatch between the models of the calculus of relations (see Defs.6 and 5) and the class of proper relation algebras (see Def. 3), evidenced by Lyndon in [Lyndon, 1950, 1956], by constructing a finite, non-simple and non-trivial relation algebra that is not representable as a proper relation algebra, results in a major drawback for its adoption as a specification language and formal development tool.

The study of the relational reduct of fork algebras, started and promoted by Paulo A.S. Veloso in [Veloso, 1996b,a], is of great interest for the community of applied relational methods in computer science as fork algebras, thought of as the models of the calculus for fork algebras (see Defs. 17 and 16), are representable in proper fork algebras (see Def. 13), a class of algebras whose carrier is formed by binary relations.

In this paper we summarised some of Velosos's results in this field, like the construction of explosive relation algebras, by controlling the fixpoints of $\star : U^2 \to U$. Our contribution is twofold; on the one hand, a generalisation of such a construction by introducing the notion of t-controlled fixpoints of $\star : U^2 \to U$, where t is a term induced by a tree-like structure and, on the other hand, the controlling technique based on the use of the pseudo-projections π and ρ, as an alternative to the $\triangledown$-controlled one, introduced by Veloso. Finally, we generalise the technique by introducing the notion of s-controlled fixpoints of $\star : U^2 \to U$, where s is a term induced by a sequence-like structure.

References

Burris, S. & Sankappanavar, H. P. (1981), *A course in universal algebra*, Graduate Texts in Mathematics, Springer-Verlag, Berlin, Germany.

Enderton, H. B. (1972), *A mathematical introduction to logic*, Academic Press.

Frias, M. F. (2002), *Fork algebras in algebra, logic and computer science*, Vol. 2 of *Advances in logic*, World Scientific Publishing Co., Singapore.

Frias, M. F. & Lopez Pombo, C. G. (2003), Time is on my side, *in* R. Berghammer & B. Möller, eds, 'Proceedings of the 7th. Conference on Relational Methods in Computer Science (RelMiCS) - 2nd. International Workshop on Applications of Kleene Algebra', Malente, Germany, pp. 105–111.

Frias, M. F. & Lopez Pombo, C. G. (2006), 'Interpretability of first-order linear temporal logics in fork algebras', *Journal of Logic and Algebraic Programming* **66**(2), 161–184.

Frias, M. F. & Orlowska, E. (1998), 'Equational reasoning in non-classical logics', *Journal of Applied Non-classical Logics* **8**(1–2), 27–66.

Frias, M. F., Baum, G. A. & Haeberer, A. M. (1997a), 'Fork algebras in algebra, logic and computer science', *Fundamenta Informaticae* **32**, 1–25.

Frias, M. F., Baum, G. A. & Maibaum, T. S. E. (2002), Interpretability of first-order dynamic logic in a relational calculus, *in* H. de Swart, ed., 'Proceedings of the 6th. Conference on Relational Methods in Computer Science (RelMiCS) - TARSKI', Vol. 2561 of *Lecture Notes in Computer Science*, Springer-Verlag, Oisterwijk, The Netherlands, pp. 66–80.

Frias, M. F., Baum, G. A., Haeberer, A. M. & Veloso, P. A. (1995a), 'Fork algebras are representable', *Bulletin of the Section of Logic* **24**(2), 64–75.

Frias, M. F., Galeotti, J. P., Lopez Pombo, C. G. & Roman, M. (2005), Fork algebra as a formalism to reason across behavioral specifications (extended abstract), *in* I. Düntsch & M. Winter, eds, 'Proceedings of the 8th. Conference on Relational Methods in Computer Science (RelMiCS) - 3nd. International Workshop on Applications of Kleene Algebra', St. Catharines, Ontario, Canada, pp. 61–68.

Frias, M. F., Haeberer, A. M. & Veloso, P. A. (1995b), 'A finite axiomatization for fork algebras', *Bulletin of the Section of Logic* **24**(4), 193–200.

Frias, M. F., Haeberer, A. M. & Veloso, P. A. (1997b), 'A finite axiomatization for fork algebras', *Logic Journal of the IGPL* **5**(3), 311–319.

Gyuris, V. (1997), 'A short proof of representability of fork algebra', *Theoretical Computer Science* **188**(1–2), 211–220.

Haeberer, A. M. & Veloso, P. A. (1991), Partial relations for program derivation: adequacy, inevitability and expressiveness, *in* 'Proceedings of IFIP

TC2 working conference on constructing programs from specifications', IFIP TC2: Software: Theory and Practice, North Holland, pp. 310–352.

Haeberer, A. M., Baum, G. A. & Schmidt, G. (1993a), Dealing with nonconstructive specifications involving quantifiers, Monografias en Ciências da Computação 4/93, Departamento de Informatica, Pontifícia Universidade Católica do Rio de Janeiro.

Haeberer, A. M., Baum, G. A. & Schmidt, G. (1993b), On the smooth calculation of relational recursive expressions out of first-order nonconstructive specificationes involving quantifiers, in D. Bjørner, M. Broy & I. V. Pottosin, eds, 'International Conference on Formal Methods in Programming and Their Applications', Vol. 735 of Lecture Notes in Computer Science, Springer-Verlag, Academgorodok, Novosibirsk, Russia, pp. 281–298.

Huntington, E. V. (1904), 'Sets of independent postulates for the algebra of logic', Transactions of the American Mathematical Society 5(3), 288–309.

Jónsson, B. & Tarski, A. (1951), 'Boolean algebra with operators, part I', American Journal of Mathematics 73, 891–939.

Jónsson, B. & Tarski, A. (1952), 'Boolean algebra with operators, part II', American Journal of Mathematics 74(3), 127–162.

Lopez Pombo, C. G., Frias, M. F. & Maibaum, T. S. E. (2020), 'On the construction of explosive relation algebras', On-line. Available at https://arxiv.org/abs/2009.02720.

Löwenheim, L. (1915), 'Uber Möglichkeiten im Relativkalkul', Mathematische Annalen 76, 447–470. See [?, pp. 228–251] for an english account, and translation, of this work.

Lyndon, R. C. (1950), 'The representation of relation algebras, part I', Annals of Mathematics (series 2) 51(2), 707–729.

Lyndon, R. C. (1956), 'The representation of relation algebras, part II', Annals of Mathematics (series 2) 63(2), 294–307.

Mikulás, S., Sain, I. & Simon, A. (2015), 'Complexity of equational theory of relational algebras with standard projection elements', Synthese 192(7), 2159–2182.

Monk, J. D. (1964), 'On representable relation algebras', Michigan Mathematical Journal 11, 207–210.

Schöder, F. W. K. E. (1895), *Algebra und Logik der Relative, der Vorlesungen über die Algebra der Logik*, Vol. 3, Abt. 1, Teubner, Leipzig.

Tarski, A. (1941), 'On the calculus of relations', *Journal of Symbolic Logic* 6(3), 73–89.

Tarski, A. (1943–1945), Untitled book manuscript containing some of Tarski's early contributions to the theory of relation algebras, written during the period 1943 to 1945, The book was never published, but most of the results in the book were later included in Tarski & Givant [1987].

Tarski, A. (1955), 'Contributions to the theory of models. III', *Indagationes Mathematicae (Proceedings)* **17**, 56–64. Also in ?.

Tarski, A. (1956), 'Equationally complete rings and relation algebras', *Indagationes Mathematicae (Proceedings)* **18**, 39–46. Also in ?.

Tarski, A. & Givant, S. (1987), *A formalization of set theory without variables*, Vol. 41 of *Colloqium Publications*, American Mathematical Society, Providence, RI, USA.

Veloso, P. A. (1996a), On finite and infinite fork algebras, Monografias en Ciências da Computação 05/96, Departamento de Informatica, Pontifícia Universidade Católica do Rio de Janeiro. See also Veloso [1996b].

Veloso, P. A. (1996b), On finite and infinite fork algebras and their relational reducts: classification and examples, Technical Report ES-418-96, Programa de Engenharia de Sistemas e Computação, COPPE, Departamento de Computação, Instituto de Matemática, Universidade Federal do Rio de Janeiro. See also Veloso [1996a].

Veloso, P. A. (1996c), Some connections between logic and computer science, *in* W. A. Carnielli & I. M. D'Ottaviano, eds, 'Proceedings of Eleventh Brazilian Conference on Mathematical Logic', Vol. 235 of *Advances in Contemporary Logic and Computer Science*, American Mathematical Society, pp. 187–260.

Veloso, P. A. & Haeberer, A. M. (1991), 'A finitary relational algebra for classical first-order logic', *Bulletin of the Section of Logic* **20**(2), 52–62.

Veloso, P. A., Haeberer, A. M. & Baum, G. A. (1992), On formal program construction within an extended calculus for binary relations, Monografias en Ciências da Computação 19/92, Departamento de Informatica, Pontifícia Universidade Católica do Rio de Janeiro.

Dedicated to Paulo A.S. Veloso in his 70th. birthday

I (Tom Maibaum) am writing this dedication on behalf of the two other authors and the ghost in the room, Armando Haeberer. Of the co-authors and the ghost, I have known Paulo by far the longest. We met in early 1977 when I spent 4 months at PUC in Rio and, unexpectedly, formed the longest research collaborations of my career. To this saintly group were added, in due course, Armando, Gabriel Baum, Marcelo and Charlie. What we all benefited from enormously was Paulo's combination of extensive general knowledge of logic and his fundamental understanding of and intuition for the basics of computing. He, like me, understood intuitively that the algebraic specification program was fundamentally misconceived because it used a model of software engineering that bore no relation to the fundamentals of the discipline. This fact was tacitly established when, in 1989, Broy and Wirsing published a paper replacing the initial algebraic approach to specification with the FOL based framework that Paulo and I had developed. The approach has not been heard of since. (Of course, people who know me might suspect that I tend to exaggerate just a little bit!)

The work on things relational and, in particular, fork algebras, followed the arrival of Armando in Rio. Paulo was Armando's PhD supervisor and Armando was what is euphemistically called a "mature" student. (That is, he was "old"!) The contrast in characters was of Grand Canyon proportions. Paulo was a quiet, reserved person, whereas Armando was the polar opposite: dramatic and flamboyant. Together, amongst other things, they set out to solve a longstanding problem in relation algebra, the finite axiomatisability of relation algebra, the so called Monk/Tarski problem. They did this by introducing an extra operator called the fork (see below). Unfortunately for them, this seemed to have caused offence to a group led by Andreka and Nemeti, who argued that because of various technical minutiae the problem was not actually solved. They became known in our circles as the "evil Hungarians". (The fact that I was also a Hungarian mathematician did not act as a deterrent.)

Marcelo then arrived at PUC to pursue a PhD under the supervision of Armando. His book on fork algebra and programming was a very good reflection of his work in this period. After Marcelo's return to Buenos Aires, Charlie became his PhD student at UBA. Together they put fork algebra in the center of the discussion about heterogeneous software specification, and the need for the glueing formalism to have good mathematical properties, a finite and complete calculus for an easy to understand semantics, providing the foundation for building tools supporting the process of software design and construction, a topic on which we have worked together for many years now.

Thus, the circle is complete: starting with my collaboration with Paulo, through his collaboration with Armando, then through Gabriel, Marcelo and Charlie and back to me. Always supported by our interaction with and motivation by Paulo. Always informed by his wisdom, knowledge and good humour.

Euclidean Machines: General Theory of Problems[‡]

Wagner de Campos Sanz

* Faculdade de Filosofia,
Universidade Federal de Goiás,
Campus Samambaia, Goiânia–GO, 74690-900, Brazil
wsanz@ufg.br

Abstract

Our subject is the Euclidean Geometry approached via one of the most successful mathematical models of XXth century: Turing machines. The comparison is made through the concept of problem, in the hope that it becomes an important trend in contemporary epistemology, connected to mathematical logic. There are some reasons for seeing analogies between Turing machines and constructions in *propositiones* of plane geometry. This essay is our humble way of doing homage to our colleague and fellow logician Professor Paulo Veloso who gave important contributions for the development of a General Theory of Problems. The paper suggests a way of interpreting Euclidean geometry with an ontology of actions by means of problems.

Dedicated to Professor Paulo Augusto Silva Veloso

1 Introduction

According to the Church Thesis,[1] Turing machines compute every intuitively calculable function. It is only a thesis, no proof can be afforded for it since its statement involves and intuitive non-formal concept. However, there are evidences giving it some support. One is the equivalence among distinct theoretical formal models—recursion, lambda-calculus, etc.—, all

---

[‡]We express our gratitude to Dr. P. Viana and Dr. M. de Castro for valuable comments and suggestions. We also thank professor L.C.D.Pereira for the discussion of an earlier version of this paper in his online seminar at PUC-Rio.

[1]Some call it Church-Turing Thesis. See p.317 of [Kleene, 1952].

them proposed as formal counterparts of the intuitive notion of calculability. Turing machines compared to other models are appealing in a specific way. They involve *anima*, that is, movement. The movements can be compared to the procedures employed by human calculators when using paper, pencil and rubber.

In what follows, we point parallels between those theoretical constructs called Turing machines and one of the component structures in *propositiones*, or *protases*, asking a construction in Euclids' geometry. Two objectives are envisaged here. First, to propose a way of looking to computability without restricting it to syntax, using diagrams instead. If sucessfull, this might imply a revision of Church thesis. Second, to sketch an alternative way of looking into mathematical ontology by using the concepts of problem and action as grounds through a discussion of constructions.

Propositiones in geometry are since long ago divided in two kinds: problems and theorems. But from XIXth century on, Euclidean geometry presentations have transformed it into a purely theorematic set of statements like Hilbert's Foundations of Geometry. Problems then seem to disappear by magic. But, historically, problems have been at the heart of mathematics. Any philosopher knowing a modest portion of history of science has noticed it. And they are in good company. Making reference to ancient geometry, Kant (Jäsche Logic, § 38) observes that:

> A postulate is a practical, immediately certain proposition, or a principle that determines a possible action, in the case of which it is presupposed that the way of executing it is immediately certain. Problems (problemata) are demonstrable propositions that require a directive, or ones that express an action, the manner of whose execution is not immediately certain. [. . .] Note 2: A problem involves (1.) the question, which contains what is to be accomplished, (2.) the resolution, which contains the way in which what is to be accomplished can be executed, and (3.) the demonstration that when I have proceeded thus, what is required will occur.

For a presentation concerning the role of problems in Kant's philosophy of mathematics see [Lassalle-Casanave, 2019].

More recently, at the beginning of XXth century, intuitionistic logic, and thus logic, received a problem interpretation for the first time, as far as we know. [Kolmogorov, 1932] problem interpretation contains the basis for a more comprehensive theory of problems.

A few years ago, [VonPlato & Mäenpää, 1990] used the problem approach to examine Euclids' geometry. Actually, they relied on Martin-Löf's intuitionist type theory in their analysis, assuming this to be a natural development of Kolmogorov's original idea concerning problems and logic. In what concerns geometry and its virtual relation to logical inference rules, they state (*ibid.*,p.281):

> The construction postulates lay down the permitted means of producing finite straight-lines, ... The functionality of postulates suggests a way of rendering them into the general pattern of natural deduction rules used in intuitionistic type theory. Its inferences may be viewed as functions from premises to conclusions. This proof functionality is explicitly recorded in proof objects, that is, in the objects given in the left side of judgements of the form a:A. It is judgements, not propositions, which figures as premises and conclusion in an inference rule.

But, this problem analysis of geometry based on intuitionistic type theory assumes a standpoint that seems alien to ancient geometry. The ontology of types is a set ontology and there are reasons to believe that an ontology of actions fits better Euclids. Additionaly, the conception of what is a hypothesis/assumption for intuitionists, and for Martin-Löf in particular, might not correspond with the way hypotheses are used in Euclidean reasonings. These two items turn to be central for an examination of the concept of geometrical construction. For a discussion concerning the difficulties in the approach of Von Plato and Mäenpää, see [Naibo, 2018].

There is a simpler way of using the concept of problem for interpreting *propositiones* in geometry. This is going to be developed bellow, although it is not our main concern here. Our main objective is to analyze *propositiones* by comparing their solutions to Turing machines appealing to the problem interpretation. And, in order to formulate the basics of a General Theory of Problems, we rely on Kolmogorov interpretation (*ibid.*) as also [Veloso, 1984] analysis of problem resolution strategies.

The assumption that *propositiones* in Euclids are problems specially those asking a construction, and its solution an algorithm, is not new as we can see in Kant's remark. The idea is somehow present in Von Plato and Mäenpää (*ibid.*) as also p.20 of [Beeson, 2010]:

> The basis for the work described here is the idea that in geometry, we can take "algorithm" in the restricted sense of "geometric construction". That is, we pursue the analogy
>
> formal number theory / Turing computable functions
>
> =

intuitionistic geometry / geometric constructions

Actually, Beeson proposes a formal syntactical system for Euclidean geometry. His system is closely analyzed in [Naibo, 2018] who raises some doubts concerning Beeson's formal system adequacy.

As we see it, the contemporary recast introduces ontological commitments that might be alien to Euclidean geometry. Hilbert's geometry, for example, transforms postulates into existential assertions, changing their original nature. The same happens with syntactical formal systems, in the measure that all *propositiones* are reinterpreted as assertions, not problems.

Among recent scholarship, [Sidoli, 2018] examines Euclids Elements and discusses what could be its underlying notion of construction. Parts of this paper are going to be closely inspected here. The author acquaints the influence of Von Plato and Mäenpää (*ibid.*) as also Beeson (*ibid.*) on his own work. He closes his exposition by saying, (*ibid.*,p.449):

> Hence, the statement of the postulates, and their articulation in the problems of the Elements, can be taken as a foundational project, the goal of which was to provide a set of tools for demonstrating that certain objects can be produced as the result of an effective–that is, a well-ordered and finite–procedure.

We think his point of view worth of appraisal and we suspect that the foundational project mentioned concerns problems. But we have two criticisms of Sidoli's perspective. They concern the question of canonical means for introducing points and the interpretation of geometry as the theory of straightedge and compass. Both issues involve the crucial question of interpreting postulates as mathematical constructive functions.

The exegetical interpretation of Euclids Elements is not our objective here, although some few remarks might go in that direction. We want rather to develop a set of theoretical tools for analyzing problems such that they make sense when applied to ancient geometry, thus creating room for an ontology of actions in mathematics. The focus here lies over the concepts of problem, solution and construction, and it is carried together with an examination of issues in Euclids' geometry as a historical document.

Some approaches mentioned above have as focus the concept of construction. Curiously, they seem to assume that the intuitionistic conception of construction should preside the examination of ancient geometry. From our point of view, this choice turns the historical perspective upside down. The Euclidean text is a first-hand historical document for considering the issue. Hence, it should be taken inversely as a document. It is the examination of intuitionistic claims about constructions that has to undergo a critical analysis using the Euclidean text as a given evidence. The concept of mathematical constructions should be investigated with one foot in the history of the discipline.[2]

Acquaintance with Turing machines is here presupposed, since they are well known in general. Further details can be found in the literature, like §67 of [Kleene, 1952].

2 Euclidean geometry

As already pointed, *propositiones* of plane geometry can be divided in two groups. Those we call problem *propositiones* where it is asked a geometrical construction–usually finalized with the expressions "as it was to be build"–and the property *propositiones* where it is asked a demonstration of a geometrical property–finalized with the expression "as it was to be proved".

Problem *propositiones* can be analyzed with the concept of Turing machines in the following sense. Seen from the point of view of action, the statement of a problem formulates an action to be accomplished (not a proposition in the contemporary sense), as for example: to build an equilateral triangle given a line segment, or to sum two given numbers. Therefore, in the same way a sum Turing machine is a machine summing two numerals according to an algorithm, then the machine solving a geometrical problem according to an algorithm as described in the *kastaskeue* of the problem *propositio* is an Euclidean machine, by analogy.

An Euclidean machine acts over a geometrical configuration, not over geometric concepts. One of the main elements of a Turing machine is its tape. The tape is divided in squares and each square contains one symbol. Each square can be read, erased and written with one symbol. The set of words that can be written in a tape are the possible configurations a tape can assume. By analogy, an Euclidean machine also assumes distinct configurations. The diagrams are examples of such configurations in this case. The exposition here is supposed to cover constructions from book I to VI.

---

[2]Hardly there would be a definitive concept of construction with which all mathematicians would be satisfied for eternity. Nonetheless, contemporary proposals concerning what should be understood as a construction must be subjected to a critical examination on the background of what has been done historically in mathematics.

2.1 Problem resolution in Euclidean geometry

Kant's conceptual analysis of problems quoted above presents the essentials for conceiving what is a problem. First, there is the task to be done (which includes the case of giving an answer to a question); second, the solution of the task; and third, this is optional (but not in the case of mathematics), the argumentative regimentation of evidences guaranteeing that the solution fits the task intended.[3]

We assume the concept of problem as being pivotal for the epistemology going to be developed bellow. A problem is a task to be effected. The solution is the complex action that accomplishes the task, but it can also be understood as the recipe for accomplishing the task.[4] In a General Theory of Problems we are interested in recipes as solutions.

According to the above division, problems are going to be differentiated into concrete problems and type problems. For a concrete problem we expect the accomplishment of a task. For a type problem we expect an algorithm to be provided describing the actions that would solve that kind of problems. A type subsumes a collection of concrete problems and subordinates a collection of subtype problems.[5] Ou focus lies over types for which the actions can be described, as in a recipe, for obtaining the solution.

Concrete problems and type problems can be categorized under different type problems. To draw a rectangle and to draw a square are both types subordinated to a more general type, for example, that of drawing a parallelogram. A concrete specific drawing of a square is by its turn subsumed under the more general problem, or type, of drawing squares, but it is also subsumed under the type of drawing rectangles, of drawing parallelograms, etc. The relation between types is complex.

By definition, in a General Theory of Problems we focus on type problems and their solutions. Euclidean geometry problem *propositiones* become then an object of investigation for such a theory. In each problem *propositio*, the parcel of the exposition that corresponds to the exhibition of a solution, that is of an algorithm or construction, is contained in the so called *kataskeue*. The verification of the solution, in its turn, is contained

---

[3]One of the essential characteristics of mathematical knowledge is the requirement of this third component which in practical situations can be dismissed. But, this characteristic is not exclusive of mathematics.

[4]The first sense can be assimilated to the second if the description of the acts accomplished is taken as the recipe.

[5]One example of concrete problem is that of finding four numbers in Fermat's last theorem.

in the so called *apodeixis*.[6]

2.1.1 A case study

In this section we analyze *propositio* III.1 of Euclids Geometry. We follow Sidoli's (*ibid.*) careful examination. It is a construction problem asking, for a given circle ABC, to find its center. The center exists by definition. We apply and expose the tools of the Problem Theory on this this example. In particular we discuss at length the counterfactual hypothesis employed in the *apodeixis* and the notion of construction that could be adequated for interpreting the use of such hypothesis.

The diagram accompanying the *propositio* is the following:

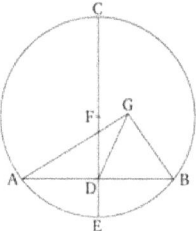

Figure 1: Diagram for *Propositio* III.1

It contains the drawings produced by the solution and, this is important, those necessary for its verification. The starting diagram would be like the following:

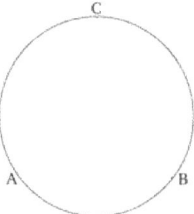

Figure 2: Initial Diagram for *Propositio* III.1

A circle must have a center, by definitions I.15 and I.16. Hence, it is virtually impossible to say that problem III.1 would be about proving its

---

[6]However there might be some disagreement of where exactly starts one and ends the other.

existence.[7] It is then merely the problem of singularizing the center. This observation has important consequences for understanding the role of definitions in ancient geometry. They establish which elements should go hand in hand. They serve to point over a diagram what is what and, many times, what it is not: what is a point, a straight-line, a circumference, etc. A point cannot be a straight-line since it has no part and no dimension. It is not clear that the terms and their statement should be understood as complete identity criteria, i.e., sufficient and necessary conditions for identification. If they can serve the purpose of distinguishing and naming elements in a diagram, this is enough. The postulates could eventually complete the distinguishing task.

That the required solution of III.1 be that of just singularizing something that must be already there in the diagram, by definition, makes doubtful any interpretation of ancient geometry where this *propositio* is transformed into an existential assertion concerning the referred point when all *propositiones* become theorems. It cannot be uniformly assumed that the solutions offered to geometrical problems constitute a demonstration of existence of an element, and neither that postulates are existential claims.

Propositio III.1 brings to the forefront the recognition question. Given a diagram, if an element is recognized as being a circle, then it is inescapable that according to the definition there must be a central point although not singularized in the diagram. The recognition also seems to require the assumption that the element was obtained through a procedure in line with postulate I.3.

It is not difficult to imagine a situation where a drawing is not recognized as being a circle. For example, if it is recognized as being an ellipsis. Then, there would be no point in trying to find its center.

The solution to III.1, its recipe, consists in: (i) to draw or produce a chord AB;[8] (ii) next, to find the medium point D of AB; (iii) next, to draw the straight-line CE perpendicular to AB, passing through D; (iv) in the sequel, to find the medium F of CE. F is then said to be the center of the circle. The actions are to be done exactly in the order they were given, on pain of not properly solving the problem.

Sidoli (*ibid.*,p.408) designates this parcel of the *propositio* treatment as **problem-construction**. It belongs to what is called the *kataskeue*. This parcel of the exposition is followed by another where the solution is proved

---

[7]However, it can be understood as the proof of existence of an algorithm. I own this observation to prof. Petrúcio.

[8]The concept of a chord is many times used inside Euclidean geometry, but no specific word is given in the definitions. They are non-deterministically described in the *kataskeue*, as is the case of AB in III.1.

to be correct, that is, one in which F is proven to be the center of the circle. The Greek term for the second is *apodeixis*. But, between the solution to the problem and the proof of correcteness there are some drawings involving point G and three lines.

Now, how should we interpret this last drawings? Is it part of the *kataskeue* or part of the *apodeixis*? The new elements do not make part of the solution properly speaking, they are envisaged in order to illustrate the problem of proving that F has the desired property. Sidoli calls such addition of elements a **proof-construction**. But what would be such a "construction"?

The "construction" is actually anticipated by a linguistic-argumentative move, a counterfactual supposition:[9] *suppose that F were not the center.* It seems natural to understand that the counterfactual supposition starts the *apodeixis*, the correctness proof.

Now comes the question of how to understand the counterfactual supposition. According to intuitionists, a constructive assumption involves the supposition of having a proof of an assertion.[10] Hence, if we were to assume that the supposition in question means (A) *to suppose the possession of a proof that F were not the center*, then we would be pressuposing the possession of a construction of (at least) two straight-lines from F to points H and I on the circumference of ABC and such that FH is not equal to FI.[11] But then, since FE and FC are equal by construction, CE has to be merely a chord of ABC but not the diameter. In that case, D would not be the middle point of the chord AB perpendicular to CE, contrary to what has been done (and proved as so) through the use of *propositio* I.10 in the *kataskeue*. Therefore the construction of FH and FI is impossible. However, the argument in the *apodeixis* does not even resemble this argument. Does it mean that the reasoning is non-constructive?[12] We don't think so, let's see why.

The supposition that F were not the center is immediately turned into the statement that a G different of F would then be the center. And this move is correct.[13] Now, should this be considered an equivalent hypothesis

---

[9]In the text that follows the counterfactual nature of the supposition is presented through the use of the past subjunctive.

[10]See [Matin-Löf, 1996].

[11]This supposition is weird, after all the *kataskeue* just gave an algorithm for finding F as the center. In case there were another construction showing that F is not the center of ABC, we would have two algorithms, each supporting one statement of a pair of contradictories.

[12]Remember that intuitionists tend to consider invalid any *reductio ad absurdum* reasoning starting with a negation $\neg p$ as the hypothesis and concluding p.

[13]That a G different of F were the center of the circle is a statement equivalent to the

or merely a conclusion extracted from the former counterfactual hypothesis? Even in case that the second reading is the one to be chosen as the correct interpretation, not much will be changed in the way we should interpret the whole argument, only its final step which becomes a classical *reductio* concluding that F is the center. But, if it is an equivalent sentence, then the supposition is indeed displaced over the new sentence. And again, there are two ways of understanding this equivalent supposition. Either it is understood as (B) *the supposition of possessing a construction/algorithm singularizing a G different of F as the center* or (C) *pretending that a G different of F were the center, without assuming the possession of any construction/algorithm.*

It is not difficult to realize that from an intuitionist point of view the supposition (B) is completely different from supposition (A) since the construction that directly proves one is not a construction that could prove the other. For proving (B) it is required to show a construction guaranteeing that any line segment from G to the circumference is equal to another. (C) is our preferred interpretation for what is being accomplished in the argument.

Where is G? We think that it does not make sense to try to answer this question with precision, once G has been **introduced** hypothetically. At best, we could say that point G should be inside the circle, in accordance with definitions I.15 and I.16. However, observe that in the diagram there is (i) a point G singularized and distinct of F. From this point are drawn lines (ii) GA, (iii) GD and (iv) GB, respectively, in accordance with postulate I.1. With these four steps what Sidoli called as a proof-construction is completely finished in this case. Thus, if the supposition were meant as a supposition of having a construction of point G, we should conclude that the proof-construction developed after this supposition just arrived safely to its final destination with triangles ADG and BDG. That is, nothing barred the accomplishment of such a construction here. This observation is relevant, since intuitionists usually claim that a *reductio* argument starting with a truly false assumption must stop at some point since it cannot go on any longer with the construction.[14] This is clearly not the case here!

The Euclidean proof continues linguistically and it involves a "game" with properties. The argumentation that follows will arrive at an absurd[15],

---

statement that F were not the center of the circle ABC. Since F is inside the circle, if a G different of F were the center, then F would not be the center because there is only one center in the circle, by definition. By another side, if F inside circle were not the center, as there must be a center inside the circle, let's say G, then G must be different of F.

[14] See §2.3 of [VanAtten, 2017].

[15] The argumentation consists basically in saying that triangles ADG and BDG should be equal if G were the center of the circle. Necessarily, GA and GB should be equal if

then refuting the counterfactual equivalent sentence that a different G were the center of circle ABC. If a refuting argumentation is producible for any point G distinct of F inside the circle, then any of these points would not be the center of the circle. Hence, constructively speaking, F has to be the center, by exclusion. At least this is how a constructive rendering of the argument can be presented, using *reductio* over the non-negative hypothesis, i.e., that G were the center and distinct of F. The same reasoning classically interpreted starts with the counterfactual supposition of a negation: that F were not the center of the circle. Once an absurd is reached, the positive conclusion can be asserted: F is the center of the circle. In any case, point G is being used as a generic object in the process of refutation.

Sidoli (*ibid.*,p.443) employs the terms "semi-constructive" and "non-constructive" for qualifying the elements added to the diagram after the solution was found, point G in particular. Basically, he says, these elements were not introduced by the rules he understands to be the right constructive interpretation of the postulates.

From our perspective, the author assumed uncritically a notion of constructivity that is alien to ancient geometry. The postulates, mainly the first three, are being interpreted as introduction rules for the geometrical objects following the path of Von Plato and Mäenpää (*ibid.*). This solution, in its turn, is umbilically linked to Martin-Löf's type theory. It treats the three first postulates as canonical constructive functions. For example, Sidoli interprets postulate I.1 as a routine–or a mathematical (introduction) function–that receives two points as parameters and gives back the straight-line between them, then assimilating the postulate to a kind of inference rule defining terms.

Nonetheless, it is debatable that in Euclidean geometry there are canonical ways of introducing points. In general, the intersection of lines singularizes a point: straight and straight-line, straight and circumference, etc. But it is not clear that in order to drawn or to produce a straight-line there should be two points identified and singularized beforehand. And this would be necessary if the first postulate is interpreted as an introduction rule depending on two points.

The act of drawing or producing a line can well be an act of will that creates its own starting and stopping points, reason why they are essentially distinct. The points are singularized in function of the action of producing

---

they were radius of the same circle. As the other sides of the triangle should be equal by construction, then angles ADG and BDG should be equal, according to *propositio* I.8. They would be right-angles. But FDB is a right-angle. Since all right-angles are equal according to postulate I.4, DGB would be equal to FDB. But it's only a (proper) part of FDB. Impossible.

the line, one as a stop point and the other as an origin point, it does not matter which is which. The problem with Sidoli's, Von Plato and Mäenpää, interpretation becomes visible when we realize that the routine associated with this introduction rule only works under the presupposition that the two points are distinct, otherwise the result would not be a straight-line[16], requiring then that a previously defined relation of apartness among points be presupposed here. But there is nothing like that in the Euclidean text. Did Euclids assume it tacitly? We doubt.

Now, how to explain that the final diagram of III.1 under our eyes singularizes a point G? As suggested, it seems that the objective of taking one G whatever exemplifies the action of taking an object in the role of a generic object. The point G singularized is different from F and it is inside the circle, in accordance with the counterfactual hypothesis. However, G could be in any place inside the circle. Hence, in order to fully work as a generic object, the two triangles obtained by drawing the three lines from G to A, B and D, respectively, should allow the development of the *reductio* argument no matter which G was picked, giving then a true generality to the whole argument. In this way, G is then simply picked up different of F, and there is no need to presuppose any construction singularizing it as the center, after all this is what was accomplished in the *kataskeue* but with F.

How did the absurd spring? The absurd is the result of some illations made over the triangles formed by the lines GA, GB and GD. It consists in saying that the angles FDB and GDB would be equal since we had concluded that GDB is a right-angle triangle–and here is the error, since one (GDB) is part of the other (FDB). Notice that if point G were picked up on the other side of CE something would change in the discourse, although the construction and the structure of the argumentation would remain almost the same. In such case we still arrive at an absurd, but now FDB would be part of GDB.

As the discourse has an important role for reaching the absurd, it seems more appropriated to say that the straight-lines GA, GD and GB are–in the discourse–hypothetically produced between the three points effectively singularized in the diagram by the construction of the solution–i.e., A, D and B–and a hypothetical generic point G whatever, different of F. This way, the triangles ADG and BDG should also be considered discursively hypothetical, even if they are singularized in the diagram. The singularization of point G is made at random in order to represent the counterfactual center, but any other could have been taken.

---

[16]See on this respect [Beeson, 2010] and [Naibo, 2018]

From our perspective, G's singularization serves the purpose of illustrating the argument of reduction to absurd. Different choices change details of the argument, but not the hypothetical construction of the triangles, with one exception, as we are going to see in the next paragraph. The absurd springs from, and among, "sayings". It involves the statement that FDB would be equal to GDB. The statement that one of them would be part of the other is a fact, once given an arbitrary point G different of F inside the circle. Concretely, the illustrating drawing shows that GDB is part of FDB. And the verification is inescapably based on the diagram, it is perceptual. But this does not change the fact that G is being supposed to be the center of circle. As the contradiction arises from sayings, had we not supposed that G were the center, no contradiction would have been obtained. The diagram itself is not contradictory, neither impossible nor absurd. Impossible is to maintain at the same time the two statements, one of them directly extracted from the counterfactual hypothesis adopted, the other obtained by visually inspecting the diagram.

Finally, notice that the way point G was picked up leaves gaps in the argumentation and in the notion of generic object. The counterfactual hypothesis implies that G be different of F and that G be inside the circle, nothing else. But G could then be equal to D. In that case, the triangles would not be obtainable by tracing straight-lines from G to A, B and D, respectively, thus apparently breaking down the argument. Also, if G were chosen in any other place over CE, the triangles would be obtained but not the absurd.[17] So, strictly speaking, the G picked up is the most generic object for the argumentation if we exclude points on the straight-line CE. Observe, however, it is absolutely clear that only F could be the center over CE since FE and FC are equal straight-lines once CE was divided in the middle according to *propositio* I.10.

Sidoli (*ibid.*,p.444) considers two possibilities concerning the counterfactual above: either it is a supposition of existence of such a G or it is a supposition that the property of being the center holds for G. We gave above the reasons why we choose the second alternative. But we see with surprise the author saying that the introduction of point G is non-constructive. For sure, it was taken almost randomly. But, the meaning the author seems to give to the term "constructive" is historically misleading, as already pointed. Although there are distinct ways of singularizing a point in geometry, a choice by will like that of a hypothesis cannot be dismissed as non-constructive or semi-constructive. In any case, it is doubtful

---

[17]v.II, p.8 of [Heath, 1956], resolves the problem by saying that the counterfactual hypothesis is used to show that point G is neither in one side nor in the other side of CE. In the case it were over CE, he says, it is trivial that the center should be the middle point.

that there is a canonical way of introducing points in geometry. And, finally, a hypothesis does not introduce anything, it only invites us to do a conjecture, to consider something, to pretend something, and there is no a priori reason to consider this activity as non-constructive.

2.1.2 Reduction and decomposition of problems

In the attempt at formulating a General Theory of Problems, [Veloso, 1984] stresses that the resolution of a problem involves globally two main strategies: reduction of a problem to others and decomposition of problems.[18]

Examining the solution of III.1 it can be observed that: step (i) employs postulate I.1 (to draw a straight-line from any point to any point); step (ii) the *propositio* I.10 (to cut a given finite straight-line in half); step (iii) the *propositio* I.11 (to draw a straight-line at right-angles to a given straight-line from a given point on it) and postulate I.2 (to produce a finite straight-line continuously in a straight-line); and (iv) the *propositio* I.10 again.

Problem III.1 is decomposable into three problems that must be solved in the sequence: (a) to find a diameter of the circle; (b) next, to find the middle point F of this diameter; (c) the last step, to prove that F is indeed the center.[19] It is obvious that a change in the sequence won't solve the problem correctly. Additionally, it can be said that the problem of finding the center of the circle reduces to the problems of finding a diameter and of finding the middle point of this diameter.[20] Reduction and decomposition are used concomitantly.

In steps (i) to (iii) of the above recipe the problem of finding a diameter is solved. This is not a deterministic step in the sense that the result of the action over the configuration is not uniquely determined; distinct diameters of ABC could be used. And, as the diameter in the configuration is obtained after drawing chord AB, not any straight-line in step (i) serves the purpose but still a a huge number of them would: all chords of ABC.

An Euclidean machine solving a problem, like that of finding a diameter of ABC, might resolve it in different ways and in different places inside the configuration.[21] This is an example of a non-deterministic solution. In

---

[18] We repute the notion of reduction in [Veloso, 1984] to be more adequate than that in [Kolmogorov, 1932].

[19] The third step does not make part of the construction solution, properly speaking. But it is part of a mathematical theory of geometric constructions.

[20] The correctness proof can be left out.

[21] Interpreting the plane as a set of points, each straight-line corresponds to "one action" and now the comparison with non-deterministic Turing machines is tighter. But, in this particular case, there is an infinite number of straight-lines, or actions, among which a choice is to be made. Seen from this perspective, a simple Euclidean machine becomes

contrast, the problem in step (iv) of finding the middle point of a given diameter has only one final result, only one element that is considered the solution over the diagram, it is deterministic.[22] We clearly see that the problem of finding a diameter is itself further decomposed into other problems, until the decomposition stops at postulates.

In one movement, both strategies are used: "divide and conquer" and reduction to other problems considered already solved. In the example, drawing the diameter is a problem that is solved by appealing to problems I.10 e I.11. If those two problems were already reduced to postulates, then the problem of tracing the diameter is also reduced to the postulates. Decomposition, i.e. "divide and conquer", and reduction are then intertwined. What Sidoli sees as the calling of routines and subroutines in the solution of a geometrical problem are truly the iterated application of those two strategies.

Ideally, all construction problems in the Euclidean geometry are reducible to a few initial postulates. Problems, postulates and solutions are, in essence, actions, even if a little bit distinct among themselves, because they are taken from different intentional perspectives. A problem is an action we propose ourselves to solve. The actions in the Euclidean geometry leave traces on the plane, that is, they transform plane configurations into other plane configurations.

More important, from our standpoint, the postulates must be interpreted as problems. This is clear for the three first postulates and not so easy to establish for the last two–but not impossible too. Actually, by homogeneity, this is so because reduction is a transformation between problems. Postulates can profitably be seen as problems of a special nature. They are problems assumed to be solvable and this is why they constitute the basis of the theory.

In order to realize that the treatment of postulates as problems makes sense, it is relevant to ask how the actions described by the postulates are supposed to be performed. Some would pretend that they are "performed" in imagination. Others that they are performed in a pure space of perception. In that last case the crucial question is: how would we differentiate a straight-line from one that is not straight? From the practical point of view, there are different ways of acting for the consecution of what has

---

an infinite machine. However, there is nothing mysterious about making a choice at will of how to draw a chord AB over the circle ABC.

[22]The word solution has been used in the text primarily to mean algorithm. The central point of the circle can be found by different algorithms. In other words, there are different solutions for finding the center. However, the word solution can also be used to refer to an element of the diagram that is being looked for, in the literature this has been many times assumed to be the primary use of the word.

been postulated, different patterns, although all those actions are no more than approximations to what has been postulated.

A straight-line can be traced over a piece of paper with the aid of a ruler. The ruler constitutes then the pattern of straightness. But it can also be produced over the ground with the aid of a nylon thread, as it is usually done when establishing the marks for the foundations of a building. In this last case the tension gives the thread its pattern of straightness. In the imagination, the straight-lines seem to depend on actions experienced before by the subject who is imagining them. By contrast, in a metaphysical pure space the straight-line is straight by fiat, with no other pattern.

Much has been debated about postulates. Different authors have observed that the three first are formulated with action verbs and should then be considered principles of a practical nature, as in Kant's words. Sidoli (ibid.,p.418,note 33), also examines the question concerning the statement of postulate I.1, exegetically. Should it be translated as *to draw a straight-line from any point to any point* or as *to produce a straight line from any point to any point*? From our perspective, the most important fact is that an action verb is employed in both interpretations. And different actions can be taken as approximations of this postulate, but none can be identified with it. It does not make sense to try to say that a perfect metaphysical action corresponds to it.

Different instruments and actions can be employed in order to perform the action postulated in I.1. It does not matter which approximation is taken, certain consequences will follow. It is assured that we have a theoretical know how for producing an equilateral triangle according to *propositio* I.1 once the instruments are chosen.

For further details about logical operations, decomposition and reduction of problems see [Sanz, 2019] and [Sanz, 2020]. There it is pointed that Kolmogorov's problem interpretation can be defined by means of a semantical relation of reduction between problems such that each logical propositional constant usual in intuitionism becomes a problem decomposition constant. However, a new constant involving the order of actions has to be added to the set of logical intuitionistic constants meaning roughly *solve x and, after, solve y*. It is revealing that Von Plato and Mäenpää (ibid.) didn't notice it.

Composition of functions is ordinarily used for obtaining this "temporal sequence". For example, in order to find the center of circle ABC, we could use the two–as many would like to say–functions: $diameter\_of(ABC)$ and $middle\_point\_of(CE)$.[23] Therefore, by composition, an ordered sequence

---

[23]Strictly speaking, $diameter\_of(ABC)$ cannot be a function. It is at best a family of

of "functions", $middle\_point\_of(diameter\_of(ABC))$, is established and the composed functions delivers what was desired. In this spirit, we can understand why the first three postulates became functions constructively interpreted in the hands of Von Plato and Mäenpää (*ibid.*) as also Sidoli (*ibid.*). However, now reinterpreted, postulate I.1 would produce a straight-line only if two distinct points were previously given as arguments. But, in regard of ancient geometry, this is a disputable interpretation.

To draw a line is an act that might create the stopping and the starting points, the action postulated can be assumed to be non-deterministic. And it is doubtful that we could find canonical ways of introducing points. At some stage there must be an unconditioned action. Some might object that unconditioned actions can be represented by constant functions, but this introduces a complexity not present in the original idea. We would need an infinite family of constant functions for representing an unconditioned action since the unconditioned action of drawing a straight-line is from any point to any (distinct) point. This should be enough for perceiving that actions and functions are distinct, conceptually speaking. The parameter of a constant function is irrelevant in such action.

The consequence of leaving unnoticed the logical constant for a sequence of actions is that Martin-Löf's type theory has a double formalism with formation rules for types and formation rules for lambda-terms.[24] As a consequence, hypotheses become the assumption of a type with an unknown variable x as a lambda-term. That is, all hypotheses are the supposition of possessing a term, a construction, but this is a distortion of the act of making hypotheses, as we already argued above.

Reductions should be much better known by logicians and philosophers. The term has been used since long ago in history. About reductions, [Heath, 1956], v.I, p.135, makes the following remark:

6. Reduction.

This is again an Aristotelian term, explained in the Prior Analytics. It is well described by Proclus in the following passage: "reduction (*apagoge*) is a transition from one problem or theorem to another, the solution or proof of which makes that which is propounded manifest also. For example, after the doubling of the cube had been investigated, they transformed the investigation into another upon which it follows, namely the finding of the two means; and from that time forward they inquired

---

functions since there are infinite distinct diameters that can be offered as the result of the "function".

[24] See [VonPlato & Mäenpää, 1990].

how between two given straight-lines two mean proportionals could be discovered. And they say that the first to effect the reduction of difficult constructions was Hippocrates of Chios, who also squared a lune and discovered many other things in geometry, being second to none in ingenuity as regards constructions.

It is not a minor issue to observe that under such a conception reduction can be applied to both, problems and theorems. Theorems are interpreted as demonstration problems according to a suggestion by Kolmogorov (*ibid.*).

Finally, the concept of construction in geometry can be used either to refer to the algorithm for producing a configuration or used to refer to the configurations obtained through postulated actions. In the second case, there is a clear way of saying that geometry produces its artifacts by using ruler and compass. If all problems in the ancient geometry reduce in the end to postulates, once we interpret the postulates as problems–i.e., assume them to be immediately solvable problems–, then any approximative actions taken as the concrete solutions for the postulates will offer a concrete reduction basis for problem *propositiones*. In this sense, in a certain measure, Euclids' geometry can be said to be a geometry of straightedge and compass, the main instruments of this approximation in the case where configurations are build over a sheet of paper. That is, this is simply one species of physical realization for Euclidean machines in which the natural geometer plays an important role.

2.2 Euclidean configurations

Euclidean machines and Euclidean configurations require a deep look, this section is partially dedicated to it, specially the connection of problems and geometric constructions. In the next paragraphs we employ an idea present in Kolmogorov (*ibid.*) concerning conditional problems and their solution. Here, the notion of supposing the possession of a solution comes into play.

Book I of Euclids Geometry presents items in the following order: (i) terms (*oroi*) or definitions ; (ii) what is asked (*aitemata*) or postulates; (iii) common notions (*koinai ennoiai*) ; (iv) propositiones (*protasis*). The other books contain expositions that keep the order but are restricted only to: (i) terms (excepting books VIII, IX, XII and XIII); and next (iv) *propositiones*. Hence, the principles of book I have apparently as their kingdom all the treatise.

In the books for plane geometry, the terms introduced give us the faculty of naming, pointing and distinguishing elements over the plane: points, straight-lines, circles, angles, etc. But not all of them are terms given previously as *oroi*. The term *parallelogrammon*, for example, is not characterized in the definitions. It appears for the first time in *propositio* I.34, being used for making reference to figures formed by the intersection of two parallel lines with other two.

2.2.1 Another case study

Considering the elements being produced or traced in plane, that is, the elements perceptually identifiable in an Euclidean plane–complexes of lines and points basically–we are calling configuration to sets of such perceptual elements some of them designated as a unity by the terms (*oroi*) introduced, for example, those in *propositio* I.22 (lines, circles, triangles, etc.)[25]:

Figure 3: Diagram for *Propositio* I.22

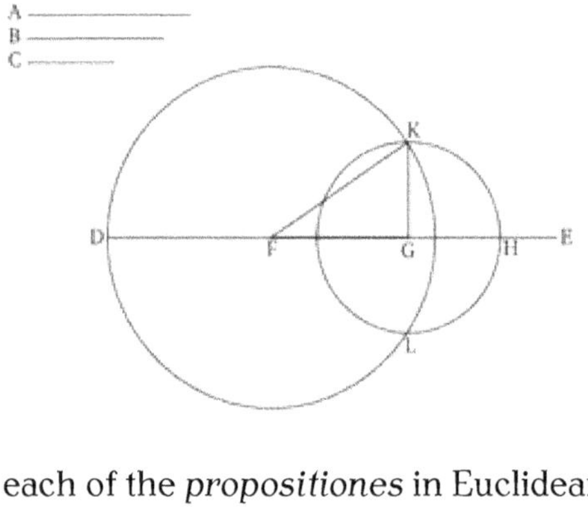

All diagrams in each of the *propositiones* in Euclidean geometry, books I-VI, correspond to a final configuration of the respective *propositio*, be it a problem or a theorem.

As said before, it is not our intention to make an exegesis of the ancient text. We aim at developing instruments for an epistemological analysis. With the concept of configuration we intend to capture the diagram visualized but not only. Since a problem involves an action to be done, there must be an initial configuration and a final configuration related to each problem. This is also the case for theorems in the measure that they are

---

[25]It is reasonable to consider that in the figure the position of the segments A, B and C, could be any other and, thus, magnitude is what really matters here. But if they were not given in some position or localization in the configuration, it would be impossible to trace the triangle.

demonstration problems, although not always the initial and the final configurations will be different, in this case. Naming them "configurations" is a way of underlining the analogy between diagrams and the states a tape may assume in a Turing machine, as already pointed.

The initial configuration of *propositio* I.22 should be the following:

Figure 4: Initial Diagram for *Propositio* I.22

This is an interesting case. A, B and C cannot be lines of any magnitude. They must be such that the sum of any two of them is greater than the third, according to the enunciation of the *propositio*. It is an important restriction and it cannot be expressed graphically. It can be expressed conditionally like: *to build a triangle with three given lines for which the sum of any two of them is greater than the third*. Although being debatable if this truly express the original formulation, at least it captures and gives proeminence to an aspect of geometric problems.

If the sum of B and C were lesser than A, the algorithm provided in the *kataskeue* will stop without providing the required triangle. In order to obtain triangle FKG it is necessary that circumferences DLK and HLK intersect each other in two distinct points K and L. If the sum is lesser, this would not happen. And the procedure could not go forward. The construction would be impossible. The procedue stops in failure.[26]

The problem interpretation proposed by Kolmogorov (*ibid.*) when making room for conditional problems fits perfectly here.[27]

The diagram of problem I.22 presents the three segments in their order

---

[26] We remind here that one of the descriptions by Brouwer - see §2.3 of [VanAtten, 2017] - of what would be a proof of a negation consists in saying that at a certain point it becomes impossible to go forward with the construction that started with the assumption. But, of course, not all situations of impossibility in a construction are relative to the proof of a negation, and this is the case here.

[27] The author makes it clear already in the first page of [Kolmogorov, 1932] that there are conditional problems, through examples.

of magnitude from the biggest to the smallest. If that were necessary for developing the solution, then the comparison of magnitudes two by two would be required. But the truth is that they can be taken in any order, the *kataskeue* works adequately no matter the position and no matter the order they are taken.[28] Magnitude is the important data for this *propositio*. The expressions *given in position* and *given in magnitude* have an specific meaning which are lenghty examined in Euclids *Data*. But, with a certain dose of freedom in using these expressions, we can say that the three straigh-lines can be given in any position inside the initial configuration once their magnitudes are in the required relation. Hence, of this case, we can say that the given is assumed to be so mainly in magnitude.

When solving a conditional problem, the fact that the condition does not hold might be considered as a solution to the problem, according to Kolmogorov's perspective. Therefore, the simple supposition that the straight-lines in I.22 are such that the sum of any two of them is greater than the third does not suffice to guarantee that the resolution will result a triangle. Here, it is necessary to presuppose more: to presuppose the possession of a proof that this is the case.[29] In case some lines are given and the construction stops before the right moment–as it is going to occur if the sum of B and C is lesser than A–, then we can say that the assumption that the lines were in that relation of magnitudes is false, which is then a solution to the problem. In other words, the supposition that there is a proof showing the line segments to be in the required relation produces a failure in the development of the algorithm which then stops. If the machine stops, some kind of solution has then been reached: either the triangle or nothing.

Propositio I.22 can be used in the solution of a future problem only if in the concrete case under examination there is a guarantee that the sum of any two straight-lines is bigger than the third. This might not appear explicitly in the *kataskeue* of the future problem, but it must be there, as it is the case of problem I.23, for example.

2.2.2 Diagrams and configurations

Let´s return to the concept of tape configuration. It is reasonable to say that modern computers are capable of some prowess in terms of image processing and recognition. Hence, it would be strange if we could not

---
[28]After *propositio* I.2 any line segment can be "transferred" to a given point.

[29]By contrast, when considering the proof that the solution in the *kataskeue* fulfills the intended role, a mere hypothesis that the lines given have the mentioned property is enough for proving what is at stake concerning the correctness of the solution.

consider more complex spatial configurations similar to that of the tape (or plane) in Euclidean machines as a possibility, barring a necessary *aggiornamento* of the theoretical model developed in the 30's of XXth century. It should be reminded that the model is older than the electronic computer devices developed during World War II. Of course, there are many questions concerning the conceptual and epistemological nature of configurations and Euclidean machines that need to be examined. We'll analyze a few in what follows.

In general, as input of a *propositio* the elements of a configuration are given in position.[30] The construction modifies this initial configuration. However, it frequently occurs that the final configuration does not represent all possible cases, like in I.1 in which two intersections might be considered for building the equilateral triangle. Also, not always the magnitude of the elements is determined in the initial configuration, like the case of the straight-line AB in I.12 that has indefinite magnitude.[31]

All the above cases can be explained appealing to a desire of simplification in the exposition that might be motivated by concision and due maybe to costs of copying.[32] Hence, the exposition continues without major considerations for minor details whose treatment would be lengthy and/or easy. For example, if the straight-line AB of *propositio* I.12 were given in a position and with a specific magnitude such that the point C were far away from it, it would be necessary to extend the line by postulate I.2 until somehow "passing" the point C. From the perspective of practical action the procedure is not difficult. From an ontological standpoint the question is important since it can be viewed from very different sides.

A geometer operates over a configuration space similar to the tape in a Turing machine. Diagrams just pick up the essential elements of a configuration. That those spaces have similar role in Turing and Euclidean machines might be a crucial theoretical observation. If this is the case, then many philosophical questions that can be asked of one can be put to the other and, perhaps, philosophical solutions to one of them could in principle be carried to the other.

The initial diagrams we mentioned before, and usually absent in the problem *propositiones*, constitutes the initial configuration of an Euclidean machine. Lots of operations are performed over this configuration, but the

---

[30] See [Taisback, 2003] about Euclids *Data* and the discussion concerning the concept of *given in position*. Also Sidoli (*ibid.*,p.408) about *propositiones* in the Elements on this respect.

[31] [Fitzpatrick, 2008] translates *apeiron* by infinite, but the term can be also employed with the sense of boundless, indefinite, that doesn't stop.

[32] The medium for registering ancient books were rare and expensive.

diagram we called final, i.e. the diagram that goes together with the *propositio* in the text, only highlights the elements necessary for illustrating the *propositio*. Since a solution like that of III.1 employs other former *propositiones*, all operations over the configuration should leave a trace although theses traces are not represented over the final diagram, as noticed by Sidoli (*ibid.*).

In the case of *propositio* I.22 the final diagram presents what would be the final configuration of the Euclidean machine, i.e. the solution algorithm, if the three lines are indeed in the required relation. If they were not, then the diagram does not represent the final configuration of the machine that stopped in failure, according to the problem interpretation of Kolmogorov.

3 Conclusion

Propositiones that ask a construction were compared to Turing machines, giving rise to what is here being called an Euclidean machine. Those machines have a peculiar characteristic that Turing machines do not possess. In the ancient text, there is regularly a proof that the machine is in accordance with its specification.

In order to speak of "specification" it is necessary to make the concept of problem to intervene between *propositiones* and machines. The concept of problem is as old as that of theorem. The question concerning which one would have preeminence over the other is also as old. The concept of problem appears sufficiently characterized in Kant, as we pointed. Nonetheless, contemporarily, we believe that two contributions for the General Theory of Problems gave it deepness and amplitude. One is Kolmogorov's interpretation of intuitionistic logic through the problem interpretation. The other is the analysis of problems carried by Veloso, in which two strategies of problem-solving are considered as principal: decomposition and reduction. We tried to show above how those two are profitably intertwined in ancient geometry, illustrating it with examples taken from Euclids Elements, in special *propositio* III.1.

In the above investigation, Sidoli's interpretation of books I-VI of Euclidean geometry was extendedly used. We disagree with him in some minor points, mainly in his full adoption of Von Plato and Mäenpää interpretation of postulates as introduction rules. As indicated, there is a direct connection of this interpretation to Martin-Löf's intuitionistic type theory. This "intuitionistic chain" of authors keeps in storage some presuppositions that we see as controverse, in particular when it comes to ancient geometry.

The concept of construction received a close analysis as it appears in Euclids and, above all, how the constructions are described in the *kataskeue* of problem *propositiones*. The word construction is itself ambiguous. It can be used to either designate the solution algorithm or to designate the drawing executed by the algorithm. We think that the first is the principal use.

Intuitionists have used the word construction for making reference to proofs, which should probably include logical steps. Intuitionists introduce an ambiguity in the concept of construction, fact that we tried to make visible by examining the Euclidean text structure as a historical document. This has been done with respect to *propositio* III.1 and in a smaller degree with I.22. The first illustrates the use of a hypothesis in the *apodeixis* from which an absurd is derived but in which the construction does not stop. The second illustrates a construction that would stop if the problem-condition is not fulfilled, although no negation is being considered. The first contains a simple supposition used with the purpose of proving something, the second contains a supposition that a property concerning the elements in the initial configuration has been proved and it concerns the *kataskeue*. This evidently are two different kinds of suppositions.

The above investigation showed how to extend the concept of computability, by extending the notion of what we understand by a computing machine, if Euclidean machines are accept as such. Euclidean machines should in fact be viewed as a more general notion. Turing machines would then be restricted Euclidean machines where the the set of tape squares constitutes the configuration with each square containing only one element.

References

Beeson, M. (2010), Constructive Geometry, *in* 'Proceedings of the Tenth Asian Logic Colloquium, Kobe, Japan, 2008', pp. 19–84.

Fitzpatrick, R. (2008), *Euclid's Elements of Geometry (translated from the Greek text compiled by J.L.Heiberg)*, University Texas, Austin. URL http://farside.ph.utexas.edu/Books/Euclid/Euclid.html.

Heath, T. (1956), *Euclid: The thirtheen Books of The Elements, V.I,II and III*, Dover, New York.

Kleene, S. (1952), *Introduction to Metamathematics*, North-Holland, Amsterdam.

Kolmogorov, A. (1932), 'Zur Deutung der Intuitionistischen Logik', *Mathematische Zeitschrift* **35**(1), 58–65. Pagination from the English edition in ?.

Lassalle-Casanave, A. (2019), *Por Construção de Conceitos: Em Torno da Filosofia Kantiana da Matemática*, Loyola, Rio de Janeiro.

Matin-Löf, P. (1996), 'On the meanings of the logical constants and the justifications of the logical laws', *Nordic Journal of Philosophical Logic* **1**(1), 11–60.

Naibo, A. (2018), Constructivity and Geometry, *in* G. et alli, ed., 'From Logic to Practice', Springer, Heidelberg, pp. 123–161.

Sanz, W. d. (2019), Hypo: A simple constructive semantics for intuitionistic sentential logic; soundness and completeness, *in* 'Proof-Theoretic Semantics: Assessment and Future Perspectives. Proceedings of the Third Tübingen Conference on Proof-Theoretic Semantics, 27–30 March 2019', pp. 153–178. URL http://dx.doi.org/10.15496/publikation-35319.

Sanz, W. d. (2020), *Kolmogorov and the General Theory of Problems*, unpublished yet.

Sidoli, N. (2018), 'Use of Constructions in Problems and Theorems in Euclid´s Elements', *Arch. Hist Exact Sci.* **72**(4), 403–452.

Taisback, C. M. (2003), *Euclid's Data: The Importance of Being Given (ACTA Historica Scientiarum Naturalum Et Medicinalium)*, Museum Tusculanum, Copenhagen.

VanAtten, M. (2017), The development of intuitionistic logic, *in* 'Stanford Encyclopedia of Philosophy'. URL https://plato.stanford.edu/entries/intuitionistic-logic-development/.

Veloso, P. (1984), 'Aspectos de uma teoria geral de problemas', *Cadernos de História e Filosofia da Ciência* **7**, 21–42.

VonPlato, J. & Mäenpää, P. (1990), 'The Logic of Euclidean Construction Procedures', *Acta Phenica* **49**, 275–293.

When databases roamed computing: Formal database specification revisited

Ionuț Țuțu*‡   Claudia E. Chiriță†   José L. Fiadeiro‡

* Simion Stoilow Institute of Mathematics of the Romanian Academy
ittutu@gmail.com

† School of Informatics, University of Edinburgh
claudia.elena.chirita@gmail.com

‡ School of Science and Engineering, University of Dundee
jfiadeiro@dundee.ac.uk

Abstract

We revisit an early paper by Marco Casanova, Paulo Veloso, and António Furtado where database specifications are developed by stepwise refinement within a heterogeneous conceptual-design framework, recasting some of their original results using concepts and techniques from institution theory and modern algebraic specification that are more adept at handling heterogeneity. Three levels of specification are considered, corresponding to information representation, database querying and manipulation, and database procedures – thus gradually progressing from high-level descriptions to implementations of database applications. Each level is grounded in a logical system formalized as an institution, while connections between levels are given in terms of institution semi-morphisms and heterogeneous refinements.

Keywords: Formal database development, Algebraic specification, Stepwise refinement, Institution theory, Kripke semantics.

1 Introduction

A significant body of work conducted by Paulo Veloso and his collaborators in the early 1980s deals with database specifications and formal program development [Veloso & Furtado, 1982; Veloso, 1983; Casanova et al., 1984; Maibaum et al., 1985]. The paper [Casanova et al., 1984] is particularly striking because it features concepts that are much more advanced than what was typical at that time and whose understanding was fully realized only several decades later. More specifically, [Casanova et al., 1984] builds on the principle of developing software applications from logical specifications through a series of refinement steps – what became known as *stepwise refinement*. Three logical levels of database development are considered therein, corresponding to information representation, algebraic specification, and database programming; each level is equipped with its own logical system and with an appropriate axiomatization of a concrete database. The main challenge addressed in that paper concerns the fact that two significant refinement steps in this process are heterogeneous, meaning that they bridge specifications written over different formalisms: one takes place between the information-representation level and the algebraic-specification level, and the other between the algebraic-specification level and the database-programming level.

The approach proposed by Casanova et al. [1984] was developed at a time when an abstract mathematical notion of logical system, later known as *institution*, was only starting to emerge, and when there was no general consensus on what refinement and mapping between logical systems really meant. After more than 30 years of continuous development, Joseph Goguen and Rod Burstall's theory of institutions [Goguen & Burstall, 1992] allows us to view Paulo Veloso's work from a different, much clearer, perspective. Similarly to [Casanova et al., 1984], the concept of *institution* originated in the 1980s within the theory of algebraic specification, one of the earliest publications being [Goguen & Burstall, 1983]. Institutions arose from the general concept of language [Burstall & Goguen, 1979] as a means to cope with the increasing number of specification formalisms in use at the time (which has since continued to increase). They are a clear category-theoretic concept that formalizes the intuitive notion of logical system as a balanced interaction between its syntax and its semantics. Virtually all logical systems used in model theory [Diaconescu, 2008] and formal specification [Sannella & Tarlecki, 2011] can be formalized as institutions. In fact, two of the logical systems used in Veloso's work, namely temporal and equational first-order logic, are some of the earliest examples of institutions [Goguen & Burstall, 1992; Fiadeiro & Costa, 1996].

Reframing the database specification method proposed by Casanova et al. [1984] in an institutional context has several key advantages:

(i) It allows us to treat the logics used at each of the three levels of formal development as concrete mathematical objects – i.e. institutions.

(ii) Building on (i), the links between the three levels can be described in terms of various notions of mapping between institutions [Goguen & Roşu, 2002]. Moving between the information-representation and the algebraic-specification level requires special attention because it involves two distinct kinds of mapping. First, it makes use of the well-known standard translation of modal (and, in particular, temporal) logics into first-order logic (see e.g. [Blackburn et al., 2001]), which was recently formalized as an *institution comorphism* by Diaconescu & Madeira [2016]. Further, it relies on an encoding of predicates as Boolean-valued functions – and also of Boolean connectives as algebraic operations – as it has been done for the OBJ family of formal-specification languages by Goguen et al. [2000], and along the lines of [Diaconescu, 2010].

(iii) These logics and encodings define a *heterogeneous logical environment* [Mossakowski & Tarlecki, 2008], which can then be flattened by means of a Grothendieck construction [Diaconescu, 2002; Mossakowski, 2002] to an ordinary institution, thus providing support for heterogeneous specification.

(iv) Lastly, we can instantiate a well-established notion of *refinement* from formal specification [Astesiano et al., 1999; Lopes & Fiadeiro, 2002; Mossakowski et al., 2004; Codescu et al., 2017] for the Grothendieck institution obtained at (iii) in order to capture the actual refinement steps defined by Casanova et al. [1984] for database development.

A tribute to Veloso More than showing how institution-based formal specification can provide a useful blueprint for applications such as formal development of databases in the sense of [Casanova et al., 1984], this paper is a tribute to Veloso (as he is known in Brazil), one of the giants on whose shoulders the third author developed his research career (initially in formal aspects of information systems, including databases), influencing early papers such as [Fiadeiro & Sernadas, 1988, 1986] and, through Tom Maibaum, becoming a most treasured colleague and friend. Visiting Veloso at PUC – Rio de Janeiro – was always inspiring; the memories of going into his office, a treasure-trove of manuscripts from which he would occasionally emerge holding a piece of paper with hand-written notes that would shatter established theories or provoke intense debate, will always remain indelible.

Growing up into research on formal specification in the 1980s was nothing less than exciting. Nothing short of a 'holy war' was raging across the community where the 'orthodoxy' defended that specifications should denote algebras (split into two camps: initial vs final algebras) and the 'dissidents' (among whom was Veloso) proclaimed that specifications should denote axiomatic theories (or theory presentations) as, like programs, they are syntactic objects. The point was not so much about the denotations of specifications (the duality between theories and classes of models levelling up the dispute) but the operations through which specifications should be structured and stepwise refinement should be framed (where the duality broke). The most heated debate was perhaps around the role of interpolation and conservative extensions in modularity; although not closing the debate, Veloso et al. [2002] made some contributions to it.

An important lesson learnt from working with Veloso was that, to be a good researcher, one does not need to follow the crowds, or to attract a crowd, or to be loud or a bully; but one should know well one's subject, always strive for elegance and simplicity, and keep one's scalpel well sharpened. We hope that traces of some of those attributes can be recognised in this paper.

2 Logical framework

In what follows, we recast the formal database specification from [Casanova et al., 1984] in the framework of institutions using modal-logic formalisms that we have been studying in the past couple of years. In doing so, and to make the paper as accessible as possible despite the fact that we are working with formal specifications, we keep the institution-theoretic preliminaries and the degree of formality of the presentation at a bare minimum. All we assume is basic familiarity with first-order and modal logic. From institution theory, we only import four key concepts whose definitions we distil into the following somewhat naive description:

Logical systems By *logical system* we mean a collection (class) Sig of *signatures*, or vocabularies (which, in practice, are structured sets of symbols), typically denoted as Σ, each of which is equipped with a class $\text{Mod}(\Sigma)$ of *models* (providing interpretations for the symbols in Σ), a set $\text{Sen}(\Sigma)$ of *sentences* (built from symbols declared in Σ), and a *satisfaction relation* $\vDash_\Sigma$ between Σ-models and Σ-sentences. When the signature Σ can be easily inferred, we may drop the subscript and denote the Σ-satisfaction relation simply by $\vDash$. The class Sig may be partially ordered according to a subsignature relation; in that case, if Σ is a subsignature of Ω, which we denote by $\Sigma \subseteq \Omega$, then every Σ-sentence can also be regarded as an Ω-sentence and every Ω-model M can be reduced to a Σ-model $M\!\restriction_\Sigma$ — which is usually obtained from M by 'forgetting' the interpretation of all symbols from Ω that do not belong to Σ. An important consistency condition is that *the meaning of sentences does not depend on the context in which they are considered*; in other words, for every signature inclusion $\Sigma \subseteq \Omega$, every Ω-model M and every Σ-sentence e, $M \vDash e$ if and only if $M\restriction_\Sigma \vDash e$.[1]

The prominent role of signatures is a hallmark of institution theory and one of the main reasons why institutions proved to be so useful in formal specification, especially when tackling issues related to modularity and heterogeneity (see e.g. [Diaconescu, 2014] and the monographs [Diaconescu, 2008; Sannella & Tarlecki, 2011]). This relativization of the semantics and syntax of logics has also profound methodological implications in the way we work with concrete logical systems. For instance, it provides a better understanding of the conventional concepts of variable and quantified sentence [Tarlecki, 1985], which we use extensively in this paper.

In most logics, a *variable* for a signature Σ is simply a constant-operation symbol x distinct from those in Σ. This allows us to extend Σ by adding x as a new constant symbol, thus obtaining a larger signature $\Sigma[x]$ – or, more generally, $\Sigma[X]$ when X is a set of variables. In such an extension, a *valuation* of X over a Σ-model M is nothing more than a model N for the extended signature $\Sigma[X]$ such that $N\restriction_\Sigma = M$. In that case, we say that N is an X-*expansion* of M. Moreover, Σ-formulae containing free variables from X are simply sentences over the extended signature $\Sigma[X]$. Variables can be bound by quantifiers in familiar fashion, yielding quantified sentences over Σ, whose semantics can be defined in terms of expansions; e.g., $M \vDash \forall X \cdot e$ if and only if $N \vDash e$ for all X-expansions N of M.

Specifications over a logical system $\mathcal{L}$ are defined as *theory presentations* $\langle \Sigma, E \rangle$, where Σ is a signature of $\mathcal{L}$ and E is a set of Σ-sentences. The semantics of $\langle \Sigma, E \rangle$ is given by the class of all Σ-models M that satisfy all sentences in E. In that case, we say that M is a model of $\langle \Sigma, E \rangle$, or simply of E. In symbols: $\mathrm{Mod}(\Sigma, E) = \{M \in \mathrm{Mod}(\Sigma) \mid M \vDash E\}$.

Moving between logical systems Suppose that $\mathcal{L}$ and $\mathcal{L}'$ are two logical systems with components Sig, Mod and Sig', Mod', respectively, such that $\mathcal{L}$ is intuitively richer than $\mathcal{L}'$. A *semi-morphism* [Sannella & Tarlecki, 1988] from $\mathcal{L}$ to $\mathcal{L}'$ consists of a subsignature-preserving map $\Phi \colon \mathrm{Sig} \to \mathrm{Sig}'$ together with a family of model-reduction functions $\beta_\Sigma \colon \mathrm{Mod}(\Sigma) \to \mathrm{Mod}'(\Phi(\Sigma))$ indexed by signatures $\Sigma \in \mathrm{Sig}$, such that the following naturality property holds: for every signature inclusion $\Sigma \subseteq \Omega$ in Sig and every Ω-model M, $\beta_\Omega(M)\restriction_{\Phi(\Sigma)} = \beta_\Sigma(M\restriction_\Sigma)$. That is, 'forgetting' the interpretation of symbols that belong only to the richer signature Ω commutes with 'forgetting' the additional structure defined by the richer logical system $\mathcal{L}$.

---

[1] Readers familiar with institution theory will recognize a simplified form of institution where Sig is a poset category and, for every signature Σ, the category $\mathrm{Mod}(\Sigma)$ is discrete.

Specification refinement Consider a semi-morphism $\langle \Phi, \beta \rangle$ between logical systems $\mathcal{L}$ and $\mathcal{L}'$, and let $\langle \Sigma', E' \rangle$ be an $\mathcal{L}'$-specification. An $\mathcal{L}$-specification $\langle \Sigma, E \rangle$ is a *refinement* of $\langle \Sigma', E' \rangle$ along $\langle \Phi, \beta \rangle$ if $\Sigma' \subseteq \Phi(\Sigma)$ and $\beta_\Sigma(M)\!\restriction_{\Sigma'} \vDash E'$ for all models M of $\langle \Sigma, E \rangle$.

We are now ready to present the three logical systems that we use to showcase the approach to formal database development presented by Casanova et al. [1984]. We list them in increasing order of complexity, from $\mathcal{L}_1$ to $\mathcal{L}_3$, maintaining a close correspondence with the notations used in that paper. We deviate in that, where $\mathcal{L}_1$ and $\mathcal{L}_3$ are modal logics, and $\mathcal{L}_2$ is equational, the logics we consider herein are all modal. To that end, in the role of $\mathcal{L}_2$ we use a logical system that we have recently developed in the context of *dynamic networks of interactions* [Țuțu et al., 2020], and which stems from previous research on actor-network theory [Fiadeiro et al., 2019; Țuțu et al., 2019]. This has the advantage of being sufficiently expressive to capture the constraints necessary at the second level of specification, and at the same time it allows us to remain within the realm of modal logics. We thus avoid a detour to first-order equational logic, which, in effect, leads to simpler mappings (semi-morphisms) to $\mathcal{L}_1$ and from $\mathcal{L}_3$. All three logics are extensions of many-sorted first-order logic as follows: $\mathcal{L}_1$ is a basic temporal extension adding modal operators for necessity and possibility; $\mathcal{L}_2$ adds parametric modalities to support query and update functions; and lastly, $\mathcal{L}_3$ introduces database procedures and relational statements.

Throughout the rest of the paper we fix a many-sorted first-order signature Σ, which serves as a common denominator of the signatures we consider at the three levels of specification. We use the signature Σ to capture the basic information structure to be implemented in a database – bearing some similarities to the entity-relationship model. Intuitively, its sorts correspond to the entities and data types of the database, its function symbols match entity attributes and elementary data operations (addition, multiplication, etc.), and its predicate symbols capture properties of entities and relations between them.

3 The use of a temporal-logic formalism

The basic temporal language The logical system $\mathcal{L}_1$ used for specifying databases at an informational level inherits the signatures of many-sorted first-order logic. That is, a *signature* of $\mathcal{L}_1$ is a plain first-order signature Σ including sorts, operations, and predicates as discussed above. The modal nature of the logic is however reflected in its Kripke semantics; an $\mathcal{L}_1$-model for Σ is a Kripke structure $\langle W, R, M \rangle$ consisting of:

- a set W of *possible worlds*, also called *states*;
- a reflexive & transitive *accessibility relation* $R \subseteq W \times W$ on states;
- a family M of first-order Σ-structures M_w, indexed by states $w \in W$, that share the same domain D – meaning that they have the same carrier sets and interpret the operation symbols in Σ in the same way. In other words, sorts and operations are *rigid*.

The sentences of $\mathcal{L}_1$ are obtained from equational and relational atoms through the application of ordinary Boolean connectives and first-order quantifiers, together with a modal operator $\Box$ of *necessity*, whose intended interpretation is "always in the future", and its dual, the *possibility* operator $\Diamond$, whose intended interpretation is "sometime in the future".

The satisfaction of sentences is standard for modal logic, and builds on the concept of satisfaction in first-order logic. For example, a Kripke structure $\langle W, R, M \rangle$ *satisfies* an atomic sentence e *at a state* $w \in W$, which we write $\langle W, R, M \rangle \vDash^w e$, if and only if the first-order structure M_w satisfies e. In regard to the two added modal operators, we define:

- $\langle W, R, M \rangle \vDash^w \Box e$ if $\langle W, R, M \rangle \vDash^z e$ for all *transitions* $(w, z) \in R$; and
- $\langle W, R, M \rangle \vDash^w \Diamond e$ if $\langle W, R, M \rangle \vDash^z e$ for some *transition* $(w, z) \in R$.

This notion of satisfaction is *local* as it depends on the possible worlds where sentences are evaluated. To formalize $\mathcal{L}_1$ as a logical system (institution), we take the usual route of defining *global satisfaction relations*: $\langle W, R, M \rangle \vDash e$ when $\langle W, R, M \rangle \vDash^w e$ for all $w \in W$.

Temporal specifications The logic $\mathcal{L}_1$ serves as a high-level description language for specifying which states of the database are consistent and which transitions are acceptable. Both kinds of criteria can be formally expressed using temporal-logic sentences: the consistency of states is addressed by *static constraints* – sentences that do not involve modal operators – while the acceptable transitions are defined by *transition constraints* – sentences that make use of the necessity or possibility operators.

To illustrate, following the lines of [Casanova et al., 1984], we consider a simple $\mathcal{L}_1$-specification $\langle \Sigma_1, E_1 \rangle$ of a database of courses offered at a university and of the students enrolled there. The signature Σ_1 consists of two sorts, Student and Course, and two predicate symbols, takes : Student Course and offered : Course, whose names are self-explanatory. The enrolment of students in courses is regulated by two constraints: a static constraint according to which *students can only take courses that are being offered by* the university, and a transition constraint enforcing that, *once enrolled in a course, a student cannot drop out of all courses*. These are formalized in $\mathcal{L}_1$ as the sentences (1) and (2) listed below, which define the axioms of $\langle \Sigma_1, E_1 \rangle$. To keep the notation short, we avoid the explicit typing of variables, and generally use symbols such as s and s' for variables of sort Student, and c and c' for variables of sort Course.

$$\forall \{s, c\} \cdot \text{takes}(s, c) \Rightarrow \text{offered}(c) \tag{1}$$
$$\forall \{s, c\} \cdot \Box \big(\text{takes}(s, c) \Rightarrow \Box \exists \{c'\} \cdot \text{takes}(s, c') \big) \tag{2}$$

4 The use of a querying & manipulation formalism

Parametric modalities The logic we use at the second level of specification, $\mathcal{L}_2$, is a multi-modal extension of first-order logic whose modalities are parameterized by first-order elements. Specifically, the signatures of $\mathcal{L}_2$ are triples $\langle \Sigma, N, \Lambda \rangle$, where Σ is a first-order signature, N is a set of *nominals*, or possible-world designators, and Λ is a family of sets Λ_u of *modalities* of type u, which are indexed by sequences u of sorts in Σ. The semantics of $\mathcal{L}_2$ is defined analogously to that of the basic temporal language, with four notable exceptions:

- every nominal $n \in N$ is interpreted as a possible world / state $W_n \in W$;
- for each modality $\lambda \in \Lambda_u$, $R_\lambda \subseteq W \times W$ is an arbitrary, unconstrained accessibility relation on states – not necessarily a preorder;
- every possible world is reachable from some W_n by chasing the accessibility relations;
- for each modality $\lambda \in \Lambda_u$, where $u = s_1 \cdots s_n$, every transition $(w, z) \in R_\lambda$ determines a non-empty subset $M_{w\lambda z} \subseteq D_{s_1} \times \cdots \times D_{s_n}$ of *parameters* of λ under which the transition (w, z) can take place.

In regard to sentences, we replace the modal operators of $\mathcal{L}_1$ with parameterized variants: given a modality $\lambda \in \Lambda_u$ of type $u = s_1 \cdots s_n$ and a sequence of Σ-terms t_i of sort s_i, for $i \in \{1, \ldots, n\}$, for every $\mathcal{L}_2$-sentence e, we have that $[\lambda\ t_1, \ldots, t_n]e$ and $\langle \lambda\ t_1, \ldots, t_n \rangle e$ are $\mathcal{L}_2$-sentences as well. The first one corresponds to the notion of λ-necessity, while the second captures λ-possibility. The semantics of these upgraded modal operators is defined as follows: for every $\mathcal{L}_2$ Kripke structure $\langle W, R, M \rangle$, every state $w \in W$, and every sentence e:

- $\langle W, R, M \rangle \vDash^w [\lambda\ t_1, \ldots, t_n] e$ when $\langle W, R, M \rangle \vDash^z e$ for all transitions $(w, z) \in R_\lambda$ such that the interpretations of the terms t_i in the domain D, denoted D_{t_i}, form a tuple in $M_{w\lambda z}$;

- $\langle W, R, M \rangle \vDash^w \langle \lambda\ t_1, \ldots, t_n \rangle e$ when $\langle W, R, M \rangle \vDash^z e$ for some transition $(w, z) \in R_\lambda$ such that the interpretations of the terms t_i in D form a tuple in $M_{w\lambda z}$.

To make full use of nominals, $\mathcal{L}_2$ also defines *nominal sentences* and *local-satisfaction operators*, much like hybrid logics do (see e.g. [Blackburn, 2000]). That is, any nominal $n \in N$ can also be regarded as a sentence and, for every nominal $n \in N$ and every $\mathcal{L}_2$-sentence e, $@n \cdot e$ is an $\mathcal{L}_2$-sentence as well. The semantics of these hybrid-logic constructs is as follows:

- $\langle W, R, M \rangle \vDash^w n$ if $w = W_n$; and

- $\langle W, R, M \rangle \vDash^w @n \cdot e$ if $\langle W, R, M \rangle \vDash^z e$, where $z = W_n$.

Therefore, we can use nominal sentences to identify certain states of a Kripke structure and local-satisfaction operators to change the state where a sentence is being evaluated.

Database queries and manipulations Intuitively, a *database query* is a simple expression (often Boolean) whose meaning varies from one state of the database to another. Following Casanova et al. [1984], we formalize queries as sentences of $\mathcal{L}_2$ that are free from the modal- and hybrid-logic operators mentioned above. In other words, queries are plain first-order Σ-sentences – just like the static constraints we considered in Section 3. The main reason we use a different terminology is that queries are meant to be evaluated at given states of a database, whereas static constraints are meant to hold globally, at all states of the database.

The chief semantic distinction between $\mathcal{L}_2$ and $\mathcal{L}_1$ is that the transitions we consider at the second level of specification are labelled and, moreover, that the labels comprise both modalities and suitable domain elements as parameters. We use these labels as abstractions of *database manipulations*. More precisely, given a Kripke structure $\langle W, R, M \rangle$ of a database, we say that a state z is obtained from another state w under the action of a manipulation operation λ (which may, but needs not be deterministic) with parameters $p_1, \ldots, p_n$ when $(w, z) \in R_\lambda$ is a λ-transition and $(p_1, \ldots, p_n) \in M_{w\lambda z}$.

As an example, the $\mathcal{L}_2$-specification $\langle \Sigma_2, E_2 \rangle$ of our university-courses database has a signature $\Sigma_2 = \langle \Sigma_1, N, \Lambda \rangle$ that consists of the same signature Σ_1 as in Section 3, a nominal init denoting an initial state of the database, and four modalities: (*i*) offer of type Course for adding a new course to those offered by the university; (*ii*) cancel of type Course for dropping a course from the university's curriculum; (*iii*) enrol of type Student Course for enrolling a student in a given course; and (*iv*) transfer of type Student Course Course for transferring a student from one course to another.

The sentences in E_2 are written according to the *structured-description* approach outlined by Casanova et al. [1984]. We use simple observations of the form offered(c) and takes(s, c) to describe the initial state of the database and modal operators to define the preconditions, the intended effects, and the potential side-effects of each manipulation operation.

When writing $\mathcal{L}_2$-sentences, we maintain the same notational convention in regard to variables as in Section 3; in addition, we use the symbol n as a nominal variable. Initially, the university offers no courses (formalized in sentence 3), and no students are taking courses (4).

$$@\text{init} \cdot \neg \exists \{c\} \cdot \text{offered}(c) \tag{3}$$

$$@\text{init} \cdot \neg \exists \{s, c\} \cdot \text{takes}(s, c) \tag{4}$$

Concerning manipulation operations, we focus on the axiomatization of the modality cancel. The other modalities can be treated in a similar manner, as in [Casanova et al., 1984]. Cancelling a course can happen only if no students are enrolled in it; we formalize this requirement as a precondition in sentence (5). The effects of cancel are straightforward: a course c' is offered after cancelling c if and only if it was already part of the curriculum and is different from c – formalized in (6); furthermore, cancel has no effect on takes-observations – according to sentence (7).

$$\forall \{c\} \cdot \langle \text{cancel } c \rangle \, true \Rightarrow (\neg \exists s \cdot \text{takes}(s, c)) \tag{5}$$

$$\forall \{n, c\} \cdot n \Rightarrow [\text{cancel } c] \forall \{c'\} \cdot \text{offered}(c') \Leftrightarrow @n \cdot \text{offered}(c') \wedge c' \neq c \tag{6}$$

$$\forall \{n, c\} \cdot n \Rightarrow [\text{cancel } c] \forall \{s, c'\} \cdot \text{takes}(s, c') \Leftrightarrow @n \cdot \text{takes}(s, c') \tag{7}$$

The first to second-level refinement In order to integrate the informational and the querying-and-data-manipulation levels, we define a semimorphism $\langle \Phi, \beta \rangle : \mathcal{L}_2 \to \mathcal{L}_1$. The signature component Φ maps every $\mathcal{L}_2$-signature $\Omega = \langle \Sigma, N, \Lambda \rangle$ to its underlying $\mathcal{L}_1$ (first-order) signature Σ, forgetting both nominals and modalities. In regard to the model-reduction functions, every Ω-structure $\langle W, R, M \rangle$ is mapped under β_Ω to the Σ-structure $\langle W', R', M' \rangle$ that has the same possible worlds and local models as $\langle W, R, M \rangle$, and for which R' is defined as the reflexive and transitive closure of the relation $\bigcup \{R_\lambda \mid \lambda \in \Lambda\}$.[2] This allows us to restate the first main

---

[2] Attentive readers may notice that, unless special care is taken when defining second-level signature extensions, neither $\mathcal{L}_2$ nor $\langle \Phi, \beta \rangle$ satisfy the consistency conditions given

result of [Casanova et al., 1984] as follows:

$$\langle \Sigma_2, E_2 \rangle \text{ is a } \langle \Phi, \beta \rangle\text{-refinement of } \langle \Sigma_1, E_1 \rangle.$$

Clearly, the axioms (3–7) alone cannot ensure the correctness of the refinement because they do not account for the remaining three modalities. This can be remedied with ease by adding precondition and effect axioms for offer, enrol, and transfer as in [Casanova et al., 1984, Section 4.2].

5 The use of a programming-language formalism

Database procedures The purpose of the third level of specification is to bring us closer to programming languages used for database applications. To that end, the logical system we introduce in this section, $\mathcal{L}_3$, is once again an extension of first-order logic, but this time with database procedures. A signature for $\mathcal{L}_3$ is a pair $\langle \Sigma, \Pi \rangle$, where Σ is a first-order signature and Π is a family of sets Π_u of *procedure names* of type u, where u is a sequence of sorts in Σ. That is, from a syntactic perspective, procedure names are indistinguishable from the typed modalities of $\mathcal{L}_2$. The difference between the two concepts becomes apparent when looking at their semantics; a $\langle \Sigma, \Pi \rangle$-structure is a pair $\langle U, P \rangle$ consisting of:

- a *universe* U of *database states*, i.e. a maximal set U of Σ-models that share the same domain D (meaning that states differ only in the interpretation of predicate symbols, and any variation in the interpretation of a predicate symbol yields a model in U);
- for every procedure π in $\Pi_{s_1 \cdots s_n}$, a function $P_\pi \colon D_{s_1} \times \cdots \times D_{s_n} \times U \to U$.

The sentences of $\mathcal{L}_3$ are universally quantified *relational equations* $\forall X \cdot L = R$, where L and R are *relational statements* defined according to the following inductive rules:

- $\pi(t_1, \ldots, t_n)$ is an (atomic) *procedure call*, where $\pi \in \Pi_{s_1 \cdots s_n}$ is a procedure name and, for every $i \in \{1, \ldots, n\}$, t_i is a Σ-term of sort s_i;

---

for the notions of logical system and semi-morphism in Section 2. That is because the concept of reachability of a state – which underlies both satisfaction relations and the naturality property of β – depends on the nominals and modalities of the signature under consideration. Therefore, one needs to impose encapsulation constraints on signatures extensions – as it has been done in the past in regard to reachability and observability [Kurz, 2002; Bidoit & Hennicker, 2002; Diaconescu & Țuțu, 2014] – to ensure that computing reachable parts of Kripke structures commutes with model-reduction functions.

- $r := \{(x_1, \ldots, x_n) \mid e\}$ is an (atomic) *relational assignment statement*, where r is a relation symbol of arity $s_1 \cdots s_n$, x_i is a variable of sort s_i, and e is a first-order sentence over the extended signature $\Sigma[x_1, \ldots, x_n]$ – i.e. with free variables $x_1, \ldots, x_n$;

- e? is an (atomic) *test statement*, where e is a Σ-sentence; and

- if p and q are relational statements, then so are $p \cup q$ (referred to as the *union* of p and q), $p \,\S\, q$ (corresponding to the *sequential composition* of p and q), and p^* (called the *iteration* of p).

Relational statements R are interpreted as accessibility relations $[\![R]\!] \subseteq U \times U$ on states: for atomic statements, details are given below, while for the union, composition, and iteration operations we follow the familiar route of dynamic logics [Harel et al., 2001] and interpret $\cup$ as the union of relations, $\S$ as diagrammatic composition, and $*$ as reflexive-and-transitive closure.

- $(A, B) \in [\![\pi(t_1, \ldots, t_n)]\!]$ if and only if $B = P_\pi(D_{t_1}, \ldots, D_{t_n}, A)$;

- $(A, B) \in [\![r := \{(x_1, \ldots, x_n) \mid e\}]\!]$ if and only if A and B agree on $\Sigma \setminus \{r\}$ and the interpretation of r in B consists of all valuations of $x_1, \ldots, x_n$ for which e holds;

- $(A, B) \in [\![e?]\!]$ if and only if $A = B$ and $A \vDash_\Sigma e$.

The interpretation of relational statements serves as a basis for defining the satisfaction of relational equations: $\langle U, P \rangle \vDash \forall X \cdot L = R$ if $[\![L]\!] = [\![R]\!]$ for all uniform valuations of X in U – i.e., from an institution-theoretic perspective, for all X-expansions of $\langle U, P \rangle$.

Database specifications as schemas The main benefit of building the third level of database specification, which has a programming-language flavour, on top of a proper logical system is that it enables us to make use of theory presentations over $\mathcal{L}_3$ as a logical counterpart of database schemas. In consequence, we can further rely on familiar concepts and methods from formal specification (such as that of refinement) to reason about schemas.

Before delving back into our running example, we recall from [Veloso et al., 1984] a couple of elementary constructs that will help improve the readability of the specification that follows:

Macro	Relational statement
IF e THEN p	$(e?\,\mathring{,}\,p) \cup \neg e?$
INSERT $r(t_1,\ldots,t_n)$	$r := \{(x_1,\ldots,x_n) \mid r(x_1,\ldots,x_n)$
	$\qquad \vee\, (x_1 = t_1 \wedge \cdots \wedge x_n = t_n)\}$
DELETE $r(t_1,\ldots,t_n)$	$r := \{(x_1,\ldots,x_n) \mid r(x_1,\ldots,x_n)$
	$\qquad \wedge\, \neg(x_1 = t_1 \wedge \cdots \wedge x_n = t_n)\}$

For the university-courses database we define a signature Σ_3 that extends the first-order signature Σ_1 from Section 3 with four procedure names – offer, cancel, enrol, and transfer – with the same typing as the homonymous modalities in Section 4. They correspond to transitions between database states, but unlike the transitions at the second level of specification, they are defined using an equational formalism instead of modal-logic operators. Our final database specification, denoted $\langle \Sigma_3, E_3 \rangle$, is given by the following four equations.

$$\forall \{c\} \cdot \mathsf{offer}(c) = \text{INSERT offered}(c) \tag{8}$$

$$\forall \{c\} \cdot \mathsf{cancel}(c) = \text{IF } \neg \exists \{s\} \cdot \mathsf{takes}(s,c) \text{ THEN DELETE offered}(c) \tag{9}$$

$$\forall \{s,c\} \cdot \mathsf{enrol}(s,c) = \text{IF offered}(c) \text{ THEN INSERT takes}(s,c) \tag{10}$$

$$\forall \{s,c,c'\} \cdot \mathsf{transfer}(s,c,c')$$
$$= \text{IF takes}(s,c) \wedge \neg\mathsf{takes}(s,c') \wedge \mathsf{offered}(c')$$
$$\text{THEN } \big(\text{DELETE takes}(s,c)\,\mathring{,}\,\text{INSERT takes}(s,c')\big) \tag{11}$$

The second to third-level refinement The connection between the querying-and-data-manipulation and the database-programming level is established by a semi-morphism $\langle \Psi, \gamma \rangle : \mathcal{L}_3 \to \mathcal{L}_2$ whose signature component Ψ maps every $\mathcal{L}_3$-signature $\Delta = \langle \Sigma, \Pi \rangle$ to $\langle \Sigma, \{\mathsf{init}\}, \Pi \rangle$. That is, it preserves all first-order symbols, introduces a single nominal, init, and repurposes procedure names as modalities (while maintaining their typing).

Concerning the semantic part of the semi-morphism, every Δ-model $\langle U, P \rangle$ is reduced along γ_Δ to the $\langle \Sigma, \{\mathsf{init}\}, \Pi \rangle$-structure $\langle W, R, M \rangle$ where:

- W is the least subset of U that contains the database state whose predicates are all $\emptyset$ (which is identified as W_{init}) and is closed under procedure applications;
- for every $\pi \in \Pi_{s_1 \cdots s_n}$, $R_\pi = \{(A, P_\pi(a_1,\ldots,a_n,A)) \mid A \in W, a_i \in D_{s_i}\}$;

- for every $(A, B) \in R_\pi$, $M_{A\pi B} = \{(a_1, \ldots, a_n) \mid P_\pi(a_1, \ldots, a_n, A) = B\}$;
- for every database state $A \in W$, $M_A = A$.

With all this groundwork prepared, we can now recast the second main result of [Casanova et al., 1984]:

$$\langle \Sigma_3, E_3 \rangle \text{ is a } \langle \Psi, \gamma \rangle\text{-refinement}^3 \text{ of } \langle \Sigma_2, E_2 \rangle.$$

The proof is similar to the one found in Casanova, Veloso, and Furtado's paper. We leave out the details but encourage readers to retrace the steps.

6 Conclusions

Building on [Casanova et al., 1984; Veloso et al., 1984], in this paper we have outlined three logic-based languages for the formal development of database specifications through stepwise refinement. We have also taken this opportunity to highlight more recent advances in algebraic specification and formal system development that are specific to our institution-theoretic background. More specifically, the formalisms that we have presented are not merely "related to logic", but are proper instances of a rigorous concept of logical system (institution). That is the case especially for the third level of specification, which we have adapted from [Veloso et al., 1984] in order to allow database schemas to be captured as theory presentations instead of programming-language descriptions, all while preserving their original semantics in terms of transitions between database states.

This institutional reframing is more than a mere change of context; it enables us to integrate the formalisms used in a heterogeneous framework by means of institution semi-morphisms and to use a single, unified notion of refinement instead of separate notions, one for each change of formalism. Importantly, these developments are achieved through a high-level refactoring without sacrificing specific techniques or details related to the three concrete levels of specification. The very possibility of such a refactoring is a testament of the fact that the ideas underlining the results reported by Casanova et al. [1984] were already of an institution-theoretic nature, even though they were not formalized as such.

---
[3]Modulo encapsulation constraints similar to the ones indicated in Footnote 2.

References

Astesiano, E., Kreowski, H. & Krieg-Brückner, B., eds (1999), *Algebraic Foundations of Systems Specification*, IFIP State-of-the-Art Reports, Springer.

Bidoit, M. & Hennicker, R. (2002), On the integration of observability and reachability concepts, *in* M. Nielsen & U. Engberg, eds, 'Foundations of Software Science and Computation Structures, 5th International Conference, FOSSACS 2002. Grenoble, France, April 8–12, 2002, Proceedings', Vol. 2303 of *Lecture Notes in Computer Science*, Springer, pp. 21–36.

Blackburn, P. (2000), 'Representation, reasoning, and relational structures: a hybrid logic manifesto', *Logic Journal of IGPL* **8**(3), 339–365.

Blackburn, P., de Rijke, M. & Venema, Y. (2001), *Modal Logic*, Vol. 53 of *Cambridge Tracts in Theoretical Computer Science*, Cambridge University Press.

Burstall, R. M. & Goguen, J. A. (1979), The semantics of Clear, a specification language, *in* D. Bjørner, ed., 'Abstract Software Specifications', Vol. 86 of *Lecture Notes in Computer Science*, Springer, pp. 292–332.

Casanova, M. A., Veloso, P. A. & Furtado, A. L. (1984), Formal data base specification – An eclectic perspective, *in* D. J. Rosenkrantz & R. Fagin, eds, 'Proceedings of the Third ACM SIGACT-SIGMOD Symposium on Principles of Database Systems, April 2–4, 1984, Waterloo, Ontario, Canada', ACM, pp. 110–118.

Codescu, M., Mossakowski, T., Sannella, D. & Tarlecki, A. (2017), 'Specification refinements: Calculi, tools, and applications', *Science of Computer Programming* **144**, 1–49.

Diaconescu, R. (2002), 'Grothendieck institutions', *Applied Categorical Structures* **10**(4), 383–402.

Diaconescu, R. (2008), *Institution-Independent Model Theory*, Studies in Universal Logic, Birkhäuser.

Diaconescu, R. (2010), 'Quasi-Boolean encodings and conditionals in algebraic specification', *Journal of Logic and Algebraic Programming* **79**(2), 174–188.

Diaconescu, R. (2014), From universal logic to computer science, and back, *in* G. Ciobanu & D. Méry, eds, 'Theoretical Aspects of Computing – ICTAC

2014 – 11th International Colloquium, Bucharest, Romania, September 17–19, 2014. Proceedings', Vol. 8687 of *Lecture Notes in Computer Science*, Springer, pp. 1–16.

Diaconescu, R. & Madeira, A. (2016), 'Encoding hybridized institutions into first-order logic', *Mathematical Structures in Computer Science* 26(5), 745–788.

Diaconescu, R. & Țuțu, I. (2014), 'Foundations for structuring behavioural specifications', *Journal of Logical and Algebraic Methods in Programming* 83(3–4), 319–338.

Fiadeiro, J. L. & Costa, J. F. (1996), 'Mirror, mirror in my hand: a duality between specifications and models of process behaviour', *Mathematical Structures in Computer Science* 6(4), 353–373.

Fiadeiro, J. L. & Sernadas, A. (1986), 'The INFOLOG linear tense propositional logic of events and transactions', *Information Systems* 11(1), 61–85.

Fiadeiro, J. L. & Sernadas, A. (1988), 'Specification and verification of database dynamics', *Acta Informatica* 25(6), 625–661.

Fiadeiro, J. L., Țuțu, I., Lopes, A. & Pavlovic, D. (2019), 'Logics for actor networks: A two-stage constrained-hybridisation approach', *Journal of Logical and Algebraic Methods in Programming* 106, 141–166.

Goguen, J. A. & Burstall, R. M. (1983), Introducing institutions, in E. M. Clarke & D. Kozen, eds, 'Logic of Programs', Vol. 164 of *Lecture Notes in Computer Science*, Springer, pp. 221–256.

Goguen, J. A. & Burstall, R. M. (1992), 'Institutions: abstract model theory for specification and programming', *Journal of the ACM* 39(1), 95–146.

Goguen, J. A. & Roșu, G. (2002), 'Institution morphisms', *Formal Aspects of Computing* 13(3–5), 274–307.

Goguen, J. A., Winkler, T., Meseguer, J., Futatsugi, K. & Jouannaud, J.-P. (2000), Introducing OBJ, in J. A. Goguen & G. Malcolm, eds, 'Software engineering with OBJ: algebraic specification in action', Advances in formal methods, Kluwer Academic.

Harel, D., Kozen, D. & Tiuryn, J. (2001), 'Dynamic logic', *SIGACT News* 32(1), 66–69.

Kurz, A. (2002), Notions of behaviour and reachable-part and their institutions, in M. Wirsing, D. Pattinson & R. Hennicker, eds, 'Recent Trends in Algebraic Development Techniques, 16th International Workshop, WADT 2002, Frauenchiemsee, Germany, September 24-27, 2002, Revised Selected Papers', Vol. 2755 of *Lecture Notes in Computer Science*, Springer, pp. 312–327.

Lopes, A. & Fiadeiro, J. L. (2002), 'Superposition: Composition vs refinement of non-deterministic action-based systems', *Electronic Notes in Theoretical Computer Science* **70**(3), 282–296.

Maibaum, T., Veloso, P. A. & Sadler, M. (1985), A theory of abstract data types for program development: bridging the gap?, in H. Ehrig, C. Floyd, M. Nivat & J. W. Thatcher, eds, 'Theory and Practice of Software Development', Vol. 185 of *Lecture Notes in Computer Science*, Springer, pp. 214–230.

Mossakowski, T. (2002), Comorphism-based Grothendieck logics, in K. Diks & W. Rytter, eds, 'Mathematical Foundations of Computer Science 2002', Vol. 2420 of *Lecture Notes in Computer Science*, Springer, pp. 593–604.

Mossakowski, T. & Tarlecki, A. (2008), Heterogeneous logical environments for distributed specifications, in A. Corradini & U. Montanari, eds, 'Recent Trends in Algebraic Development Techniques', Vol. 5486 of *Lecture Notes in Computer Science*, Springer, pp. 266–289.

Mossakowski, T., Sannella, D. & Tarlecki, A. (2004), A simple refinement language for CASL, in J. L. Fiadeiro, P. D. Mosses & F. Orejas, eds, 'Recent Trends in Algebraic Development Techniques, 17th International Workshop, WADT 2004, Barcelona, Spain, March 27–29, 2004, Revised Selected Papers', Vol. 3423 of *Lecture Notes in Computer Science*, Springer, pp. 162–185.

Sannella, D. & Tarlecki, A. (1988), 'Toward formal development of programs from algebraic specifications: implementations revisited', *Acta Informatica* **25**(3), 233–281.

Sannella, D. & Tarlecki, A. (2011), *Foundations of Algebraic Specification and Formal Software Development*, Monographs in Theoretical Computer Science. An EATCS Series, Springer.

Tarlecki, A. (1985), Bits and pieces of the theory of institutions, in D. H. Pitt, S. Abramsky, A. Poigné & D. E. Rydeheard, eds, 'Category Theory and Computer Programming, Tutorial and Workshop, Guildford,

UK, September 16–20, 1985 Proceedings', Vol. 240 of *Lecture Notes in Computer Science*, Springer, pp. 334–365.

Țuțu, I., Chiriță, C. E., Lopes, A. & Fiadeiro, J. L. (2019), Logical support for bike-sharing system design, in M. H. ter Beek, A. Fantechi & L. Semini, eds, 'From Software Engineering to Formal Methods and Tools, and Back', Vol. 11865 of *Lecture Notes in Computer Science*, Springer, pp. 152–171.

Țuțu, I., Fiadeiro, J. L. & Chiriță, C. E. (2020), 'Dynamic reconfigurations through hybrid lenses', Presentation given at the 25th International Workshop on Algebraic Development Techniques – WADT 2020.

Veloso, P. A. (1983), Problems as abstract data types: Applications to program construction, in M. Broy & M. Wirsing, eds, 'Proceedings 2nd Workshop on Abstract Data Type, 1983. University of Passau, Germany', University of Passau.

Veloso, P. A. & Furtado, A. L. (1982), Stepwise construction of algebraic specifications, in H. Gallaire, J. Nicolas & J. Minker, eds, 'Advances in Data Base Theory, Vol. 2, Based on the Proceedings of the Workshop on Logical Data Bases, December 14–17, 1982, Centre d'études et de recherches de Toulouse, France', Advances in Data Base Theory, Plemum Press, pp. 321–352.

Veloso, P. A., Casanova, M. A. & Furtado, A. L. (1984), Formal data base specification – An eclectic perspective, Technical Report 1/84, Pontificia Universidade Catolica do RJ.

Veloso, P. A., Fiadeiro, J. L. & Veloso, S. R. (2002), 'On local modularity and interpolation in entailment systems', *Information Processing Letters* **82**(4), 203–211.

Living and working with Paulo A. S. Veloso[‡]

Sheila Regina Murgel Veloso*

* Computing and System Engineering Dept., Engin. Fac.
State University of Rio de Janeiro (UERJ)
Rio de Janeiro RJ, Brazil
sheila.murgel.bridge@gmail.com

Abstract

This is a testimony of fifty years of shared life with a very special human being named Veloso. The facts reported here are a partial and personal view of his creativity and enthusiasm for understanding and clarifying matters in many aspect of Computer Science, Logic, Mathematics and Philosophy. I have focused mainly on the work we have done together, and I am happy to share this view to take part in his Festschrift.

This is a testimony of fifty years of living and working with Veloso (Paulo Augusto, to me). This year we celebrated our fiftieth anniversary, and I am very happy to participate in this volume, giving a particular view of our life in common, focusing on the work we have made together.

We got married on April 28th,1970, one day after the presentation of his Master's degree dissertation. We both knew that our time in Rio after the wedding would be short: time for me to finish my undergraduate course in Chemical Engineering in December and for us to decide where to go so that he could make his PhD studies and I could get my Master's degree. The choice was between Berkeley and Waterloo, but I voted for Sunny California: the thought of spending 4 years in the chilling winter of Canada sounded too terrifying to me.

Back then, Paulo was teaching at Coppe, at the Systems Engineering Program. At that time, Coppe had hired their best former Master's students (the so called "golden boys") and encouraged them to take their PhD abroad and come back to Brazil afterwards. Many of Paulo's former Master's colleagues were already at Berkeley when we arrived there.

I think we made a good choice, not just regarding the weather. We both loved Berkeley, not only because it was a beautiful place, but our

---

[‡]I gratefully acknowledge Paula Murgel Veloso's help in editing this article in LaTeX, and in reviewing the English writing and the bibliographic citations.

stay there has opened our minds to a new life in many ways. The University was one of the best in the world in Computer Science, and also the students were allowed to take courses in the most diverse areas, from Humanities to Technical fields. This environment provided us the opportunity of attending courses in Math and Philosophy, being taught by the top professors in these areas.

Paulo decided to obtain a second Master's degree, in Logic at the Math department, working under the supervision of Leon Henkin. His work, on the history of the development results on representation of relation algebras, during the period of 1940–1970, focused mainly on the results presented by Lyndon, Tarski, Jónsson and Monk on the error and correction of Lyndon's characterization of representability [Lyndon, 1950, 1956], [Tarski, 1955], [Jónsson, 1959], [Monk, 1964].This work has awakened Paulo's interest on the theory of relation algebras, which has accompanied him for the rest of his life. Also, his work under the supervision of Arthur Gill, on network of finite-state machines, had an algebraic flavor. Logic and algebra became the foundation of Paulo's approach to Computer Science.

A private curiosity fits in here: my mother, visiting us at Berkeley, saw Paulos's sketch papers over his work table and was very surprised with what she saw: lots of little circles attached one to the other by straight lines. "Was this part of his thesis?!", she has asked me. It was a transition states diagram of finite automata.

Where do I enter in this world?

Well, during the first semester of 1971, I took some classes in the Math department, as I was inclined to Math during my undergraduate course in Chemical Engineering at UFRJ. When I started my Master's course in the Chemical Engineering department at Berkeley, in the fall semester of 1971, I realized that my interest was much more in the Math realm than in the Chemical Engineering one. So, after many considerations, I decided to quit the Chemical Engineering Master's course and apply for the Math department. Paulo was a good advisor and companion in this changing process, and, after that decision, we had one more interest in common that resulted in many hours of conversation on math problems over a capuccino at the Caffe Mediterraneum, after having browsed over books at Shakespeare & Co. Bookstore on Telegraph Avenue.

In June 1975, we came back to Brazil, having achieved many realizations: his Master's degree in Math (Logic) and his PhD degree in Electrical Engineering and Computer Science, and my Master's degree in Math (Algebra), and a production made in the U.S.A., to be born in Brazil: Paula, our eldest daughter.

One of Paulo's main characteristics is his scientific curiosity, not only in the main fields of his research, but whenever he is interested in something he goes deep inside. So, we brought in our luggage, back from Berkeley, more than fifty boxes of books and xerox copies of many books on foundations, including foundations of Biology, of Zoology and many other subject matters to be read when he would have time in the future.

Paulo went back to Coppe at PESC (Systems Engineering and Computer Science Program), and I started looking for a place to finish my PhD studies, which I started in the Math department at Berkeley (by that time my interest had turned towards algebra and logic as well). After an intense search I finally registered at PESC to work with Sueli Mendes, who had experience in Logic and Computing.

We have not worked at the same place for long, for the so called "golden boys" dispersed into different "Programas" at Coppe, according to their newly acquired interests, and Paulo found himself a little alone, sharing more affinity with the group recently formed by the "Founding Fathers" (Carlos José Lucena, Antonio Luz Furtado, Arndt von Staa, and Roberto Lins de Carvalho) of the Department of Informatics at PUC-Rio. So, after Lins' invitation in 1976, Paulo started working at PUC, continuing to work part time at Coppe.

At PUC, he found a fruitful environment to develop his work on the theory of problems, logical specifications and relation algebra etc., interacting with Lins, Furtado, Lucena and students, such as Tarcisio Pequeno, Markus Endler, Atendolpho Pereda, Armando Haeberer, Marcelo Frias, Mario Benevides, Hermann Haeusler, among others, who became his collaborators, some of them still working with him. Many researchers were invited to visit the newly created department and projects of collaboration with Tom Maibaum (Imperial College), Jean-Luc Rémy (Nancy-University) and Luiz Fiadeiro (Imperial College and Lisbon New University), provided a rich creative and productive life to Paulo.

In the meantime, I finished my PhD, our second daughter, Flávia, was born, I entered UFRJ (Federal University of Rio de Janeiro) to teach at the Math department and was also invited to teach at PESC. Rubens Mello, a member of my PhD thesis' board, who was also a colleague of Paulo at PUC, told me at the end of my defense: "Now you are free to work with Paulo". And that was true, because we had lots of things to work together, but I felt more at ease to work with him as a colleague (even though I have always had a lot to learn from him).

The work with Tom Maibaum and José Fiadeiro was a very productive one, for they both had good intuitions and practical feelings about program development while, at the same time, they used logic as a common language to talk about specifications (of abstract data types, programs, databases, ...).

In this context, the so called Modularization Property captured most of their attention. This property plays a special role in composing implementation steps for formal development of specifications and programs in a step-by-step fashion. In a nutshell, when working with logical specifications, presented by first-order axioms, one has a specification M that one wishes to implement on a (more concrete) specification P. For this purpose, one has to provide, on top of P, some support for the abstract concepts of M. One can do that by extending the concrete specification P to a specification Q, by adding symbols to correspond to the abstract ones in M, not disturbing the given concrete specification P. This involves two logical concepts: this extension Q of P should be conservative, and the translation of M to Q is an interpretation of theories, in the sense that it translates each consequence of M into a consequence of Q. Considering two consecutive implementation steps: a first implementation of M on P (with mediating specification Q) and a second implementation of P on T (with mediating specification R), one would wish to compose these two implementations, so as to obtain a composite implementation of M directly on T. This is where the Modularization Property comes into play. For it will allow one to obtain such a mediating specification S together with an interpretation Q into S and a conservative extension R into S.

Many, many discussions over this subject took place, both here in Brazil and during some visits of Paulo to the Imperial College from 1984 until the early 90's. I also took part in some of these discussions and have some articles on it with Paulo and José Fiadeiro, considering the Modularizaition Property in modular software development of labelled families [Veloso et al., 1998], in the context of π-institutions [Veloso & Veloso, 2001; Veloso et al., 2002] and the preservation of conservativeness in the introduction of new operations into logical specifications [Veloso & Veloso, 1997].

Paulo also had a fruitful collaboration with Walter Carnielli (Unicamp, CLE), working on an alternative view of "most and generally", other than the usually non-monotonic one. These ideas arose from the valuable collaboration of Antonio Mario Sette (Unicamp, CLE).

A basis for generic reasoning, reasoning about intuitive ideas as "most", "in general", "typically", was settled by means of generalized quantifiers over ultrafilters. This results in a conservative extension of classical first-order logic. To express assertions and reasoning about vague notions ("most", "generally", "several", "tipically") an operator ∇ (and axioms to characterize the vague notion expressed by ∇) is added. So, one can express, for instance, "Birds generally fly" by $\nabla v F(v)$. For reasoning about generic objects, "generic" individuals were introduced as those possessing the properties that most individuals have. For instance, in the case of birds, we select a new constant symbol c and express that c is generic (with respect to flying, i.e a "generic" bird as one that exhibits exactly the properties that birds generally posses) by $\nabla v F(v) \leftrightarrow F(c)$. The non-monotonic approach, via defaults for example, takes a negative view, in the sense of "in the absence of information to the contrary", while the generalized quantifiers' approach has a positive interpretation. This approach lead to many philosophical and technical discussions specially among the non-monotonic community, which usually wanted to establish a correspondence between a typical object (an ideally abstracted one, which has a property P exactly when P holds generically), with a particular object having property P by default. The generalized approach can be thought of as complementary to the non-monotonic one. Also, some issues may have been left somewhat unclear, leading to misunderstandings, and some objections against the approach have been raised, such as lack of intuitive justification. Thus, we have revisited and discussed some issues concerning logics for "generally" and "rarely", with the aim of clarifying the approach, its underlying motivations, intuitions and justification [Veloso & Veloso, 2011].

Meanwhile, we have been working on some issues of qualitative logic for "generally", such as obtaining a framework for theorem proving in logics for "generally". By introducing special functions, generic functions (which are similar to Skolem functions, and coherent functions) one can reduce proof procedures in logic for "generally" to proof procedures and theorem provers for classical logic [Veloso & Veloso, 2002, 2004].

Also, we considered filter logic to address some versions of "most", showing that they can also be embedded into classical first order theory of certain special predicates, instead of functions as in the above case [Veloso & Veloso, 2005]. With Leonardo Vana, our DSc student at PESC, now teaching at UFF (Fluminense Federal University), we introduce deductive systems for "generally" in sequent and natural deduction styles [Vana et al., 2008, 2014].

In 1995, Paulo took a leave of absence from PUC-Rio to enter as a fulll-time professor at UFRJ. This provided a bigger interaction of Paulo with graduate students at PESC [1] and a fruitful collaboration with Mario Benevides. An instance of this collaboration is the the work we have done involving Mario Benevides, his student Carla Delgado, Petrucio Viana and Renata de Freitas, our students at that time, treating vague notions in a modal environment [Veloso et al., 2010], and the works on the relations of fork algebra with arrow and modal logics [Veloso et al., 2007].

---

[1] While writing this article, Leonardo Vana, sent me some of his impressions on working with Paulo, which I briefly report in the appendix.

Petrucio's research for his PhD degree introduced us to graphical deduction systems for binary relations. Graphical reasoning became our favorite and passionate research theme since then. Petrucio and Renata were our students since their Master's studies and worked very closely. Supervising them and collaborating with them has been very pleasant and fruitful, rendering very interesting discussions and publications [de Freitas et al., 2009, 2010], to mention a few. Our collaboration has surpassed academia, we worked as a family.

Speaking on working within the family, we had a collaboration with Paula, our daughter, who is a mathematician, teaching at UFF, who helped us in the algebraic approach of formalizing a tool to analyse and compare logics based on some ideas introduced by J. Piaget [Veloso et al., 2011b,a]. I would say that this was an exciting and grateful experience for Paulo and me.

As I mentioned above, graphical reasoning became our favorite research theme. The formalism of graph calculus was originally conceived to handle graphically binary relations and their operations [Curtis & Lowe, 1996; de Freitas et al., 2009]. It can also be used as graphical (meta-)language for modal logics, providing an intuitive and natural way of handling modal formulas. This was the approach we have taken, working until recently with Mario Benevides, in a very pleasant collaboration environment. Mario's interest in modal logics was very helpful in providing a general framework for expressing modal formulas and frame properties [Veloso et al., 2014].

A collaboration that must be recalled is the one with Armando Haeberer. Armando was a passionate person and used to dive deeply into whatever he was interested in. Back in the late eighties, Paulo and him started a close exchange on fork algebras. He used to make phone calls from Argentina, late at night, and Paulo and him discussed about it for hours. His coming to Brazil, working at PUC-Rio, resulted into a very rich and fruitful interchanging of ideas relating fork algebras both to logic and to computer science.

Last, but not least, the longterm collaboration with Luiz Carlos Pereira (PUC-Rio, Philosophy Dept.) and Hermann (PUC-Rio, Dept. of Informatics) should be mentioned. I think that it all started because Paulo's interest in logic naturally drove him to look for cooperation at the Philosophy Department. It was through Oswaldo Chateaubriand that he began exchanging ideas with the Philosophy Department's academic staff. Nowadays, his involvement with philosophy takes most of his attention. Luiz Carlos Pereira's dynamism and ability to organize and gather people, make Paulo fell very comfortable to transit through and collaborate with people who participate at the Conesul Colloquium of Philosophy of Formal Sciences (Abel Lasalle-Casanave, Eduardo Giovanini, Wagner Sanz, among other philosophers) and with Jean-Baptiste Joinet (Université Jean-Moulin Lyon 3, Faculté de Philosophie). Hermann too transits through these areas very easily, and Paulo and him became a bridge from computer science to mathematics and logic.

As a conclusion, I am grateful to say that in preparing this article, many, many good memories of these fifty years of living together were redeemed. Maybe I have forgotten to mention some important people who have worked with Paulo or some areas in which Paulo has been interested, but my intention was mainly to focus on the work we both have done together. I have to say that I have had the privilege of witnessing his creative process in searching to understand and clarify matters in logic and foundations of computing, and also of sharing with him common interests in movies, music, the french language and traveling, specially to Paris. Even the not so nice time we have spent together doing bureaucratic work may also be regarded as a collaboration task. So, Paulo, I can state "we have came a long way, baby", and it was worth every minute of it.

A Leonardo's account: A special mind

In 2003, at PESC, a group of graduate students and professors interested in logic (among whom Leonardo and Paulo Augusto) used to attend a reading seminar. At the end of Leonardo's presentation, some points in his reading assignment remained unclear to him. Paulo asked him whether these points could be translated into "a several-minute-long talk with many questions left unanswered" (in Leonardo's own words). Leonardo left the seminar feeling bad for not having understood Paulo's comments. Talking to his colleagues, he was comforted by "Don't worry; the questions proposed by Veloso take 6 months to be understood and another 12 months to be answered". This made him change his point of view of his expectations from his DSc studies: he had to review his approach on reading articles. He thanks Paulo for this challenge.

In many other occasions, Paulo's cultural profile made Leonardo realise that, as an aspiring DSc, he was not only expected to know some specific subject, but a vast general culture might also help a researcher to correlate theories and come up with new ideas. This realisation helped Leonardo finish his DSc studies.

References

Curtis, S. & Lowe, G. (1996), 'Proofs with graphs', *Science of Computer Programming* **26**(1-3), 197–216.

de Freitas, R., Veloso, P. A., Veloso, S. R. & Viana, P. (2009), 'On graph reasoning', *Information and Computation* **207**(10), 1000–1014.

de Freitas, R., Veloso, P. A., Veloso, S. R. & Viana, P. (2010), A calculus for graphs with complement, *in* 'International Conference on Theory and Application of Diagrams', Springer, pp. 84–98.

Jónsson, B. (1959), 'Representation of modular lattices and of relation algebras', *Transactions of the American Mathematical Society* **92**(3), 449–464.

Lyndon, R. C. (1950), 'The representation of relational algebras', *Annals of Mathematics* pp. 707–729.

Lyndon, R. C. (1956), 'The representation of relation algebras, ii', *Annals of Mathematics* pp. 294–307.

Monk, D. (1964), 'On representable relation algebras', *The Michigan Mathematical Journal* **11**(3), 207–210.

Tarski, A. (1955), 'Contributions to the theory of models, iii', *Indagationes Mathematicae* **58**, 56–64.

Vana, L. B., Veloso, P. A. & Veloso, S. R. (2008), 'Sequent calculi for 'generally'', *Electronic Notes in Theoretical Computer Science* **205**, 49–65.

Vana, L. B., Veloso, P. A. & Veloso, S. R. (2014), On the structure of natural deduction derivations for 'generally', *in* 'Advances in Natural Deduction', Springer, pp. 103–128.

Veloso, P. A. & Veloso, S. R. (1997), 'On methods for safe introduction of operations', *Information Processing Letters* **64**(5), 231–238.

Veloso, P. A. & Veloso, S. R. (2001), 'On local modularity variants and π-institutions', *Information Processing Letters* **77**(5-6), 247–253.

Veloso, P. A. & Veloso, S. R. (2004), 'On ultrafilter logic and special functions', *Studia Logica* **78**(3), 459–477.

Veloso, P. A. & Veloso, S. R. (2005), 'On 'most' and 'representative': Filter logic and special predicates', *Logic Journal of the IGPL* **13**(6), 717–728.

Veloso, P. A. & Veloso, S. R. (2011), Revisiting 'generally' and 'rarely', *in* J.-Y. Beziau & M. E. Coniglio, eds, 'Logic without Frontiers: Festschrift for Walter Alexandre Carnielli on the occasion of his 60th Birthday', Vol. 17 of *Tributes series*, College Publications, pp. 183–208.

Veloso, P. A., De Freitas, R. P., Viana, P., Benevides, M. & Veloso, S. R. (2007), 'On fork arrow logic and its expressive power', *Journal of philosophical logic* **36**(5), 489–509.

Veloso, P. A., Fiadeiro, J. L. & Veloso, S. R. (2002), 'On local modularity and interpolation in entailment systems', *Information processing letters* **82**(4), 203–211.

Veloso, P. A., Veloso, S. R. & Benevides, M. R. (2014), 'On a graph approach to modal logics', *Electronic Notes in Theoretical Computer Science* **305**, 123–139.

Veloso, P. A., Veloso, S. R., Viana, P., Freitas, R. d., Benevides, M. & Delgado, C. (2010), 'On vague notions and modalities: a modular approach', *Logic Journal of IGPL* **18**(3), 381–402.

Veloso, S. R. & Veloso, P. A. (2002), On special functions and theorem proving in logics for 'generally', *in* 'Brazilian Symposium on Artificial Intelligence', Springer, pp. 1–10.

Veloso, S. R., Veloso, P. A. & Fiadeiro, J. L. (1998), 'Labeled families in modular software development', *Journal of the Brazilian Computer Society* **5**(1), 20–31.

Veloso, S. R., Veloso, P. A. & Veloso, P. M. (2011*a*), On piaget-like monoids: monoids for logics, *in* 'Annals of the International Conference on Logic, Methodology and Philosophy of Science', pp. 102–115.

Veloso, S. R., Veloso, P. A. & Veloso, P. M. (2011*b*), 'A tool for analysing logics', *Electronic Notes in Theoretical Computer Science* **269**, 125–137.

www.ingramcontent.com/pod-product-compliance
Lightning Source LLC
Chambersburg PA
CBHW070725160426
43192CB00009B/1318